Monitoring the Comprehensive Nuclear-Test-Ban Treaty: Surface Waves

Edited by
Anatoli L. Levshin
Michael H. Ritzwoller

Springer Basel AG

Reprint from Pure and Applied Geophysics
(PAGEOPH), Volume 158 (2001), No. 8

Editors:

Anatoli L. Levshin
University of Colorado
Dep. of Physics / Seismology Group
Campus Box 390
Boulder, Colorado 80309-0390
USA
e-mail: levshin@lemond.colorado.edu

Michael H. Ritzwoller
University of Colorado
Dep. of Physics / Seismology Group
Campus Box 390
Boulder, Colorado 80309-0390
USA
e-mail: ritzwoller@lemond.colorado.edu

A CIP catalogue record for this book is available from the Library of Congress,
Washington D.C., USA

Deutsche Bibliothek Cataloging-in-Publication Data

Monitoring the comprehensive nuclear test ban treaty. - Basel ; Boston ; Berlin : Birkhäuser
   (Pageoph topical volumes)

Surface waves / ed. by Anatoli L. Levshin ; Michael H. Ritzwoller. - 2001
   ISBN 978-3-7643-6551-6   ISBN 978-3-0348-8264-4 (eBook)
   DOI 10.1007/978-3-0348-8264-4

© 2001 Springer Basel AG
Originally published by Birkhäuser Verlag Basel, Switzerland in 2001
Printed on acid-free paper produced from chlorine-free pulp

9 8 7 6 5 4 3 2 1

Contents

Pure appl. geophys. 158 (2001) 1339–1340
0033–4553/01/081339–02 $ 1.50 + 0.20/0

❚ **Pure and Applied Geophysics**

Monitoring the Comprehensive Nuclear-Test-Ban Treaty

Preface

The first nuclear bomb was detonated in 1945, thus ushering in the nuclear age. A few political leaders quickly saw a need to limit nuclear weapons through international cooperation and the first proposals to do so were made later in that same year. The issue of nuclear testing, however, was not formally addressed until 1958 when the United States, the United Kingdom, and the Soviet Union, initiated talks intended to establish a total ban on that testing (a Comprehensive Test-Ban Treaty or CTBT). Those talks ended unsuccessfully, ostensibly because the participants could not agree on the issue of on-site verification.

Less comprehensive treaties did, however, place some constraints on nuclear testing. The United States, the United Kingdom, and the Soviet Union, in 1963, negotiated the Limited Test-Ban Treaty (LTBT) which prohibited nuclear explosions in the atmosphere, outer space and under water. The Threshold Test-Ban Treaty (TTBT), signed by the United States and the Soviet Union in 1974, limited the size, or yield, of explosions permitted in nuclear tests to 150 kilotons.

Seismological observations played an important role in monitoring compliance with those treaties. Many of the world's seismologists set aside other research projects and contributed to that effort. They devised new techniques and made important discoveries about the Earth's properties that enhance our ability to detect nuclear events, to determine their yield, and to distinguish them from earthquakes. Seismologists are rightfully proud of their success in developing methods for monitoring compliance with the LTBT and TTBT.

Although seismologists have also worked for many years on research related to CTBT monitoring, events of recent years have caused them to redouble their efforts in that area. Between 1992 and 1996 Russia, France and the United States all placed moratoria on their nuclear testing, though France did carry out a few tests at the end of this period. In addition, the United States decided to use means other than testing to ensure the safety and reliability of its nuclear arsenal, and all three countries, as well as the United Kingdom, agreed to continue moratoria as long as no other country tested. Those developments, as well as diplomatic efforts by many nations, led to the renewal of multilateral talks on a CTBT that began in January 1994. The talks led to the Comprehensive Nuclear-Test-Ban Treaty. It was adopted by the United Nations General Assembly on 10 September 1996 and has since been signed by 160 nations. Entry of the treaty into force, however, is still uncertain since it requires ratification by all 44 nations that have some nuclear capability and, as of 30 April 2001, only 31 of those nations have done so.

Although entry of the CTBT into force is still uncertain, seismologists and scientists in related fields, such as radionuclides, have proceeded with new research on issues relevant to monitoring compliance with it. Results of much of that research may be used by the International Monitoring System, headquartered in Vienna, and by several national centers and individual institutions, to monitor compliance with the CTBT. New issues associated with CTBT monitoring in the 21st century have presented scientists with many new challenges. They must now be able to effectively monitor compliance by several countries that have not previously been nuclear powers. Effective monitoring requires that we able to detect and locate substantially smaller nuclear events than ever before and to distinguish them from small earthquakes and other types of explosions. We must have those capabilities in regions that are seismically active and geologically complex, and where seismic waves might not propagate efficiently.

Major research issues that have emerged for monitoring a CTBT are the precise location of events, and discrimination between nuclear explosions, earthquakes, and chemical explosions, even when those events are relatively small. These issues further require that we understand how seismic waves propagate in the solid Earth, the oceans and atmosphere, especially in regions that are structurally complex, where waves undergo scattering and, perhaps, a high degree of absorption. In addition, we must understand how processes occurring at the sources of explosions and earthquakes manifest themselves in recordings of ground motion.

Monitoring a CTBT has required, and will continue to require, the best efforts of some the world's best seismologists. They, with few exceptions, believe that methods and facilities that are currently in place will provide an effective means for monitoring a CTBT. Moreover, they expect that continuing improvements in those methods and facilities will make verification even more effective in the future. This topical series on several aspects of CTBT monitoring is intended to inform readers of the breadth of the CTBT research program, and of the significant progress that has been made toward effectively monitoring compliance with the CTBT.

The following set of papers, edited by Drs. A.L. Levshin and M.H. Ritzwoller presents research results on surface waves that are applicable for monitoring a CTBT. It is the fourth of eight topics addressed by this important series on *Monitoring the Comprehensive Nuclear-Test-Ban Treaty*. Previously published topics are Source Location, Hydroacoustics, and Regional Wave Propagation and Crustal Structure. The topics to appear in ensuing issues are Source Processes and Explosion Yield Estimation, Infrasound, Data Processing, and Source Discrimination.

Brian J. Mitchell
Saint Louis University
Series Editor

Pure appl. geophys. 158 (2001) 1341–1348
0033–4553/01/081341–8 $ 1.50 + 0.20/0

❙ Pure and Applied Geophysics

Introduction

MICHAEL H. RITZWOLLER[1] and ANATOLI L. LEVSHIN[1]

This is one of several volumes planned to appear in *Pure and Applied Geophysics* covering a range of topics related to monitoring the Comprehensive Nuclear-Test-Ban Treaty (CTBT). This volume concentrates on the measurement and use of surface waves and the papers fall into two general categories: The development and/ or application of methods to summarize information in surface waves (e.g., surface-wave tomography) or the use of these summaries to improve capabilities to monitor and verify the CTBT by advancing the art of surface-wave identification, measurement, and source characterization. Because of the emphasis here on a type of wave rather than on a specific application, the papers in this volume overlap those in the other volumes appreciably. Readers interested in the application of surface waves are encouraged also to investigate the contents of the other volumes, after thoroughly digesting the results in this volume, of course.

Surface waves compose the longest and largest amplitude parts of broadband seismic waveforms generated both by explosions and shallow earthquakes. In addition, they contain most of the low frequency information radiated by seismic sources. Measurements of the properties of surface waves have been important for evaluating source mechanisms, estimating yields, and helping to discriminate nuclear explosions from naturally occurring earthquakes, and have been widely used by national and international organizations charged with monitoring and verifying various nuclear test treaties. Under the CTBT, concentration has shifted from teleseismic monitoring of a threshold yield targeted on a few well-defined locations to identifying and characterizing signals from weak nuclear explosions and earthquakes using potentially very noisy and incomplete regional data following events that may be distributed widely in space. Concentration is no longer on yield estimation, but rather on being able to discriminate explosions from naturally occurring earthquakes and to locate small events using sparse regional networks in complex tectonic environments with the accuracy and precision demanded by the CTBT.

[1] Center for Imaging the Earth's Interior, Department of Physics, University of Colorado at Boulder, Boulder, Colorado, 80309-0390, U.S.A.
E-mail: levshin@lemond.colorado.edu, ritzwoller@lemond.colorado.edu

Within the context of the CTBT, the use and interpretation of information from surface waves has grown in significance. There are two general uses of surface waves under the CTBT. First, the comparison of the amplitudes of surface waves and body waves remains the most reliable regional discriminant, an example of which is the well known $m_b : M_s$ discriminant (e.g., STEVENS and DAY, 1985). Second, broadband surface-wave dispersion provides important information used in estimating 3-D seismic models of the crust and uppermost mantle which are necessary to obtain accurate locations of small events for which only regional data may be available. The success of both applications depends on obtaining reliable surface-wave dispersion measurements and representing these measurements in a useful form, usually as group- or phase-velocity maps.

The measurement of the group velocity of Rayleigh and Love waves is performed on the envelope of the surface-wave packet and can be robustly measured across a broad frequency band, from several seconds to hundreds of seconds period (e.g., DZIEWONSKI et al., 1969; LEVSHIN et al., 1972; CARA, 1973; KODERA et al., 1976; RUSSELL et al., 1988; LEVSHIN et al., 1989, 1992; RITZWOLLER et al., 1995). A recent experiment by a number of research groups in the U.S. revealed general agreement among the various methods and codes used to measure group velocities (WALTER and RITZWOLLER, 1998). Phase-velocity measurements are typically obtained by wave-form fitting (e.g., WOODHOUSE and DZIEWONSKI, 1984) or by differencing phase spectra obtained at adjacent stations or from nearby events. There are three key reasons why group velocities have been considered more useful in nuclear monitoring than phase velocities. First, absolute phase-velocity measurements are strongly affected by initial source phase (e.g., KNOPOFF and SCHWAB, 1968; MUYZERT and SNIEDER, 1996), which may be poorly known or completely unknown for small events. Group velocities are much less sensitive to source characteristics (e.g., LEVSHIN et al., 1999). Second, phase velocities are difficult to measure unambiguously below about 30 s period. Finally, although multi-station and multi-event differential phase measurements are largely unaffected by source phase, they are typically too sparsely distributed to be of general use in constructing tomographic maps. With a few notable exceptions surface-wave data processing for use in nuclear monitoring has concentrated on estimating velocities rather than wave amplitudes, polarizations, or scattering. If the emphasis on constructing 3-D models to improve regional location capabilities continues, it is likely that a larger share of future efforts will be devoted to short-period phase-velocity estimation and the use of more complicated wavefield effects to constrain 3-D models, such as polarization anomalies (e.g., LEVSHIN et al., 1994; LASKE, 1995) and scattering (e.g., POLLITZ, 1994).

The estimation of dispersion maps by tomography (e.g., DITMAR and YANOVS-KAYA, 1987; YANOVSKAYA and DITMAR, 1990) is now commonplace and new methods such as kriging (e.g., SCHULTZ et al., 1998) have emerged. Dispersion maps on a variety of scales have appeared in the last several years. For example, there are global phase-velocity maps (e.g., LASKE and MASTERS, 1996; TRAMPERT and WOOD-

HOUSE, 1996; ZHANG and LAY, 1996; EKSTRÖM et al., 1997; VAN DER HEIJST and WOODHOUSE, 1999) as well as regional studies across Eurasia (e.g., WU et al., 1997; CURTIS et al., 1998; GRIOT et al., 1998; RITZWOLLER and LEVSHIN, 1998; RITZWOLLER et al., 1998; YANOVSKAYA and ANTONOVA, 2000) and elsewhere (e.g., Antarctica: VDOVIN, 1999; South America: VDOVIN et al., 1999; Arctic: LEVSHIN et al., 2001). Two papers in this volume describe the application of surface-wave tomography to regions of interest for monitoring the CTBT. *Pasyanos, Walter, and Hazler* present a study of the Middle East, North Africa, southern Eurasia and the Mediterranean using Rayleigh and Love waves at periods ranging from 10 s to 60 s. *Mokhtar, Ammon, Herrmann, and Ghalib* present a tomographic inversion of Rayleigh and Love group velocities across the Arabian peninsula in the period range of 5–60 s. These and other observational efforts exemplify the advances that are emerging as data sets accumulate and, in particular, as the frequency band of observation lowers.

Advances in surface wave methodology continue to emerge both on regional (e.g., STEVENS and McLAUGHLIN, 1997) and global scales (e.g., WANG and DAHLEN, 1995; WANG et al., 1998). In this volume, *Barmin, Ritzwoller, and Levshin* discuss a tomographic method for constructing both isotropic and azimuthally anisotropic surface-wave maps. Although their algorithm is based on a regular grid, it extends naturally to irregular grids, and recent advances in the construction and use of irregular grids in tomography (e.g., SAMBRIDGE et al., 1995; SPAKMAN and BIJWAARD, 1998) are now being exploited in surface-wave tomography, as described here by *Spakman and Bijwaard*. Irregular grids are most useful when the spatial distribution of data is inhomogenous, as is common in regional surface-wave tomography. Also in this volume, *Larson and Ekström* show that at periods above about 50 s group velocity maps constructed directly with regional tomography agree well with those computed from phase-velocity maps which were themselves constructed globally. Thus, information from disparate data types appears to provide consistent constraints on the 3-D structure of the earth. Other researchers have demonstrated that broadband group- and phase-velocity maps can be simultaneously inverted for 3-D structure on both regional (e.g., VILLASEÑOR et al., 2001) and global scales (e.g., *Stevens and McLaughlin* in this volume). In addition, VILLASEÑOR et al. (2001) established that the resulting model of the mantle agrees well with a recent model constructed with teleseismic body wave travel times (e.g., SPAKMAN and BIJWAARD, 1998).

Tomographic maps have four principal applications: to detect and extract surface waves from noisy records, to help discriminate nuclear explosions from other sources of seismic energy, to characterize sources, and to be used as data in inversion for the shear-velocity structure of the crust and uppermost mantle.

First, the focus of the CTBT on small events makes the detection of seismic signals and the extraction of useful information a crucial task. The detection and extraction of surface waves is facilitated by using phase-matched filters (e.g., HERRIN and GOFORTH, 1977; HERRMANN and RUSSELL, 1990), which are designed to compensate for the dispersion of the surface wave-train. In this volume, *Levshin and*

Ritzwoller argue that to perform optimally these filters need to be tuned regionally with group velocity delays which may be efficiently summarized as group travel-time correction surfaces for each monitoring station. They and *Barmin, Ritzwoller, and Levshin* present examples in this volume of group velocity correction surfaces for a few stations in Central Asia. *Levshin and Ritzwoller* also demonstrate how these correction surfaces can be used to detect weak surface-wave signals buried in noise.

The second important application of surface-wave observations is in the discrimination of nuclear explosions from numerous other natural and human-made seismic phenomena. The surface-wave magnitude in combination with the body-wave magnitude obtained for each event is then used as part of the well known $m_b : M_s$ discriminant. After the surface wave has been extracted from the observed waveform, the amplitude is typically measured in a window centered around 20 s period from which the surface-wave magnitude M_s is inferred. The exact procedure varies depending on the monitoring agency. There has been considerable debate concerning the appropriate distance correction to use in computing M_s (e.g., MARSHALL and BASHAM, 1972; VON SEGGERN, 1975; HERAK and HERAK, 1993), and for paths less than ~20° it is common practice not to use surface-wave amplitude measurements. The effect has been to constrain M_s to relatively large events for which surface waves are well observed beyond epicentral distances of 20°. The prototype International Data Centre (PIDC) recently adopted a new M_s-distance relation of REZAPOUR and PEARCE (1998) which appears to justify the use of surface-wave amplitudes at all distances below 100 degrees and hence extends M_s to smaller events. The procedure for estimating M_s at the PIDC is thoroughly described in this volume by *Stevens and McLaughlin* who show that the automated methods that they developed and that are now in place at the PIDC demonstrate a detection threshold approximately one magnitude unit lower than those of other global networks that use visual detection of surface waves. They also argue that continuing improvements in 3-D earth models will advance surface-wave identification and reduce the magnitude threshold further. Observational efforts aimed at producing group velocity maps at periods well below 20 s, such as those of *Pasyanos, Walter, and Hazler* and *Mokhtar, Ammon, Herrmann, and Ghalib* in this volume, hold out the hope to reduce the period at which M_s is measured below 20 s. The effect envisioned will be to reduce the size of events further for which reliable M_s measurements can be obtained. *Levshin and Ritzwoller* sound a cautionary note by demonstrating how spectral amplitudes below 20 s period vary strongly on the relatively small scales across the Kirghiz Seismic Network (KNET). Also in this volume *Herak, Panza, and Costa* demonstrate how estimates of M_s depend on source depth because the maximum observed amplitude near 20 s period is a function of the excitation of overtones. They postulate a correction to M_s depending on earthquake depth, which they argue is important for calibrating the M_s scale but is of little practical significance for the $m_b : M_s$ discriminant because the correction is zero for events shallower than 20 km.

The third application of surface-wave observations relevant to monitoring the CTBT concerns source characterization, because source depth and the source mechanism may together form a useful discriminant. For example, events that are deeper than 2–3 km below the earth's surface are most probably natural phenomena. The difficulty is in discriminating very shallow natural events from explosions. In this volume, *Bukchin, Mostinsky, Egorkin, Levshin, and Ritzwoller* argue that source depth and both the isotropic and nonisotropic components of the moment tensor can be estimated if body-wave polarization data and surface-wave amplitudes are considered simultaneously. (A good introduction to moment tensor estimation is presented by DZIEWONSKI *et al.*, 1981). They test the hypothesis that events missing the isotropic component are earthquakes and near-surface events which have a significant isotropic component are explosions, by analyzing data following events on and near the Lop Nor test site in China. They argue that tests of this potential discriminant are encouraging, but a much larger data set of earthquakes must be considered to determine the false alarm rate (i.e., the percentage of earthquakes with surface-wave amplitudes consistent with a substantial isotropic component of the moment tensor).

The fourth and final application of surface-wave dispersion data regards improving and focusing regional models. The inversion of broadband regional surface-wave maps can provide a more detailed picture of the earth's lithosphere than can emerge from globally propagating surface waves or observations of teleseismic body waves alone. By raytracing through such models, it is possible to construct body wave travel-time correction surfaces for a set of monitoring stations for use in improving location estimates of weak seismic events (e.g., VILLASEÑOR *et al.*, 2000). This is especially important in regions of complex structure, where locations based on global models or coarse regional models are invariably biased in the absence of good azimuthal coverage. In this volume, *Hazler, Sheehan, McNamara, and Walter* present one-dimensional shear-velocity models for several tectonic regions of North Africa by inverting average Rayleigh wave group velocity dispersion curves from 10 s to 160 s period. This set of 1-D models is posited as a replacement for a 3-D model which, the authors argue, cannot be reliably estimated given the poor data distribution traversing most of North Africa. On a vastly larger scale, *Stevens and McLaughlin* present the results of inverting group and phase velocities on a 5° × 5° grid worldwide.

The papers presented in this volume cut across essentially all of the major applications of surface waves to monitoring the CTBT. We believe that for this reason the volume will provide a reasonable introduction to the state of research in this area and act as a guide for further exploration.

REFERENCES

BIJWAARD, H., SPAKMAN, W., and ENGDAHL, E. R. (1998), *Closing the Gap between Regional and Global Travel-time Tomography*, J. Geophys. Res. *13*, 30,055–30,078.
CARA, M. (1973), *Filtering of dispersed wave trains*, Geophys. J. R. astr. Soc. *33*, 65–80.

CURTIS, A., TRAMPERT, J., SNIEDER, R., and DOST, B. (1998), *Eurasian Fundamental Mode Surface Wave Phase Velocities and their Relationship with Tectonic Structures*, J. Geophys. Res. *103*, 26,919–26,947.

DITMAR, P. G., and YANOVSKAYA, T. B. (1987), *A Generalization of the Backus-Gilbert Method for Estimation of Lateral Variations of Surface-wave Velocity* (in Russian), Izv. Akad. Nauk SSSR, Fiz. Zeml. *6*, 30–60.

DZIEWONSKI, A., BLOCH, S., and LANDISMAN, N. (1969), *A Technique for the Analysis of Transient Seismic Signals*, Bull. Seismol. Soc. Am. *59*, 427–444.

DZIEWONSKI, A. M., CHOU, T.-A., and WOODHOUSE, J. H. (1981), *Determination of Earthquake Source Parameters from Waveform Data for Studies of Global and Regional Seismicity*, J. Geophys. Res. *86*, 2825–2952.

EKSTRÖM, G., TROMP, J., and LARSON, E. W. F. (1997), *Measurements and Global Models of Surface Wave Propagation*, J. Geophys. Res. *102*, 8137–8158.

GRIOT, D. A., MONTAGNER, J. P., and TAPPONIER, P. (1998), *Surface-wave Phase-velocity Tomography and Azimuthal Anisotropy in Central Asia*, J. Geophys. Res. *103*, 21,215–21,232.

HADOUCHE, O., and ZÜRN, W. (1992), *On the Structure of the Crust and Upper Mantle beneath the Afro-Arabian Region from Surface-wave Dispersion*, Tectonophys. *209*, 179–196.

HERAK, M., and HERAK, D. (1993), *Distance Dependence of M_s and Calibrating Function for 20 Second Rayleigh Waves*, Bull. Seismol. Soc. Am. *83*, 1881–1892.

HERRIN, E., and GOFORTH, T. (1977), *Phase-matched Filters: Application to Study of Rayleigh Waves*, Bull. Seismol. Soc. Am. *67*, 1259–1275.

HERRMANN, R. B., and RUSSELL, D. R. (1990), *Ground Roll: Rejection Using Adaptive Phase-matched Filters*, Geophysics *55*, 776–781.

KODERA, K., DE VILLEDARY, C., and GENDRIN, R. (1976), *A New Method for the Numerical Analysis of Non-stationary Signals*, Phys. Earth Planet. Int. *12*, 142–150.

KNOPOFF, L., and Schwab, F. A. (1968), *Apparent Initial Phase of a Source of Rayleigh Waves*, J. Geophys. Res. *73*, 755–760.

LASKE, G., (1995), *Global Observations of Off-great-circle Propagation of Long-period Surface Waves*, J. Geophys. Res. *90*, 605–621.

LASKE, G., and MASTERS, G. (1996), *Constraints on Global Phase Velocity Maps from Long-period Polarization Data*, J. Geophys. Res. *101*, 16,059–16,075.

LEVSHIN, A. L., PISARENKO, V. F., and POGREBINSKY, G. A. (1972), *On a Frequency-time Analysis of Oscillations*, Ann. Geophys. *28*, 211–218.

LEVSHIN, A. L., YANOVSKAYA, T. B., LANDER, A. V., BUKCHIN, B. G., BARMIN, M. P., RATNIKOVA, L. I., and ITS, E. N., *Surface waves in vertically inhomogeneous media*. In *Seismic Surface Waves in a Laterally Inhomogeneous Earth* (ed. Keilis-Borok, V. I.) (Kluwer Academic Publisher, Dordrecht, 1989) pp. 131–182.

LEVSHIN, A. L., RATNIKOVA, L., and BERGER, J. (1992), *Peculiarities of Surface-wave Propagation across Central Eurasia*, Bull. Seismol Soc. Am. *82*, 2464–2493.

LEVSHIN, A. L., RITZWOLLER, M. H., and RATNIKOVA, L. I. (1994), *The Nature and Cause of Polarization Anomalies of Surface Waves Crossing Northern and Central Eurasia*, Geophys. J. Int. *117*, 577–590.

LEVSHIN, A. L., and RITZWOLLER M. H. (1995), *Characteristics of Surface Waves Generated by Events on and near the Chinese Nuclear Test Site*, Geophys. J. Int. *123*, 131–149.

LEVSHIN, A. L., RITZWOLLER, M. H., BARMIN, M. P., RATNIKOVA, L. I., and PADGETT, C. A. (1998), *Automated surface wave analysis using phase-matched filters from dispersion maps*, Proceedings of the 20th Seismic Research Symposium on Monitoring a CTBT, 466–475.

LEVSHIN, A. L., RITZWOLLER, M. H., and RESOVSKY, J. S. (1999), *Source Effects on Surface-wave Group travel times and group-velocity maps*, Phys. Earth Planet. Int. *115*, 293–312.

LEVSHIN, A. L., RITZWOLLER, M. H., BARMIN, M. P., VILLASEÑOR, A., and PADGETT, C. A. (2001), *New Constraints on the Arctic Crust and Uppermost Mantle: Surface-wave Group Velocities, P_n, and S_n*, Phys. Earth Planet. Int. *123*, 185–204.

MARSHALL, P. D., and BASHAM, P. W. (1972), *Discrimination between Earthquakes and Underground Nuclear Explosions Employing an Improved M_s Scale*, Geophys. J. R. astr. Soc. *28*, 431–458.

MUYZERT, E., and SNIEDER, R. (1996), *The Influence of Errors in Source Parameters on Phase-velocity Measurements of Surface Waves*, Bull. Seismol. Soc. Am. *86*, 1863–1872.

POLLITZ, F. F. (1994), *Surface-wave Scattering from Sharp Lateral Heterogeneities*, J. Geophys. Res. *99*, 21,891–21,909.

REZAPOUR, M., and PEARCE, R. G. (1998), *Bias in Surface-wave Magnitude M_s Due to Inadequate Distance Corrections*, Bull. Seismol. Soc. Am. *88*, 43–61.

RITZWOLLER, M. H., LEVSHIN, A. L., SMITH, S. S., and LEE, C. S. (1995), *Making accurate continental broadband surface-wave measurements*, Proceedings of the 17th Seismic Research Symposium on Monitoring a CTBT, pp. 482–490.

RITZWOLLER, M. H., and LEVSHIN, A. L. (1998), *Eurasian Surface-wave Tomography: Group Velocities*, J. Geophys. Res. *103*, 4839–4878.

RITZWOLLER, M. H., LEVSHIN, A. L., RATNIKOVA, L. I., and EGORKIN, A. A., Jr., (1998), *Intermediate Period Group Velocity Maps across Central Asia, Western China, and Parts of the Middle East*, Geophys. J. Int. *134*, 315–328.

RODGERS, A. J., WALTER, W. R., MELLORS, R. J., AL-AMRI, A. M. S., and ZHANG, Y. S. (1999), *Lithospheric Structure of the Arabian Shield and Platform from Complete Regional Waveform Modeling and Surface-wave Group Velocities*, Geophys. J. Int. *138*, 871–878.

RUSSELL, D. W., HERRMANN, R. B., and HWANG, H. (1988), *Application of Frequency-variable Filters to Surface-wave Amplitude Analysis*, Bull. Seismol. Soc. Am. *78*, 339–354.

SAMBRIDGE, M., BRAUN, J., and McQUEEN, H. (1995), *Geophysical Parameterization and Interpolation of Irregular Data Using Natural Neighbors*, Geophys. J. Int. 837–857.

SCHULTZ, C., MYERS, S., HIPP, J., and YOUNG, C. (1998), *Nonstationary Bayesian Kriging: Application of Spatial Corrections to improve Seismic Detection, Location, and Identification*, Bull. Seismol. Soc. Am. *88*, 1275–1288.

SPAKMAN, W., and BIJWAARD, H. (1998), *Irregular Cell Parameterization of Tomographic Problems*, Ann. Geophys. *16*, 18.

STEVENS, J. L., and DAY, S. M. (1985), *The Physical Basis of the $m_b : M_s$ and Variable Frequency Magnitude Methods for Earthquake/Explosion Discrimination*, J. Geophys. Res. *90*, 3009–3020.

STEVENS, J. L., and McLAUGHLIN, K. L. (1997), *Improved methods for regionalized surface-wave analysis*, Proceedings 17th Annual Seismic Research Symposium on Monitoring a CTBT, pp. 171–180.

TRAMPERT, J., and WOODHOUSE, J. (1996), *High Resolution Global Phase-velocity Distributions*, Geophys. Res. Lett. *23*, 21–24.

VAN DER HEIJST, J. J., and WOODHOUSE, J. W. (1999), *Global High Resolution Phase-velocity Distribution of Overtone and Fundamental Mode Surface Waves Determined by Mode Branch Stripping*, Geophys. J. Int. *137*, 601–620.

VON SEGGERN, D. H. (1975), *Distance-amplitude relationships for long-period P, S, and LR from measurements on recordings of the long-period experimental stations*, Teledyne Geotech Report SDAC-TR-75-15, Defense Advanced Research Projects Agency, September, submitted.

VDOVIN, O. Y. (1999), *Surface-Wave Tomography of South America and Antarctica*, Ph.D. Thesis, Department of Physics, University of Colorado at Boulder.

VDOVIN, O. Y., RIAL, J. A., LEVSHIN, A. L., and RITZWOLLER, M. H. (1999), *Group-velocity Tomography of South America and the Surrounding Oceans*, Geophys. J. Int. *136*, 324–330.

VILLASEÑOR, A., RITZWOLLER, M. H., LEVSHIN, A. L., BARMIN, M. P., ENGDAHL, E. R., SPAKMAN, W., and TRAMPERT, J. (2001), *Shear velocity structure of Central Eurasia from inversion of surface wave velocities*, Phys. Earth Planet. Int. *123*, 169–184.

VILLASEÑOR, A., RITZWOLLER, M. H., BARMIN, M. P., ENGDAHL, E. R., and LEVSHIN, A. L. (2000), *Computation of travel times and station correction surfaces in Eurasia using three dimensional velocity models*, Proceedings of the 22nd Seismic Reasearch Symposium on Monitoring a CTBT, *II*, 453–462.

WALTER, W. R., and RITZWOLLER, M. H. (1998), *Summary report on the Workshop on the U.S. Use of Surface Waves for Monitoring the CTBT, UCRL-ID-131835*, Lawrence Livermore National Laboratory, 16 pp.

WANG, Z., and DAHLEN, F. A. (1995), *Validity of Surface-wave Ray Theory on a Laterally Heterogeneous Earth*, Geophys. J. Int. *123*, 757–773.

WANG, Z., TROMP, J., and EKSTRÖM, G. (1998), *Global and Regional Surface-wave Inversion: A Spherical-spline Parameterization*, Geophys. Res. Lett. *25*, 207–210.

WOODHOUSE, J. H., and DZIEWONSKI, A. M. (1984), *Mapping the Upper Mantle: Three-dimensional Modelling of Earth Structure by Inversion of Seismic Waveforms*, J. Geophys. Res. *89*, 5953–5986.

WU, F. T., LEVSHIN, A. L., and KOZHEVNIKOV, V. M. (1997), *Rayleigh Wave-group Velocity Tomography of Siberia, China, and the Vicinity*, Pure appl. geophys. *149*, 447–473.

YANOVSKAYA, T. B. and DITMAR, P. G. (1990), *Smoothness Criteria in Surface-wave Tomography*, Geophys. J. Int. *102*, 63–72.

YANOVSKAYA, T. B. and ANTONOVA, L. M. (2000), *Lateral Variations in the Structure of the Crust and Upper Mantle in the Asia Region from Data on Group Velocities of Rayleigh Waves*, Fizika Zemli, Izv. Russ. Acad. Sci. *36*(2), 121–128.

ZHANG, U.-S., and LAY, T. (1996), *Global Surface-wave Phase Velocity Variations*, J. Geophys. Res. *101*, 8415–8436.

Surface Wave Tomography:
Methods and Results

Pure appl. geophys. 158 (2001) 1351–1375
0033–4553/01/081351–25 $ 1.50 + 0.20/0

┃Pure and Applied Geophysics

A Fast and Reliable Method for Surface Wave Tomography

M. P. Barmin,[1] M. H. Ritzwoller,[1] and A. L. Levshin[1]

Abstract — We describe a method to invert regional or global scale surface-wave group or phase-velocity measurements to estimate 2-D models of the distribution and strength of isotropic and azimuthally anisotropic velocity variations. Such maps have at least two purposes in monitoring the nuclear Comprehensive Test-Ban Treaty (CTBT): (1) They can be used as data to estimate the shear velocity of the crust and uppermost mantle and topography on internal interfaces which are important in event location, and (2) they can be used to estimate surface-wave travel-time correction surfaces to be used in phase-matched filters designed to extract low signal-to-noise surface-wave packets.

The purpose of this paper is to describe one useful path through the large number of options available in an inversion of surface-wave data. Our method appears to provide robust and reliable dispersion maps on both global and regional scales. The technique we describe has a number of features that have motivated its development and commend its use: (1) It is developed in a spherical geometry; (2) the region of inference is defined by an arbitrary simple closed curve so that the method works equally well on local, regional, or global scales; (3) spatial smoothness and model amplitude constraints can be applied simultaneously; (4) the selection of model regularization and the smoothing parameters is highly flexible which allows for the assessment of the effect of variations in these parameters; (5) the method allows for the simultaneous estimation of spatial resolution and amplitude bias of the images; and (6) the method optionally allows for the estimation of azimuthal anisotropy.

We present examples of the application of this technique to observed surface-wave group and phase velocities globally and regionally across Eurasia and Antarctica.

Key words: Surface waves, group velocity, tomography, seismic anisotropy.

1. Introduction

We present and discuss a method to invert surface-wave dispersion measurements (frequency-dependent group or phase velocity) on regional or global scales to produce two-dimensional (2-D) isotropic and azimuthally anisotropic maps of surface-wave velocities. Such "tomographic" maps represent a local spatial average of the phase or group velocity at each location on the map and summarize large volumes of surface-wave dispersion information in a form that is both useful and easily transportable. Dispersion information in this form can be applied naturally to a number of problems relevant to monitoring the nuclear Comprehensive Test-Ban

[1] Department of Physics, University of Colorado at Boulder, Boulder, CO 80309-0390, USA.
E-mail: levshin@lemond.colorado.edu

Treaty (CTBT); For example, (1) to create phase-matched filters (e.g., HERRIN and GOFORTH, 1977; RUSSELL *et al.*, 1988; LEACH *et al.*, 1998; LEVSHIN and RITZWOLLER, 2001, this volume) designed to detect weak surface-wave signals immersed in ambient and signal-generated noise as a basis for spectral amplitude measurements essential to discriminate explosions from earthquakes (e.g., STEVENS and DAY, 1985; STEVENS and MCLAUGHLIN, 1997) and (2) in inversions to estimate the shear-velocity structure of the crust and upper mantle (e.g., VILLASEÑOR *et al.*, 2001) which is useful to improve regional event locations. The method we discuss here is designed to produce accurate and detailed regional surface-wave maps efficiently and reliably, as well as to provide information about the quality of the maps. The method may be applied, perhaps with a few extensions, to other 2-D inverse problems such as P_n and S_n tomography (e.g., LEVSHIN *et al.*, 2001).

We note, as a preface to further discussion, that the relationship between observed seismic waveforms and an earth model is not linear. Thus, the problem of using surface-wave data to constrain the structure of the crust and upper mantle is nonlinear. In surface-wave inversions, however, the inverse problem is typically divided into two parts: A nearly linear part to estimate 2-D dispersion maps and a nonlinear part in which the dispersion maps are used to infer earth structure. It is the nearly linear part that we call surface-wave tomography and that is the subject of this paper. Some surface-wave inversion methods linearize the relation between the seismic waveforms and an earth model (e.g., NOLET, 1987; SNIEDER, 1988; MARQUERING *et al.*, 1996) and iteratively estimate the earth model. Therefore, these methods do not estimate dispersion maps on the way to constructing structural models. We take the path through the dispersion maps for the following reasons.

- Surface-wave dispersion maps, like a seismic model, summarize large volumes of data in a compact form, but remain closer to the data than the models.
- They are less prone to subjective decisions made during inversion and contain fewer assumptions (both hidden and explicit).
- Because of the foregoing, dispersion maps are more likely than models to be consumed and utilized by other researchers.
- Dispersion maps are directly applicable to detect and extract surface waves from potentially noisy records, which is important in discriminating explosions from earthquakes for CTBT monitoring.

On the negative side, dispersion maps contain only part of the information concerning earth structure in the seismogram, are the products of inversions themselves, and contain uncertainties due to both observational and theoretical errors.

There are a number of surface-wave tomographic techniques currently in use by several research groups around the world. These techniques differ in geometry (i.e., Cartesian versus spherical), model parameterization (e.g., global versus local basis functions), certain theoretical assumptions (particularly about wave paths and scattering), the regularization scheme, and whether azimuthal anisotropy can be estimated simultaneously with the isotropic velocities. Because surface-wave tomographic inversions are invariably ill-posed, the regularization scheme is the focal point of any inversion method. There is a large, general literature on ill-posed linear or linearized inversions that applies directly to the surface-wave problem (e.g., TIKHONOV, 1963; BACKUS and GILBERT, 1968, 1970; FRANKLIN, 1970; AKI and RICHARDS, 1980; TARANTOLA and VALETTE, 1982; TARANTOLA, 1987; MENKE, 1989; PARKER, 1994; TRAMPERT, 1998). We do not intend to extend this literature, rather the purpose of this paper is to describe one useful path through the numerous options available to an inversion method. Our method appears to provide robust and reliable dispersion maps on both global and regional scales.

The surface-wave tomographic method we describe here has the following characteristics:

- *Geometry*: Spherical;
- *Scale*: The region of inference is defined by an arbitrary simple closed curve;
- *Parameterization*: Nodes are spaced at approximately constant distances from one another, interpolation is based on the three nearest neighbors;
- *Theoretical Assumptions*: Surface waves are treated as rays sampling an infinitesimal zone along the great circle linking source and receiver, scattering is completely ignored;
- *Regularization*: Application of spatial smoothness (with a specified correlation length) plus model amplitude constraints, both spatially variable and adaptive, depending on data density;
- *Azimuthal Anisotropy*: May optionally be estimated with the isotropic velocities.

The theoretical assumptions that we make are common in most of surface-wave seismology. The method we describe generalizes naturally to non-great circular paths, if they are known, with finite extended Fresnel zones (e.g., PULLIAM and SNIEDER, 1998). The incorporation of these generalizations into surface-wave tomographic methods is an area of active research at this time. The use of the scattered wavefield (the surface wave coda) is also an area of active research (e.g., POLLITZ, 1994; FRIEDERICH, 1998), but usually occurs within the context of the production of a 3-D model rather than 2-D dispersion maps.

The choices of parameterization and regularization require further comment.

1.1. Parameterization

There are four common types of basis functions used to parameterize velocities in surface-wave tomography: (1) Integral kernels (the Backus-Gilbert approach), (2) a truncated basis (e.g., polynomial, wavelet, or spectral basis functions), (3) blocks, and (4) nodes (e.g., TARANTOLA and NERSESSIAN, 1984). In each of these cases, the tomographic model is represented by a finite number of unknowns. Blocks and nodes are local whereas wavelets and polynomials are global basis functions. (To the best of our knowledge, wavelets have not yet been used in surface-wave tomography.) Backus-Gilbert kernels are typically intermediary between these extremes. Blocks are

2-D objects of arbitrary shape with constant velocities and are typically packed densely in the region of study. They are typically regularly shaped or sized, although there are notable exceptions (e.g., SPAKMAN and BIJWAARD, 1998, 2001). Nodes are discrete spatial points, not regions. A nodal model is therefore defined at a finite number of discrete points and values in the intervening spaces are determined by a specific interpolation algorithm in the inversion matrix and travel-time accumulation codes. Nodes are not necessarily spaced regularly. The ability to adapt the characteristics of these basis functions to the data distribution and other a priori information is a desirable characteristic of any parameterization, and is typically easier with local than with global basis functions. Blocks can be thought of as nodes with a particularly simple interpolation scheme. Thus we use nodes rather than blocks because of their greater generality.

To date, most surface-wave travel time tomographic methods have been designed for global application and have utilized truncated spherical harmonics or 2-D B-splines as basis functions to represent the velocity distribution (e.g., NAKANISHI and ANDERSON, 1982; MONTAGNER and TANIMOTO, 1991; TRAMPERT and WOODHOUSE, 1995, 1996; EKSTRÖM *et al.*, 1997; LASKE and MASTERS, 1996; ZHANG and LAY, 1996). There are two notable exceptions. The first is the work of DITMAR and YANOVSKAYA (1987) and YANOVSKAYA and DITMAR (1990) who developed a 2-D Backus-Gilbert approach utilizing first-spatial gradient smoothness constraints for regional application. This method has been extensively used in group velocity tomography (e.g., LEVSHIN *et al.*, 1989; WU and LEVSHIN, 1994; WU *et al.*, 1997; RITZWOLLER and LEVSHIN, 1998; RITZWOLLER *et al.*, 1998; VDOVIN *et al.*, 1999) for studies at local and continental scales. The main problem is that the method has been developed in Cartesian coordinates, and sphericity is approximated by an inexact earth flattening transformation (YANOVSKAYA, 1982; JOBERT and JOBERT, 1983) which works well only if the region of study is sufficiently small (roughly less than one-tenth of the earth's surface). YANOVSKAYA and ANTONOVA (2000) recently extended the method to a spherical geometry, however. The second is the irregular block method of SPAKMAN and BIJWAARD (2001).

We prefer local to global basis functions due to the simplicity of applying local damping constraints, the ability to estimate regions of completely general shape and size, and the ease by which one can intermix regions with different grid spacings. For example, with local basis functions it is straightforward to allow damping to vary spatially, but it is considerably harder to target damping spatially with global basis functions. This spatially targeted damping is a highly desirable feature, particularly if data distribution is inhomogeneous.

1.2. Regularization

The term 'regularization', as we use it, refers to constraints placed explicitly on the estimated model during inversion. These constraints appear in the "penalty function" that is explicitly minimized in the inversion. We prefer this term to 'damping' but take the terms to be roughly synonymous and will use them interchangeably. Regularization commonly involves the application of some combination of constraints on model amplitude, the magnitude of the perturbation from a reference state, and on the amplitude of the first and/or second spatial gradients of the model. It is typically the way in which *a priori* information about the estimated model is applied and how the effects of inversion instabilities are minimized. The strength of regularization or damping is usually something specified by the user of a tomographic code, but may vary in an adaptive way with information regarding data quantity, quality, and distribution and relating to the reliability of the reference model or other *a priori* information. As alluded to above, the practical difference between local and global basis functions manifests itself in how regularization constraints (e.g., smoothness) are applied as well as the physical meaning of these constraints.

As described in section 2, the regularization scheme that we have effected involves a penalty function composed of a spatial smoothing function with a user-defined correlation length and a spatially variable constraint on the amplitude of the perturbation from a reference state. The weight of each component of the penalty function is user specified, but the total strength of the model norm constraint varies with path density. Our experience indicates that Laplacian or Gaussian smoothing methods are preferable to gradient smoothing methods. The first-spatial gradient attempts to produce models that are locally flat, not smooth or of small amplitude. This works well if data are homogeneously distributed, but tends to extend large amplitude features into regions with poor data coverage and conflicts with amplitude penalties if applied simultaneously. The model amplitude constraint smoothly blends the estimated model into a background reference in regions of low data density such as the areas on the fringe of the region under study. In such areas the path density is very low, and the velocity perturbations will be automatically overdamped due to amplitude constraints. The dependence on data density is also user specified.

2. Surface-wave Tomography

Using ray theory, the forward problem for surface-wave tomography consists of predicting a frequency-dependent travel time, $t_{R/L}(\omega)$, for both Rayleigh (R) and Love (L) waves from a set of 2-D phase or group velocity maps, $c(\mathbf{r}, \omega)$:

$$t_{R/L}(\omega) = \int_p c_{R/L}^{-1}(\mathbf{r}, \omega)ds \ , \tag{1}$$

where $\mathbf{r} = [\theta, \phi]$ is the surface position vector, θ and ϕ are colatitude and longitude, and p specifies the wave path. The dispersion maps are nonlinearly related to the seismic structure of the earth, $\mathcal{M}(\mathbf{r}, z)$, where for simplicity of presentation we have assumed isotropy, and $\mathcal{M}(\mathbf{r}, z) = [v_s(z), v_p(z), \rho(z)](\mathbf{r})$ is the position dependent structure vector composed of the shear and compressional velocities and density. Henceforth, we drop the R/L subscript and, for the purposes of discussion here, do not explicitly discriminate between group and phase velocities or their integral kernels.

By surface-wave tomography we mean the use of a set of observed travel times $t^{obs}(\omega)$ for many different paths p to infer a group or phase velocity map, $c(\mathbf{r})$, at frequency ω. We assume that

$$t^{obs}(\omega) = t(\omega) + \epsilon(\omega) \ ,$$

where ϵ is an observational error for a given path. The problem is linear if the paths p are known. Fermat's Principle states that the travel time of a ray is stationary with respect to small changes in the ray location. Thus, the wave path will approximate that of a spherically symmetric model, which is a great-circle linking source and receiver. This approximation will be successful if the magnitude of lateral heterogeneity in the dispersion maps is small enough to produce path perturbations smaller than the desired resolution.

2.1. The Forward Problem

Because surface-wave travel times are inversely related to velocities, we manipulate equation (1) as follows. Using a 2-D reference map, $c_o(\mathbf{r})$, the travel-time perturbation relative to the prediction from $c_o(\mathbf{r})$ is:

$$\delta t = t - t_o = \int_p \frac{ds}{c} - \int_p \frac{ds}{c_o} = \int_p \frac{m}{c_o} \, ds \tag{2}$$

$$m = \frac{c_o - c}{c} \ , \tag{3}$$

where we have suppressed the $\mathbf{r} = [\theta, \phi]$ dependence throughout and have assumed that the ray-paths are known and are identical for both c and c_o.

For an anisotropic solid with a vertical axis of symmetry, the surface-wave velocities depend on the location $\mathbf{r}$ and the local azimuth ψ of the ray. In the case of a slightly anisotropic medium, SMITH and DAHLEN (1973) show that phase or group velocities can be approximated as:

$$c(\mathbf{r}) = c_I(\mathbf{r}) + c_A(\mathbf{r}) \tag{4}$$

$$c_I(\mathbf{r}) = A_0(\mathbf{r}) \tag{5}$$

$$c_A(\mathbf{r}) = A_1(\mathbf{r}) \cos(2\psi) + A_2(\mathbf{r}) \sin(2\psi)$$
$$+ A_3(\mathbf{r}) \cos(4\psi) + A_4(\mathbf{r}) \sin(4\psi) \ , \tag{6}$$

where c_I is the isotropic part of velocity, c_A is the anisotropic part, A_0 is the isotropic coefficient, and $A_1, \ldots, A_4$ are anisotropic coefficients. If we assume that $|c_A/c_I| \ll 1$ so that $(1 + c_A/c_I)^{-1} \approx (1 - c_A/c_I)$, and if the reference model c_o is purely isotropic, then by substituting equations (4)–(6) into (3) we get:

$$m \approx \frac{c_o - c_I}{c_I} - \frac{c_A}{c_I} - \left(\frac{c_A}{c_I}\right)\left(\frac{c_o - c}{c_I}\right)$$

$$\approx \frac{c_o - c_I}{c_I} - \frac{c_A}{c_I^2}c_o \ , \tag{7}$$

where the latter equality holds if $|c_A/c_I| \ll 1$ as before. Equation (7) can be rewritten as:

$$m(\mathbf{r}, \psi) = \sum_{k=0}^{n} \gamma_k(\psi)m_k(\mathbf{r}) \ , \tag{8}$$

where n should be 0, 2 or 4 for a purely isotropic model, a 2ψ anisotropic model, or a 4ψ anisotropic model, respectively, and γ_k and m_k are defined as follows:

$$\begin{aligned}
\gamma_0(\psi) &= 1 & m_0(\mathbf{r}) &= (c_o(\mathbf{r}) - c_I(\mathbf{r}))/c_I(\mathbf{r}) \\
\gamma_1(\psi) &= -\cos(2\psi) & m_1(\mathbf{r}) &= A_1(\mathbf{r})c_o(\mathbf{r})/c_I^2(\mathbf{r}) \\
\gamma_2(\psi) &= -\sin(2\psi) & m_2(\mathbf{r}) &= A_2(\mathbf{r})c_o(\mathbf{r})/c_I^2(\mathbf{r}) \\
\gamma_3(\psi) &= -\cos(4\psi) & m_3(\mathbf{r}) &= A_3(\mathbf{r})c_o(\mathbf{r})/c_I^2(\mathbf{r}) \\
\gamma_4(\psi) &= -\sin(4\psi) & m_4(\mathbf{r}) &= A_4(\mathbf{r})c_o(\mathbf{r})/c_I^2(\mathbf{r}) \ .
\end{aligned} \tag{9}$$

2.2. The Inverse Problem

Our goal is to estimate the vector function $\mathbf{m}(\mathbf{r}) = [m_0(\mathbf{r}), \ldots, m_n(\mathbf{r})]$ using a set of observed travel-time residuals d relative to the reference model $c_o(\mathbf{r})$:

$$d = \delta t^{obs} = t^{obs} - t_o = \int_p \frac{m}{c_o} ds + \epsilon \ . \tag{10}$$

From $\mathbf{m}(\mathbf{r})$ we can reconstruct $c_I, A_1, \ldots, A_4$ for substitution into equations (4)–(6):

$$c_I = \frac{c_o}{1 + m_0} \tag{11}$$

$$A_k = \frac{m_k}{1 + m_0}c_I \quad (k \neq 0) \ . \tag{12}$$

We define the linear functionals G_i as:

$$G_i(\mathbf{m}) = \sum_{k=0}^{n} \int_{p_i} (\gamma_k(\psi(\mathbf{r}))c_o^{-1}(\mathbf{r}))m_k(\mathbf{r})ds \ . \tag{13}$$

By substituting equations (8) and (13) into equation (10) for each path index ($1 \leq i \leq N$) we obtain the following:

$$d_i = \delta t_i^{\text{obs}} = G_i(\mathbf{m}) + \epsilon_i \ . \tag{14}$$

To estimate $\mathbf{m}$ we choose to minimize the following penalty function:

$$(\mathbf{G}(\mathbf{m}) - \mathbf{d})^T \mathbf{C}^{-1}(\mathbf{G}(\mathbf{m}) - \mathbf{d}) + \sum_{k=0}^{n} \alpha_k^2 \|F_k(\mathbf{m})\|^2 + \sum_{k=0}^{n} \beta_k^2 \|H_k(\mathbf{m})\|^2 \ , \tag{15}$$

where $\mathbf{G}$ is a vector of the functionals G_i. For an arbitrary function $f(\mathbf{r})$ the norm is defined as: $\|f(\mathbf{r})\|^2 = \int_S f^2(\mathbf{r}) \, d\mathbf{r}$.

The first term of the penalty function represents data misfit ($\mathbf{C}$ is the *a priori* covariance matrix of observational errors ϵ_i). The second term is the spatial smoothing condition such that

$$F_k(\mathbf{m}) = m_k(\mathbf{r}) - \int_S S_k(\mathbf{r}, \mathbf{r}')m_k(\mathbf{r}')d\mathbf{r}' \ , \tag{16}$$

where S_k is a smoothing kernel defined as follows:

$$S_k(\mathbf{r}, \mathbf{r}') = K_{0k} \exp\left(-\frac{|\mathbf{r} - \mathbf{r}'|^2}{2\sigma_k^2}\right) \tag{17}$$

$$\int_S S_k(\mathbf{r}, \mathbf{r}') \, d\mathbf{r}' = 1 \ , \tag{18}$$

and σ_k is spatial smoothing width or correlation length. The minimization of the expression in equation (16) explicitly ensures that the estimated model will approximate a smoothed version of the model.

The final term in the penalty function penalizes the weighted norm of the model,

$$H_k(\mathbf{m}) = \mathcal{H}(\rho(\mathbf{r}), \chi(\mathbf{r}))m_k \ , \tag{19}$$

where $\mathcal{H}$ is a weighting function that depends on local path density ρ for isotropic structure and a measure of local azimuthal distribution χ for azimuthal anisotropy. Thus, for $k = 0$, $\mathcal{H} = \mathcal{H}(\rho)$ and for $k = 1, \ldots, 4$, $\mathcal{H} = \mathcal{H}(\chi)$. Path density is defined as the number of paths intersecting a circle of fixed radius with center at the point $\mathbf{r}$. For isotropic structure we choose $\mathcal{H}$ to approach zero where path density is suitably high and unity in areas of poor path coverage. The function $\mathcal{H}(\rho)$ can be chosen in various ways. We use $\mathcal{H} = \exp(-\lambda\rho)$, where λ is a user-defined constant. An example is shown in Figure 1. To damp azimuthal anisotropy in regions with poor azimuthal coverage, we define $\chi(\theta, \phi)$ to measure the azimuthal distribution of ray paths at point (θ, ϕ). To find χ we construct a histogram of azimuthal distribution of raypaths in the vicinity of (θ, ϕ) for a fixed number n of azimuthal bins in the interval between $0°$ and $180°$, and evaluate the function

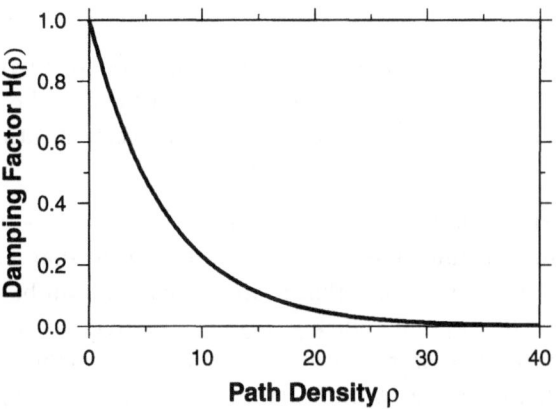

Figure 1

Example of the model norm weighting function, $\mathscr{H}(\rho)$ that we commonly use; e.g., as Figure 5. Here we choose the constant $\lambda \sim 0.147$ so that when path density (ρ) is less than about 20 paths per 50,000 km^2 damping toward the input reference model becomes increasingly strong.

$$\chi = \frac{\displaystyle\sum_{i=1}^{n} f_i}{n \max_i f_i} , \tag{20}$$

where f_i is the density of azimuths in the ith bin. Values of χ are in the range $1/n \le \chi \le 1$. $\chi \approx 1$ characterizes an almost uniform distribution of azimuths, and $\chi \approx 1/n$ is an indicator of the predominance of a single azimuthal direction (large azimuthal gap). We assume that the anisotropic coefficients cannot be determined reliably in regions where χ is less than ~ 0.3. Examples of a χ map and the histogram of azimuthal density, $f(\psi)$, are given in Section 3.4 (Figs. 8c,d).

Because $\mathbf{m}$ is a perturbation from a reference state, the effect of the third term in the penalty function is to merge the estimated model smoothly and continuously into the isotropic reference state in regions of poor data coverage. In regions of good coverage, this term has no effect so that the only regularization is the smoothness constraint represented by the second term in expression (15).

The user-supplied regularization constants, α_k and β_k, define the relative strengths of the three terms in the penalty function. The smoothing width or correlation length σ_k is also specified by the user. These parameters should be varied systematically in applying the method. In practice, we often estimate the isotropic and anisotropic maps simultaneously. In this case we normally use slightly different values of all three constants α_k, β_k, σ_k for the isotropic and anisotropic maps to make anisotropic maps more smooth. Typically $\alpha_1 = \alpha_2$ and $\alpha_3 = \alpha_4$, $\sigma_1 = \sigma_2$, $\sigma_3 = \sigma_4$, and $\beta_k = \beta$.

2.3. Discretization

The discretization of the equations in the preceding section involves two steps: (1) the formation of a discrete grid and the evaluation of the model on this grid and (2) the discretization of the penalty function. We discuss each in turn.

2.3a. Grid, Nearest Neighbors, and Interpolation

The goal is to generate a discrete grid with nodes that are approximately constantly spaced on a sphere such that nearest neighbors can be identified and the model evaluated at these points quickly. Significant advances in nonconstant grid generation have been made in recent years by several researchers (e.g., SAMBRIDGE *et al.*, 1995; SPAKMAN and BIJWAARD, 1998), particularly for application in 3-D body-wave tomography. The tomographic method that we describe is applicable irrespective of grid. However, because path coverage for regional surface waves tends to be less variable than for body waves, and surface-wave tomography is only in 2-D (e.g., BIJWAARD *et al.*, 1998; VAN DER HILST *et al.*, 1997; GRAND *et al.*, 1997; ZHOU, 1996), we find that a constant grid is sufficient for our purposes. Generalization to nonconstant grids for surface wave tomography is, however, a useful direction for future research (e.g., SPAKMAN and BIJWAARD, 2001).

We create a nearly constant grid on a sphere by performing a central projection onto the sphere of a grid on a reference cube, such that the cube and the sphere share a common center. The advantage of using this reference cube is that nearest neighbors on each of the six faces of the cube are identified trivially. Efficient neighbor identification for interpolation during model evaluation is important for travel-time accumulation, which occurs during the construction of the inversion matrix, and in the application of the smoothness constraint. Thus, this method is computationally very efficient because it imposes a natural ordering for the nodes on the sphere, which avoids the need for the creation and use of an adjacency matrix (e.g., SLOAN, 1987; SAMBRIDGE *et al.*, 1995). To ensure that the distances between nodes on the sphere are approximately constant, the grid on the reference cube must be non-constant. Without providing the details, Figure 2 demonstrates the mapping between the face of a cube and the spherical shell related to the face. The current method of grid generation guarantees that the areas defined by adjacent quadruples of nodes on the sphere differ by no more than 10% from the average area. One could produce a mapping with smaller variation in these areas, but this is good enough for our purposes.

The value of the model at each location on the sphere is evaluated from the values at the three nearest nodes. This is done by constructing Delaunay triangles from the set of nodes on the sphere (e.g., AURENHAMMER, 1991; BRAUN and SAMBRIDGE, 1997). Each triangle defines a flat plane between the three nodes at the vertices on the plane which is nearly the tangent plane to the sphere. We define a local Cartesian coordinate system on this plane and determine the distances between the point of

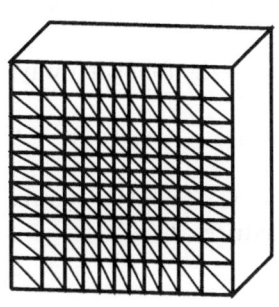 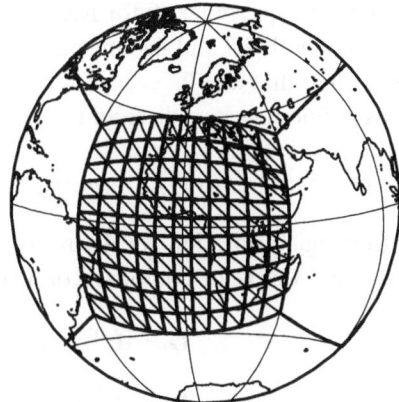

Figure 2
An example of the Delaunay triangulation on a sphere by defining a variable triangular grid on a reference cube and performing a central projection of the grid onto the sphere.

interest and the three defining nodes approximately. Typical internodal distances, even for global inversions, are 200 km or less, therefore this local Cartesian approximation is accurate enough for our purposes. Three-point linear interpolation is used to evaluate the model within each Delaunay triangle. Thus, the value of the model at some arbitrary point $\mathbf{r}$ can be expressed as the weighted sum of the values at the three neighboring nodes:

$$m_k(\mathbf{r}) = \sum_{j=1}^{M} m_k(\mathbf{r}_j) w_j(\mathbf{r}) \; , \tag{21}$$

where $\mathbf{r}_j$ are the locations of the M nodes (vortices of triangles) defining the model. The weights $w_j(\mathbf{r})$ are non-zero only inside the triangles surrounding $\mathbf{r}_j$ and linearly depend on two local coordinates inside the Delaunay triangle enclosing $\mathbf{r}$. The weights $w_j(\mathbf{r})$ form the set of the local basis functions such that the values of the basis functions range from 0 to 1 with a maximum value of 1 at the point $\mathbf{r}_j$.

2.3b. The Inversion Matrix

To construct the inversion matrix we must substitute equation (21) into expression (15). After integrating, the penalty function can be rewritten in matrix form as the sum of two quadratic forms,

$$(\mathbf{Gm} - \mathbf{d})^T \mathbf{C}^{-1}(\mathbf{Gm} - \mathbf{d}) + \mathbf{m}^T \mathbf{Qm} \; , \tag{22}$$

in which the second term is the regularization condition that includes both smoothness and model norm constraints. Let N be the number of data, n be the isotropic/anisotropic index (0 for isotropic, 2 for isotropic plus 2ψ anisotropic, 4 for isotropic plus both 2ψ and 4ψ anisotropic), and M be the number of nodes defining the model such that $k = 0, \ldots, n$, $i = 1, \ldots, N$, $j = 1, \ldots, M$.

Let us define now our discrete model as a vector $\mathbf{m}$ in the following way:

$$\mathbf{m} = (m_0(\mathbf{r}_1), m_0(\mathbf{r}_2), \ldots, m_0(\mathbf{r}_M), \ldots, m_n(\mathbf{r}_1), m_n(\mathbf{r}_2), \ldots, m_n(\mathbf{r}_M))^T .$$

Without changing notation we discretize $\mathbf{G}$ to create a $N \times (n+1)M$ matrix in the following way. Let $\mathbf{G}$ be composed of a set of n submatrices, $\mathbf{U}^k$,

$$\mathbf{G} = \begin{bmatrix} \mathbf{U}^0 \end{bmatrix} \quad (n = 0)$$

$$\mathbf{G} = \begin{bmatrix} \mathbf{U}^0 \vdots \mathbf{U}^1 \vdots \mathbf{U}^2 \end{bmatrix} \quad (n = 2) \tag{23}$$

$$\mathbf{G} = \begin{bmatrix} \mathbf{U}^0 \vdots \mathbf{U}^1 \vdots \mathbf{U}^2 \vdots \mathbf{U}^3 \vdots \mathbf{U}^4 \end{bmatrix} \quad (n = 4)$$

where $\mathbf{U}^k$ is defined as follows:

$$U_{ij}^k = \int_{p_i} \left(\gamma_k(\psi) c_0^{-1}(\mathbf{r}) \right) w_j(\mathbf{r}) ds . \tag{24}$$

The $(n+1)M \times (n+1)M$ regularization matrix $\mathbf{Q}$ is the result of discrete numerical integration of the last two terms in equation (15), and can be determined in the following way:

$$\mathbf{Q} = \mathbf{F}^T \mathbf{F} + \mathbf{H}^T \mathbf{H} , \tag{25}$$

where the smoothing constraint is incorporated within the $(n+1)M \times (n+1)M$ block-diagonal matrix $\mathbf{F}$ as follows:

$$\mathbf{F} = \begin{bmatrix} \alpha_0 \mathbf{F}^0 & \cdots & 0 & \cdots & 0 \\ \vdots & \vdots & \vdots & \vdots & \vdots \\ 0 & \cdots & \alpha_k \mathbf{F}^k & \cdots & 0 \\ \vdots & \vdots & \vdots & \vdots & 0 \\ 0 & \cdots & 0 & \cdots & \alpha_n \mathbf{F}^n \end{bmatrix} . \tag{26}$$

The $M \times M$ matrices $\mathbf{F}^k = \left(F_{jj'}^k \right)$ $(k = 0, \ldots, n; \, j, j' = 1, \ldots, M)$ are:

$$F_{jj'}^k = \begin{cases} 1 & j = j' \\ -S_k(\mathbf{r}_j, \mathbf{r}_{j'})/p_k & j \neq j', \end{cases} \quad p_k = \sum_{j'} S_k(\mathbf{r}_j, \mathbf{r}_{j'}) . \tag{27}$$

The model norm constraint is encoded within the $(n+1)M \times (n+1)M$ matrix $\mathbf{H}$ which consists of $(n+1)$ diagonal matrices $\mathbf{H}^k = \left(H_{jj'}^k \right)$:

$$\mathbf{H} = \begin{bmatrix} \beta_0\mathbf{H}^0 & \cdots & 0 & \cdots & 0 \\ \vdots & \vdots & \vdots & \vdots & \vdots \\ 0 & \cdots & \beta_k\mathbf{H}^k & \cdots & 0 \\ \vdots & \vdots & \vdots & \vdots & \vdots \\ 0 & \cdots & 0 & \cdots & \beta_n\mathbf{H}^n \end{bmatrix} \tag{28}$$

where

$$H_{jj'}^k = \begin{cases} \mathscr{H}\left(\rho(\mathbf{r}_j), \chi(\mathbf{r}_j)\right) & j = j' \\ 0 & j \neq j' \end{cases} \tag{29}$$

With these definitions the forward problem for the travel-time perturbation relative to an isotropic reference model is:

$$\delta\mathbf{t} = \mathbf{Gm} , \tag{30}$$

and the estimated model is:

$$\hat{\mathbf{m}} = \mathbf{G}^\dagger \mathbf{C}^{-1} \delta\mathbf{t} \tag{31}$$

where the inversion operator, $\mathbf{G}^\dagger$, is defined as follows:

$$\mathbf{G}^\dagger = \left(\mathbf{G}^T\mathbf{C}^{-1}\mathbf{G} + \mathbf{Q}\right)^{-1}\mathbf{G}^T . \tag{32}$$

2.4. Resolution Analysis

We agree with LEVEQUE et al. (1993) who argue that the estimation of the resolution matrix is generally preferable to checkerboard tests such as those performed by RITZWOLLER and LEVSHIN (1998). Note that from equations (30)–(32):

$$\hat{\mathbf{m}} = \mathbf{G}^\dagger \mathbf{C}^{-1} \delta\mathbf{t} = \left(\mathbf{G}^\dagger \mathbf{C}^{-1}\mathbf{G}\right)\mathbf{m} = \mathscr{R}\mathbf{m} \tag{33}$$

$$\mathscr{R} = \left(\mathbf{G}^T\mathbf{C}^{-1}\mathbf{G} + \mathbf{Q}\right)^{-1}\mathbf{G}^T\mathbf{C}^{-1}\mathbf{G} . \tag{34}$$

The matrix $\mathscr{R}$ is the resolution matrix. In this application each row of $\mathscr{R}$ is a resolution map defining the resolution at one spatial node. Thus, the resolution matrix is very large and the information it contains is somewhat difficult to utilize. We attempt to summarize the information in each resolution map by estimating two scalar quantities at each point: spatial resolution and amplitude bias.

To estimate spatial resolution we fit a cone to each resolution map. This cone approximates closely the response of the tomographic procedure to a δ-like perturbation at the target node. Figure 3a shows a δ-like input perturbation (the local basic function) at the specified spatial location. Figure 3b displays the resolution map for that spatial location for the 50 s Rayleigh wave. The cone that

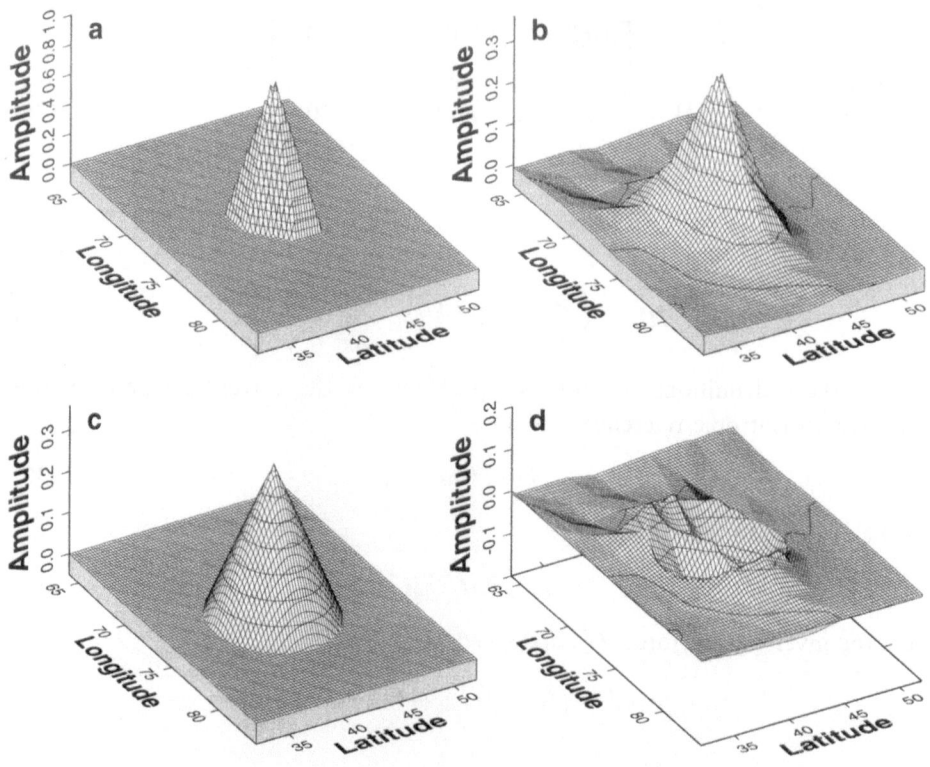

Figure 3
Graphical description of the resolution analysis. (a) Minimum sized function that can be estimated with a $2° \times 2°$ grid. The function is centered at 42°N latitude and 73°E longitude. (b) The row of the resolution matrix (a resolution map) for the point specified in (a) for the 50 s Rayleigh wave. (c) The cone that best fits the row of the resolution matrix shown in (b). A comparison of (a) with (b) and (c) demonstrates the spatial spreading produced in the tomographic procedure. (d) The difference between the resolution map and the best-fitting cone.

best fits the resolution surface for this point is shown in Figure 3c and the difference between the fit cone and the resolution map appears in Figure 3d. We define the resolution σ_R as the radius of the base of the fit cone. This value may be interpreted as the minimum distance at which two δ-shaped input anomalies (i.e., Fig. 3a) can be resolved on a tomographic map. Of course, resolution cannot be less than 2ℓ, where ℓ is the distance between the nodes. In the example in Figure 3, nodes are separated by 2 equatorial degrees ($\sim$222 km). Therefore, if σ_R is estimated to be less than 2ℓ or 444 km, we redefine resolution as $\sigma_R = 2\ell = 444$ km.

It is also useful to know how reliably the amplitude of the estimated anomalies may be determined. To do this we apply the appropriate row of the resolution matrix (eq. 33) associated with node (θ_0, ϕ_0) to a test model consisting of a cylinder of unit height with a diameter equal to $2\sigma_R$ centered at (θ_0, ϕ_0). We then define the amplitude of the fit surface as the average amplitude within σ_R of the center of the input cylinder. The relative difference between the input and estimated amplitudes is then taken as the amplitude bias estimate for this point on the map.

Examples of the estimated resolution and amplitude bias are shown in Figure 4 for the 20 s Rayleigh wave. Across much of Eurasia the 20 s Rayleigh wave data yields nearly optimal resolution for a $2° \times 2°$ grid spacing; about 450 km. Amplitude bias at the estimated resolution is typically within about ±10% at each spatial point. Near the periphery of the map where data coverage degrades, estimates of spatial resolution become unreliable but amplitude bias grows rapidly. Thus, amplitude bias is a more reliable means of estimating the reliability of dispersion maps in regions of extremely poor data coverage, using the method we describe here.

2.5. Computational Requirements

The following formulas summarize computational time (550 MHz, DEC Alpha) and memory requirements for a purely isotropic inversion:

$$t \sim 68 \left(\frac{k}{d}\right)^4 \text{ hours} \quad \text{(computational time in hours)} \tag{35}$$

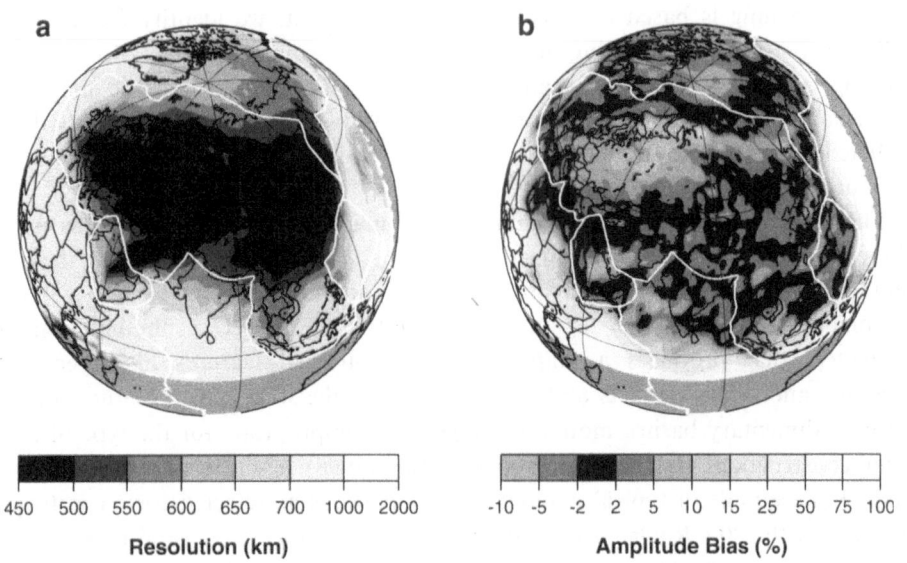

Figure 4
(a) Spatial resolution in km for the 20 s Rayleigh wave across Eurasia. Resolution depends on data coverage. In the central part of Eurasia the resolution is high ($\sim$ 450–500 km) in areas of high path density and degrades rapidly on the periphery of the region where path density (Fig. 5d) is low. (b) Amplitude bias for Rayleigh waves at 20 s period. Units of amplitude bias are percent such that 0% means that the cylindrical test function's amplitude has been fully recovered upon inversion. Amplitude bias across the region varies between about ±10% depending on path coverage.

$$M \sim 29 \left(\frac{k}{d}\right)^4 \text{Gb} \quad \text{(memory usage in Gb)} , \tag{36}$$

where k is the fraction of the earth's surface covered and d is the distance between nodesf in equatorial degrees. For Eurasian tomography about half of the earth's surface is covered ($k \sim 0.5$) and $d = 2$ degrees $= 222$ km, thus $t \sim 15$ minutes and $M \sim 115$ Mb.

3. Examples of Applications

3.1. Preliminaries

The technique described above has been extensively tested using different cell sizes, regularization parameters, and data sets from different regions of the world: Eurasia, Antarctica, South America, and the Arctic. The two conditions necessary for constructing reliable tomographic images are preliminary outlier rejection (data "cleaning") and a careful choice of regularization (or damping) parameters appropriate for a given path coverage.

Data cleaning is based on a two part process. First, we identify outliers in a preliminary way by clustering measurements into summary rays. Second, the resulting data are inverted for an overdamped, smooth tomographic map and outliers are then identified by comparing observed group travel times with those predicted from the smooth map. The usual percentage of the rejected measurements is about 2–3% of all observations.

The choice of regularization parameters is made after several iterations using different combinations of the parameters α_k, β_k, σ_k. The criteria for choosing the best combination are subjective and are based on common sense and some *a priori* information regarding the region under study. We select a combination of parameters that produces a map free from aphysical features like speckling, streaking, and other artifacts and that also reveals the well known features of the region (sedimentary basins, mountain ranges, etc.) appropriate for the type of map under construction. RITZWOLLER and LEVSHIN (1998) describe this procedure in detail. For example, a typical combination of parameters selected for an isotropic inversion of the 20 s Rayleigh wave data for Eurasia on a $2° \times 2°$ grid is: $\alpha_0 = 800$, $\beta_0 = 1$, and $\sigma_0 = 200$. The resulting maps are relatively insensitive to small (20–30%) changes in the damping parameters. Similar robustness of maps of azimuthal anisotropy to changes in the anisotropy damping parameters was demonstrated by VDOVIN (1999) for Antarctica, but in other areas of the world both the pattern and the amplitude of anisotropy change strongly with damping (e.g., Eurasia and the Arctic, LEVSHIN et al., 2001) as discussed further in section 3.4.

3.2. Regional Isotropic Group-velocity Maps

Regional group- and phase-velocity maps have been produced by a number of researchers (e.g., SUETSUGU and NAKANISHI, 1985; CURTIS et al., 1998; RITZWOLLER et al., 1999; and many others). Using the protocol described in section 3.1, we have recently constructed a set of isotropic group velocity maps of Eurasia and surrounding areas for Rayleigh and Love waves from 15 s to 200 s period. An example for the 20 s Rayleigh wave is shown in Figure 5a. As input data we used 12900 Rayleigh group velocity measurements obtained from records of both global (GSN, GEOSCOPE) and regional (CDSN, CSN, USNSN, MEDNET, Kirgiz and Kazak networks) networks. The basic characteristics of the measurement procedure, data control and weighting are described in detail in RITZWOLLER and LEVSHIN (1998). Because 20 s Rayleigh waves are most sensitive to upper crustal velocities, the corresponding group velocity map clearly shows the significant sedimentary basins across Eurasia and on the periphery of the Arctic Ocean as low velocity anomalies (e.g., Barents Sea shelf, western Siberian sedimentary complex, Pre-Caspian, South Caspian, Black Sea, Tadzhik Depression, the Tarim Basin, Dzhungarian Basin, Ganges Fan and Delta, etc.) There is qualitative agreement between the observed group velocity map and the prediction of a hybrid model composed of crustal structure from the model CRUST5.1 (MOONEY et al., 1998) and mantle velocities from the model S16B30 (MASTERS et al., 1996). The comparison is shown in Figures 5b,c. The estimated r.m.s. group velocity misfit at 20 s period is significantly less for our maps (0.08 km/s) than for the map computed from the model CRUST5.1/ S16B30 (0.14 km/s). The numbers for the 50 s Rayleigh wave are correspondingly 0.05 km/s and 0.16 km/s. Similar results are reported by RITZWOLLER and LEVSHIN (1998) which used the tomographic method of DITMAR and YANOVSKAYA (1987).

Figure 6 presents group travel-time correction surfaces for the 40 s Rayleigh wave for several stations in Central Asia. These surfaces summarize travel-time information in group velocity maps to be used to improve detection and discrimination schemes in nuclear monitoring (LEVSHIN and RITZWOLLER, 2001, this volume). LEVSHIN and RITZWOLLER (2001) also present an example of travel-time correction surfaces for the 20 s Rayleigh waves.

3.3. Global Isotropic Phase-velocity Maps

The tomographic method described above identifies the region of interest by requiring the user to define a simple closed curve on the sphere and identify a single point outside the contour that distinguishes the inside from the outside of the region of interest. If the contour is a very small circle surrounding the point, then the region of interest becomes nearly the entire sphere. In this way, our method can be used to produce global tomographic maps on a regular grid. An example is shown in Figure 7a, in which we have inverted the 100 s Rayleigh wave phase velocity data of TRAMPERT and WOODHOUSE (1995, 1996). Trampert and Woodhouse's map is shown

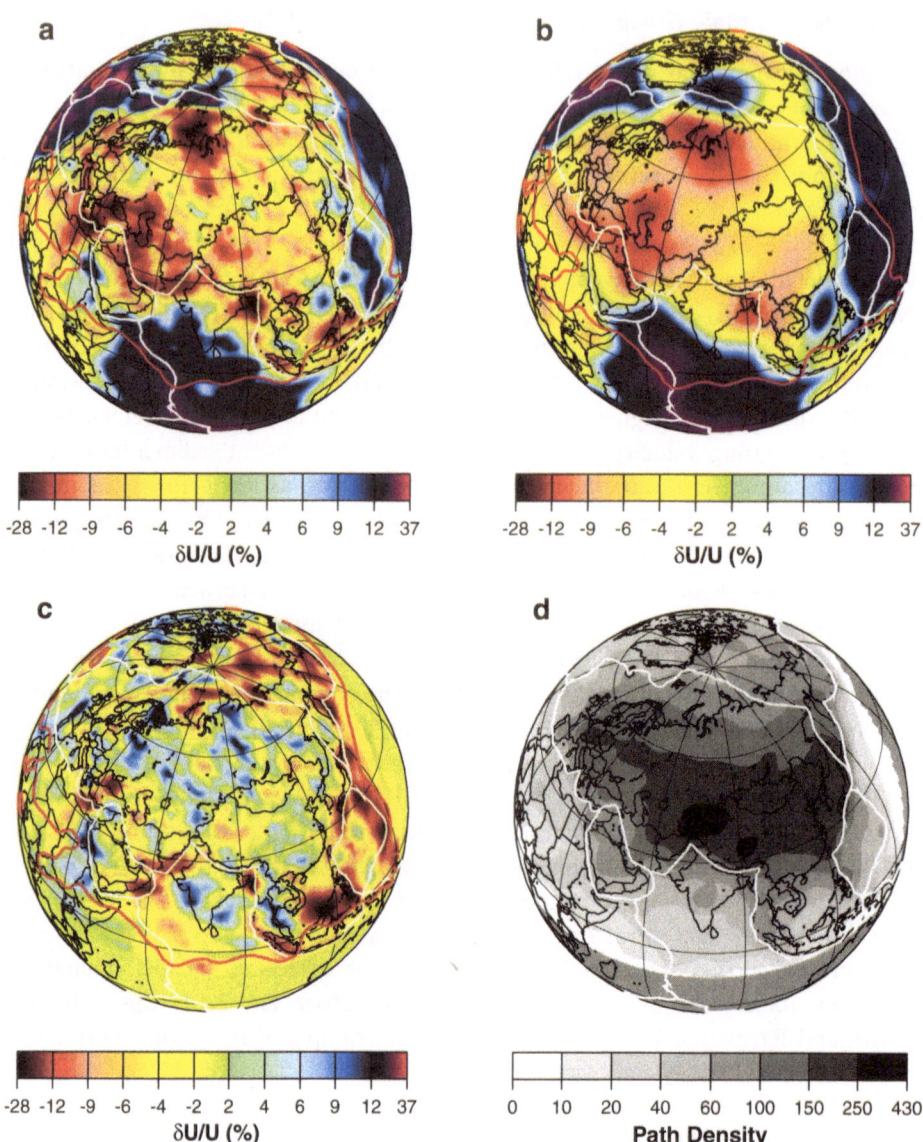

Figure 5

(a) The group-velocity map across Eurasia for the 20 s Rayleigh wave using the method described in this paper. A $2° \times 2°$ grid is used. (b) The group-velocity map computed from the smoothed version of the model CRUST5.1/S16B30. Maps (a) and (b) are plotted in percent relative to the same average velocity. (c) The difference between maps (a) and (b) relative to the same average in (a) and (b). (d) Path density, defined as the number of rays intersecting a $2°$ square cell ($\sim$50,000 km^2). White lines are plate boundaries. The red lines delineate the contour of 20 paths per 50,000 km^2. Inside this contour we have the greatest confidence in the estimated maps. Outside it, model norm damping begins to take effect.

Group Velocity Correction Surfaces

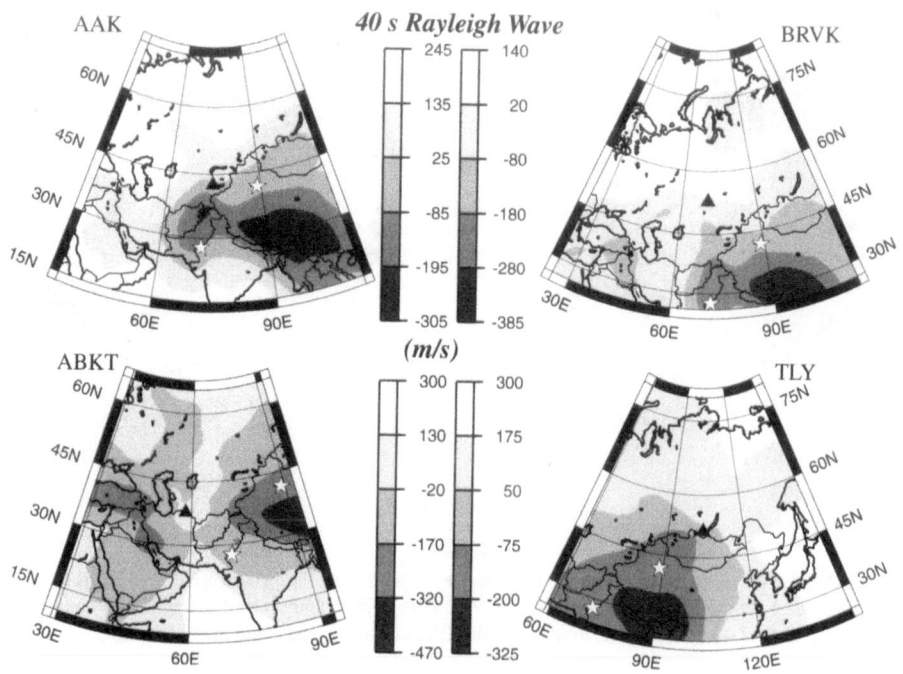

Figure 6

Group velocity correction surfaces for four stations in Central and Southern Asia for the 40 s Rayleigh wave. For each geographical point the maps define the group velocity perturbation that should be applied to a 40 s Rayleigh wave observed at a station if an event were located at the chosen point. Perturbations are relative to the group velocity at the station. Units are m/s. The locations of the Chinese and Indian test sites are indicated with stars.

in Figure 7b, where they used spherical harmonics up through degree and order 40. The major features of these maps are nearly identical. We have chosen the damping parameters, however, to accentuate smaller scale features than those apparent in the spherical harmonic parameterization. There is considerable signal remaining in the data set of Trampert and Woodhouse to be fit by smaller scale features than those apparent in Figure 7b. For example, the rms misfit to Trampert and Woodhouse's data produced by the map in Figure 7a is about 8.1 s compared with the 10.8 s produced by the spherical harmonic map in Figure 7b; about a 40% reduction in variance.

3.4. Azimuthal Anisotropy

We follow the majority of the studies of azimuthal anisotropy and our discussion above (equations (4)–(6)) by parameterizing azimuthal anisotropy for group velocity as:

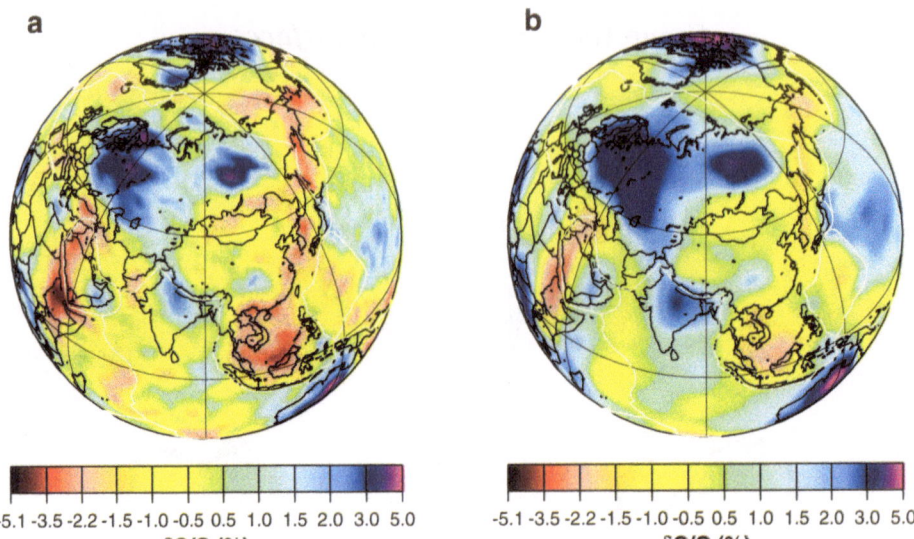

Figure 7

(a) Global 100 s Rayleigh-wave phase-velocity map estimated with the procedure described in this paper using the data of TRAMPERT and WOODHOUSE (1995, 1996). (b) Trampert and Woodhouse's map using a degree 40 spherical harmonic parameterization using the same data as in (a).

$$U(\mathbf{r}, \psi) = U_0(\mathbf{r}) + U_1(\mathbf{r})\cos 2\psi + U_2(\mathbf{r})\sin 2\psi + U_3(\mathbf{r})\cos 4\psi + U_4(\mathbf{r})\sin 4\psi \ , \quad (37)$$

where U_0 is isotropic group velocity at spatial point $\mathbf{r} = (\theta, \phi)$, U_1 and U_2 define the 2ψ part of azimuthal anisotropy, and U_3 and U_4 the 4ψ part of azimuthal anisotropy.

Figure 8 presents examples of the 2ψ component of group velocity azimuthal anisotropy for the 50 s Rayleigh wave across Antarctica and the surrounding oceans. We took the approximately 2200 observations and divided them into two separate sets of about 1100 measurements each which we then inverted separately for the two maps in Figures 8a,b. Both maps display spatially smooth anisotropy patterns, and the fast axes at many locations tend to be parallel to the directions of plate motions. The main features of the maps are similar, but there are differences in detail. In order to quantify the correlation between these two 2ψ maps we use the coherence function defined by GRIOT *et al.* (1998) which takes into account differences in the fast axes directions $(\alpha_1(\theta, \phi), \alpha_2(\theta, \phi))$ and the amplitudes $(A_1(\theta, \phi), A_2(\theta, \phi); A = (U_1^2 + U_2^2)^{1/2})$ of the two maps. The coherence K as a function of rotation angle α, varying between $-90°$ and $90°$, is defined as follows:

$$K(\alpha) = \frac{\sum_\theta \sum_\phi A_1(\theta, \phi) A_2(\theta, \phi) \sin\theta \exp\left(-\frac{(\alpha_1(\theta,\phi) - \alpha_2(\theta,\phi) + \alpha)^2}{2D_{cor}^2}\right)}{\left(\sum_\theta \sum_\phi \sin\theta A_1^2(\theta, \phi)\right)^{1/2} \left(\sum_\theta \sum_\phi \sin\theta A_2^2(\theta, \phi)\right)^{1/2}} \ . \quad (38)$$

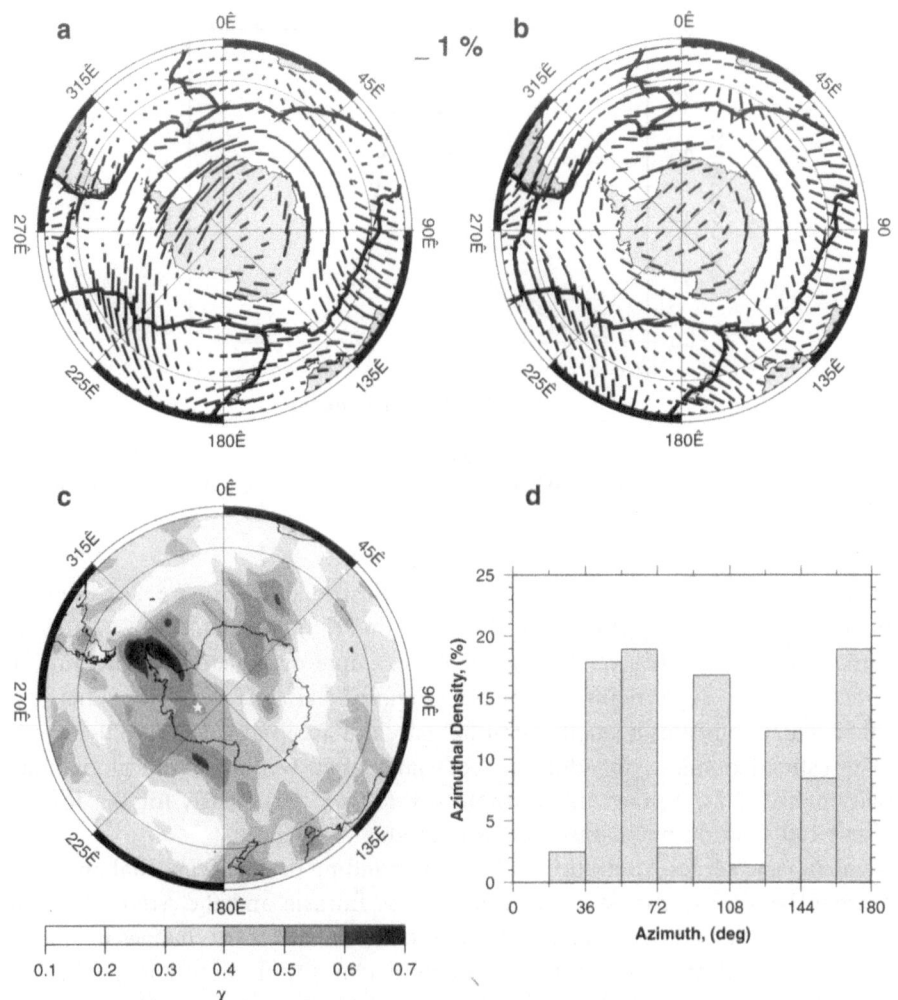

Figure 8

(a) and (b). The 2ψ component of the 50 s Rayleigh-wave group-velocity anisotropy across Antarctica and the surrounding oceans. Results from two equal data subsets of about 1100 measurements each are shown for comparison. (c) Distribution of the function $\chi(\theta, \phi)$ characterizing the azimuthal coverage for the entire set of 2200 Rayleigh-wave paths. (d) Histogram of azimuthal distribution at the fixed point $\theta = 173°$ (83°S), $\phi = 267°$ (93°W) shown by the star in (c).

Here D_{cor} is the uncertainty in the anisotropic direction, and was set to equal 10°. The resulting curve is shown in Figure 9. It is evident that the two maps are correlated, and the average absolute difference in orientation of the fast axes across the maps is less than 20°. The low value of the maximum coherence (~ 0.45) reflects differences in amplitudes of the anisotropic coefficients between the two maps.

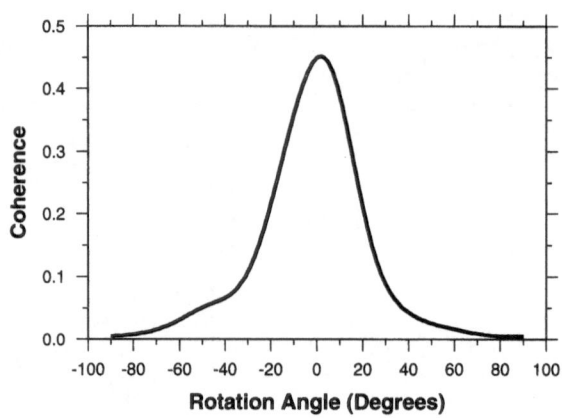

Figure 9
Coherence as defined by GRIOT *et al.* (1998) between the two 2ψ anisotropy maps shown in Figure 8.

The azimuthal coverage of the region is illustrated by Figure 8c, where the behavior of the function $\chi(\theta, \phi)$ defined by equation (20) is shown. The area in which $\chi > 0.3$ covers about 60% of the region. An example of a histogram of azimuthal distribution for a single point is shown in Figure 8d. In the vicinity of this point, $\chi = 0.53$ which indicates a small azimuthal gap and acceptable azimuthal coverage.

The estimated maps, in addition, correlate fairly well with the global phase velocity maps of TRAMPERT and WOODHOUSE (1996), particularly for the Rayleigh 50 s and 100 s period maps, but there are notable differences.

In summary, across Antarctica and the surrounding oceans azimuthal anisotropy appears to be a fairly robust observable. Across Eurasia and the Arctic this is not true, however. We find that similar data subsetting and arbitrary changes in damping and parameterization can produce substantial changes in both the pattern and amplitude of the estimated anisotropy (e.g., LEVSHIN *et al.*, 2001). Although azimuthal anisotropy can be rapidly and efficiently estimated with the algorithm described above, it remains a difficult target to estimate reliably with surface-wave data, particularly in continental regions.

4. Conclusions

We have described a rapid method for constructing surface-wave tomographic maps on local, regional or global scales. Extensive testing of this technique on data sets obtained from a variety of regions around the globe have confirmed its efficiency in producing detailed and reliable surface-wave group and phase-velocity tomographic maps together with useful measures of map quality.

Acknowledgements

We would like to thank Jeannot Trampert for contributing his measurements of surface-wave phase velocities that allowed us to construct Figure 7 and Eugene Lavely and Tatyana Yanovskaya for valuable reviews. All maps were generated with the Generic Mapping Tools (GMT) data processing and display package (WESSEL and SMITH, 1991,1995). This work was supported by the DSWA contract DSWA01-97-C-0157, DTRA contract DTRA01-99-C-0019, the Arms Control and Disarmament Agency, and NSF grants OPP-9818498 and OPP-9615139.

REFERENCES

AKI, K., and RICHARDS, P. G., *Quantitative Seismology*, vol. II, (W. H. Freeman and Co., 1980).

AURENHAMMER, F. (1991), *Voronoi Diagrams: A Survey of Fundamental Geometric Data Structure*, Assoc. Comput. Mach. Comput. Surv. *23*(3), 345–405.

BACKUS, G., and GILBERT, J. F. (1968), *Resolving Power of Gross Earth Data*, Geophys. J. R. Astron. Soc. *16*, 169–205.

BACKUS, G., and GILBERT, J. F. (1970), *Uniqueness in the Inversion of Inaccurate Gross Earth Data*, Philos. Trans. R. Soc. London, Ser. A *266*, 123–192.

BIJWAARD, H., SPAKMAN, W., and ENGDAHL, E. R. (1998), *Closing the Gap between Regional and Global Travel Time Tomography*, J. Geophys. Res. *103*, 30,055–30,078.

BRAUN, J., and SAMBRIDGE, M. (1997), *A Numerical Method for Solving Partial Differential Equations on Highly Irregular Evolving Grids*, Nature *376*, 665–660.

CURTIS, A., TRAMPERT, J., SNIEDER, R., and DOST, B. (1998), *Eurasian Fundamental Mode Surface Wave Phase Velocities and their Relationship with Tectonic Structures*, J. Geophys. Res. *103*, 26,919–26,947.

DITMAR, P. G., and YANOVSKAYA, T. B. (1987), *A Generalization of the Backus-Gilbert Method for Estimation of Lateral Variations of Surface Wave Velocity* (in Russian), Izv. Akad. Nauk SSSR, Fiz. Zeml *6*, 30–60.

EKSTRÖM, G., TROMP, J., and LARSON, E. W. F. (1997), *Measurements and Global Models of Surface-wave Propagation*, J. Geophys. Res. *102*, 8147–8158.

FRANKLIN, J. N. (1970), *Well-posed Stochastic Extensions of Ill-posed Linear Problems*, J. Math. Analysis Applic. *31*, 682–716.

FRIEDERICH, W. (1998), *Propagation of Seismic Shear and Surface Waves in a Laterally Heterogeneous Mantle by Multiple Forward Scattering*, Geophys. J. Int. *136*, 180–204.

GRAND, S. P., VAN DER HILST, R. D., and WIDIYANTORO, S. (1997), *Global Seismic Tomography: A Snapshot of Convection in the Earth*, GSA Today 7(4), 1–7.

GRIOT, D. A., MONTAGNER, J. P., and TAPPONIER, P. (1998), *Surface-wave Phase Velocity Tomography and Azimuthal Anisotropy in Central Asia*, J. Geophys. Res. *103*, 21,215–21,232.

HERRIN, E., and GOFORTH, T. (1977), *Phase-matched Filters: Application to the Study of Rayleigh Waves*, Bull. Seismol. Soc. Am. *67*, 1259–1275.

JOBERT, N., and JOBERT, G. (1983), *An Application of Ray Theory to the Propagation of Waves along a Laterally Heterogeneous Spherical Surface*, Geophys. Res. Lett. *10*, 1148–1151.

LASKE, G., and MASTERS, G. (1996), *Constraints on Global Phase Velocity Maps from Long-period Polarization Data*, J. Geophys. Res. *101*, 16,059–16,075.

LEACH, R. R., HARRIS, D. B., and WALTER, W. R. (1998), *Phase-matched filtering of after-shock sequences to detect Rayleigh waves in low SNR seismograms*, In *Proceedings of the 20th Annual Seismic Research Symposium on Monitoring a Comprehensive Test-Ban Treaty*, DoD and DoE, pp. 458–465.

LEVEQUE, J.-J., RIVERA, L., and WITTLINGER, G. W. (1993), *On the Use of Checkerboard Tests to Assess the Resolution of Tomographic Inversions*, Geophys. J. Int. *115*, 313–318.

LEVSHIN, A. L., YANOVSKAYA, T. B., LANDER, A. V., BUKCHIN, B. G., BARMIN, M. P., RATNIKOVA, L. I., and ITS, E. N. *Seismic Surface Waves in Laterally Inhomogeneous Earth* (ed. Keilis-Borok, V. I.) (Kluwer. Publ., Dordrecht, 1989).

LEVSHIN, A. L., RITZWOLLER, M. H., BARMIN, M. P., VILLASEÑOR, A., and PADGETT, C. A. (2001), *New Constraints on the Arctic Crust and Uppermost Mantle: Surface-wave Group Velocities, P_n, and S_n*, Phys. Earth. Planet. Int. *123*, 185–204.

LEVSHIN, A. L., and RITZWOLLER, M. H. (2001), *Automated Detection, Extraction, and Measurement of Regional Surface Waves*, this volume.

MARQUERING, H., SNIEDER, R., and NOLET, G. (1996), *Waveform Inversions and the Significance of Surface-mode Coupling*, Geophys. J. Int. *124*, 258–270.

MASTERS, G., JOHNSON, S., LASKE, G., and BOLTON, H. (1996), *A shear velocity model of the mantle*, Philos. Trans. R. Soc. London, Ser. A, *354*, 1385–1411.

MENKE, W., *Geophysical Data Analyses: Discrete Inverse Theory* (New York, Academic Press, 1989).

MONTAGNER, J. P., and TANIMOTO, T. (1991), *Global Upper Mantle Tomography of Seismic Velocities and Anisotropies*, J. Geophys. Res. *96*, 20,337–20,351.

MOONEY, W. D., LASKE, G., and MASTERS, G. (1998), *CRUST 5.1: A Global Crustal Model at 5 Degrees by 5 Degrees*, J. Geophys. Res. *103*, 727–747.

NAKANISHI, I., and ANDERSON, D. L. (1982), *World-wide Distribution of Group Velocity of Mantle Rayleigh Waves as Determined by Spherical Harmonic Inversion*, Bull. Seismol. Soc. Am. *72*, 1185–1194.

NOLET, G., Seismic wave propagation and seismic tomography. In *Seismic Tomography*, (Reidel, Dordrecht, 1987), pp. 1–23

PARKER, R. L., *Geophysical Inverse Theory* (Princeton, NJ, Princeton University Press, 1994).

POLLITZ, F. F. (1994), *Surface-wave Scattering from Sharp Lateral Discontinuities*, J. Geophys. Res. *99*, 21,891–21,909.

PULLIAM, J., and SNIEDER, R. (1998), *Ray Perturbation Theory, Dynamic Ray tracing and the Determination of Fresnel Zones*, Geophys. J. Int. *135*, 463–469.

RITZWOLLER, M. H., and LEVSHIN, A. L. (1998), *Eurasian Surface-wave Tomography: Group Velocities*, J. Geophys. Res. *103*, 4839–4878.

RITZWOLLER, M. H., LEVSHIN, A. L., RATNIKOVA, L. I., and EGORKIN, A. A., Jr (1998), *Intermediate Period Group Velocity Maps Across Central Asia, Western China, and Parts of the Middle East*, Geophys. J. Int. *134*, 315–328.

RITZWOLLER, M. H., BARMIN, M. P., VILLASEÑOR, A., LEVSHIN, A. L., ENGDAHL, E. R., SPAKMAN, W., and TRAMPERT, J. (1999), *Construction of a 3-D P and S model of the crust and upper mantle to improve regional locations in W. China, Central Asia, and parts of the Middle East*, Proceedings of the 21th Annual Seismic Research Symposium on Monitoring a Comprehensive Test-Ban Treaty, DoD and DoE, pp. 656–665.

RUSSELL, D. W., HERRMAN, R. B., and Hwang, H. (1988), *Application of Frequency-variable Filters to Surface-wave Amplitude Analysis*, Bull. Seismol. Soc. Am. *78*, 339–354.

SAMBRIDGE, M., BRAUN, J., and MCQUEEN, H. (1995), *Geophysical Parameterization and Interpolation of Irregular Data Using Natural Neighbors*, Geophys. J. Int. *122*, 837–857.

SLOAN, S. W. (1987), *A Fast Algorithm for Constructing Delaunay Triangulation in the Plane*, Adv. Eng. Software *9*(1), 34–55.

SMITH, M. L., and DAHLEN, F. A. (1973), *The Azimuthal Dependence of Love and Rayleigh Wave Propagation in a Slightly Anisotropic Medium*, J. Geophys. Res. *78*, 3321–3333.

SNIEDER, R. (1988), *Large-scale Waveform Inversions of Surface Waves for Lateral Heterogeneities*, J. Geophys. Res. *93*, 12,055–12,065.

SPAKMAN, W., and BIJWAARD, H. (1998), *Irregular Cell Parameterization of Tomographic Problems*, Ann. Geophys. *16*(18).

SPAKMAN, W., and Bijwaard, H. (2001), *Irregular Cell Parameterization of Tomographic Inverse Problems*, Pure appl. geophys., this volume.

STEVENS, J. L., and DAY, S. M. (1985), *The Physical Basis of m_b:M_s and Variable Frequency Magnitude Methods for Earthquake/Explosion Discrimination*, J. Geophys. Res. *90*, 3009–3020.

STEVENS, J. L., and MCLAUGHLIN, K. L. (1997), *Improved methods for regionalized surface wave analysis*, Proceedings of the 17th Annual Seismic Research Symposium on Monitoring a CTBT, pp. 171–180.

SUETSUGU, D., and NAKANISHI, I. (1985), *Surface-wave Tomography for the Upper Mantle beneath the Pacific Ocean. Part I: Rayleigh Wave Phase Velocity Distribution*, J. Phys. Earth *33*, 345–368.

TARANTOLA, A., *Inverse Problems Theory, Methods for Data Fitting and Model Parameter Estimation* (Amsterdam, Elsevier, 1987).

TARANTOLA, A., and VALETTE, B. (1982), *Generalized Nonlinear Inverse Problems Solved Using the Least-squares Criterion*, Revs. Geophys. *20*(2), a 219–232.

TARANTOLA, A., and NERSESSIAN, A. (1984), *Three-dimensional Tomography without Block*, Geophys. J. R. Astr. Soc. *76*, 299–306.

TIKHONOV, A. N. (1963), *On the Solution of Improperly Posed Problems and the Method of Regularization*, Dokl. Akad. Nauk SSSR, *151*(501).

TRAMPERT, J. (1998), *Global Seismic Tomography; The Inverse Problem and Beyond*, Inverse Problems, *14*, 371–385.

TRAMPERT, J., and WOODHOUSE, J. (1995), *Global Phase Velocity Maps of Love and Rayleigh Waves between 40 and 150 Seconds*, Geophys. J. Int. *122*, 675–690.

TRAMPERT, J., and WOODHOUSE, J. (1996), *High Resolution Global Phase Velocity Distributions*, Geophys. Res. Lett. *23*, 21–24.

VAN DER HILST, R. D., WIDIYANTORO, S., and ENGDAHL, E. R. (1997), *Evidence for Deep Mantle Circulation from Global Tomography*, Nature *386*, 578–584.

VDOVIN, O. Y., RIAL, J. A., LEVSHIN, A. L., and RITZWOLLER, M. H. (1999), *Group-velocity Tomography of South America and the Surrounding Oceans*, Geophys. J. Int. *136*, 324–330.

VDOVIN, O. Y. (1999), *Surface-wave Tomography of South America and Antarctica*, Ph.D. Thesis, Department of Physics, University of Colorado at Boulder.

VILLASEÑOR, A., RITZWOLLER, M. H., LEVSHIN, A. L., BARMIN, M. P., ENGDAHL, E. R., SPAKMAN, W., and TRAMPERT J. (2001), *Shear-velocity Structure of Central Eurasia from Inversion of Surface-wave Velocities*, Phys. Earth Planet. Inter. *123*, 169–184.

WESSEL, P., and SMITH, W. H. F. (1991), *Free Software Helps Map and Display Data*, EOS Trans. AGU *72*, 441pp.

WESSEL, P., and Smith, W. H. F. (1995), *New Version of the Generic Mapping Tools Released*, EOS Trans. AGU *76*, 329pp.

WU, F. T., and LEVSHIN, A. (1994), *Surface-wave Group Velocity Tomography of East Asia*, Phys. Earth Planet. Int. *84*, 59–77.

WU, F. T., LEVSHIN, A. L., and KOZHEVNIKOV, V. M. (1997), *Rayleigh-wave Group Velocity Tomography of Siberia, China, and the Vicinity*, Pure appl. geophys. *149*, 447–473.

YANOVSKAYA, T. B. (1982), *Distribution of surface group velocities in the North Atlantic*, Fizika Zemli, Izv. Acad. Sci. USSR, *2*, 3–11.

YANOVSKAYA, T. B., and Ditmar, P. G. (1990), *Smoothness Criteria in Surface-wave Tomography*, Geophys. J. Int. *102*, 63–72.

YANOVSKAYA, T. B., and ANTONOVA, L. M. (2000), *Lateral Variations in the Structure of the Crust and Upper Mantle in the Asia Region from Data on Group Velocities of Rayleigh Waves*, Fizika Zemli, Izv. Russ. Acad. Sci. *36*(2), 121–128.

ZHANG, Y.-S., and Lay, T. (1996), *Global Surface-wave Phase Velocity Variations*, J. Geophys. Res. *101*, 8415–8436.

ZHOU, H. (1996), *A high-resolution P-wave Model for the Top 1200 km of the Mantle*, J. Geophys. Res. *101*, 27,791–27,810.

(Received December 1, 1999, revised April 25, 2000, accepted May 15, 2000)

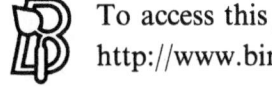

Pure appl. geophys. 158 (2001) 1377–1399
0033–4553/01/081377–23 $ 1.50 + 0.20/0

❙ Pure and Applied Geophysics

Global Models of Surface Wave Group Velocity

Erik W. F. Larson[1] and Göran Ekström[1]

Abstract—Measurements of group velocity are derived from phase-velocity dispersion curves and modeled with global laterally-varying isotropic structure. Maps for both Love and Rayleigh waves are created in the period range 35 s to 175 s. The data set of group-velocity measurements includes over 50,000 minor-arc observations and 5,000 major-arc observations. The errors in the measurements are estimated by an empirical method of comparing pairwise-similar paths, resulting in uncertainties which are 20% to 40% of the size of the typical measurement. The models are determined by least-squares inversion for spherical harmonic maps expanded up to degree 40. This parameterization allows for resolution of structures as small as 500 km. The models explain 70–98% of the variance relative to the Preliminary Reference Earth Model (PREM). For the area of Eurasia, the group-velocity maps from this study are compared with those of Ritzwoller and Levshin (1998). The results of the two studies are in very good agreement, particularly in terms of spatial correlation. The models also agree in amplitude at wavelengths longer than 30 degrees. For shorter wavelengths, the agreement is good only for models at short periods. The global maps are useful for prediction of group arrival times, for revealing tectonic structures, for determination of seismic event locations and source parameters, and as a basis for regional group-velocity studies.

Key words: Surface waves, Love wave, Rayleigh waves, group velocity, tomography.

1. Introduction

In recent years there have been parallel efforts to use surface waves to determine upper mantle seismic velocities. Regional studies usually model surface-wave group velocity, and global studies typically model phase velocity. Some groups have also modeled phase velocities on a regional scale (e.g., Curtis et al., 1998), however most regional studies model group velocity, for which Ritzwoller and Levshin (1998) provide an excellent recent history. There have been few studies of global group velocities, the most recent being from a decade ago – a study which had much lower resolution than is attainable today (Rosa, 1987). Several groups have recently published high-resolution global phase-velocity models (Trampert and Woodhouse, 1996; Zhang and Lay, 1996; Laske and Masters, 1996; Ekström et al., 1997; van Heijst and Woodhouse, 1999).

[1] Department of Earth and Planetary Sciences, Harvard University, Cambridge, Massachusetts, USA. E-mail: eriklarson@post.harvard.edu

The difference between regional and global modeling has arisen partly because of the different methods of measuring phase and group arrival times. The group arrival time is easier to determine as a function of frequency, as the effects of the source generally are ignored. Variations on the frequency-time analysis method are routinely used for this purpose (DZIEWONSKI et al., 1969; LEVSHIN et al., 1989, 1992). There are several methods for measuring phase velocity which generally require synthetic seismograms to be generated for each path. Two "automatic" methods have recently been developed to measure many paths for global tomographic studies (TRAMPERT and WOODHOUSE, 1995; EKSTRÖM et al., 1997). Also, if the phase dispersion (phase velocity as a function of frequency) has been measured for a given path, it is possible to derive the group dispersion as we show in section two.

An important difference between regional and global studies is the path coverage. In most studies all sources and stations lie within the region being modeled. For regional studies this leads to fewer data constraints near the edges of the region, and, in particular, fewer crossing paths. This can lead to large errors in the models near the edges and it is usually difficult to determine how much of the model is affected by this problem. Global models are not affected by this since all stations and earthquakes are within the region of study. Another disadvantage of the path coverage of regional studies is that all path lengths are relatively short. This can make it more difficult to measure the dispersion of long-period waves and to determine the large-scale structure.

An additional difference between measuring phase and group dispersion is the sensitivity to the source parameters. The "source phase" (due to earthquake focal geometry) must be accounted for in the measurement of phase velocities (KNOPOFF and SCHWAB, 1968). Therefore, studies measuring phase-velocity dispersion are limited to earthquakes for which source mechanisms have been determined. The Harvard centroid-moment tensor catalog (DZIEWONSKI et al., 1981) provides source mechanisms, although only for large earthquakes. Since direct group-velocity measurements are insensitive to the source phase, smaller earthquakes with undetermined focal mechanisms are often used in regional group-velocity studies. A group-velocity measurement is somewhat sensitive to the depth of the earthquake, but this effect is usually ignored (LEVSHIN et al., 1999). Our method for measuring group-velocity accounts for both source depth and mechanism.

Both phase-velocity and group-velocity measurements are affected by errors in source parameters (MUYZERT and SNIEDER, 1996; LEVSHIN et al., 1999). These errors may be significant for an individual measurement, particularly for Rayleigh waves. However, as LEVSHIN et al. (1999) point out, in the resulting maps the errors typically appear only near the edges where there is poor path coverage. Therefore, global studies with very good path coverage are less affected by such errors than regional studies.

In this work we bridge the gap between the global phase-velocity studies and regional group-velocity studies by determining new high-resolution global group-velocity models. We determine maps of surface-wave group velocities at periods ranging from 35 s to 175 s. The parameterization for the maps allows structures as small as 500 km to be modeled. These maps are directly useful in several ways: to provide group arrival-time predictions for teleseismic studies, to investigate tectonic questions with crust and upper mantle signatures, and as starting models for high-resolution regional studies. In addition, these models enable us to study carefully whether regional and global studies agree. The agreement of regional and global models has been questioned in the literature (PASSIER and SNIEDER, 1995; CHEVROT et al., 1998). The resolution in this study is sufficiently high to directly compare our global maps with a continental-scale study of Eurasian group velocities (RITZWOLLER and LEVSHIN, 1998). We compare the maps both in the spatial and spectral domains. We choose the RITZWOLLER and LEVSHIN (1998) study because it is based on more data and covers a larger region than any other work, allowing the maximum area for comparison.

We make several standard assumptions in the determination of the group-velocity maps. First, we ignore the deviation of the raypaths from the great circle. For smooth models this effect has been shown to be small when the data are phase- or group-dispersion curves (WANG and DAHLEN, 1995; WANG et al., 1998). The assumption of a "smooth model" is more questionable for the maps in this study with the largest amplitude anomalies (the shortest period models). Second, we do not allow for azimuthal anisotropy in our model parameterization. In most areas of the world anisotropy has been shown to be of small magnitude compared to isotropic structure, particularly at long wavelengths (NISHIMURA and FORSYTH, 1988; MONTAGNER and TANIMOTO, 1990; TRAMPERT and WOODHOUSE, 1996). For the scale and resolution of this study, the effects of anisotropy will not bias the isotropic results (LARSON et al., 1998), except possibly in the eastern Pacific, where the anisotropy may be large. Third, we do not account for mislocation of earthquakes or errors in station time. These effects have been shown to be negligible in other works (RITZWOLLER and LEVSHIN, 1998; EKSTRÖM et al., 1997).

2. Method and Data

The group velocity of a wave, U, is defined by the angular frequency, ω, and the wavenumber, k, as

$$U = \frac{d\omega}{dk} = c + k\frac{dc}{dk} \ , \tag{1}$$

where $c = \omega/k$ is the phase velocity. We can substitute $k = \omega/c$ and

$$\frac{dc}{dk} = \frac{dc}{d\omega}\frac{d\omega}{dk} = \frac{dc}{d\omega}U \ . \tag{2}$$

Then, at a particular frequency, we can express the group velocity by

$$U = \frac{c}{(\omega/c)(dc/d\omega) - 1}.$$ (3)

Therefore, if we know the phase-velocity dispersion, $c(\omega)$, across some frequency range, we can easily calculate the group velocity $U(\omega)$ across that range.

For our purposes we wish to know the perturbation of group velocity away from some reference model, $\delta U = U - U_0$. For a surface wave recorded on a seismogram, the measurable datum is the average perturbation of the group velocity along the path from the earthquake to the station, $\overline{\delta U}$ (where the bar indicates an average). In terms of the reference phase velocity, c_0, and path-average perturbed phase velocity, $\overline{\delta c}$,

$$\overline{\delta U} = \frac{c_0 + \overline{\delta c}}{\omega/(c_0 + \overline{\delta c})((dc_0/d\omega) + (d\overline{\delta c}/d\omega)) - 1} - U_0.$$ (4)

If the phase-velocity dispersion of a surface wave has been measured, the average group velocity for the path from the earthquake to the station can be easily determined. For our analysis we use $\overline{\delta U}/U_0$ as the datum, with the reference velocities c_0 and U_0 from PREM (DZIEWONSKI and ANDERSON, 1981).

We measure phase-velocity dispersion using the frequency-band expansion method (EKSTRÖM et al., 1997), and then calculate the frequency-dependent group-velocity perturbation. The principle of the method is to estimate the optimal phase-matched filter to fit the data. For a given path, this phase-matched filter provides the apparent average phase-velocity perturbation as a function of frequency. A key part of the method is that it expands the range of frequencies to fit, beginning with the long periods and gradually widening the range to include higher frequencies. This method has the advantage of being entirely automatic with no user interaction, allowing for the measurement of a great many paths. In addition, the method provides an estimate of the quality of each measurement (A, B, or C), which is based on the fit to the data. We sample the phase-dispersion curves, which are represented by six B-splines, at a number of frequencies and use the method above (Eq. 4) to extract the group-velocity perturbation. The range of periods for minor-arc paths is 35 s to 175 s and for major-arc paths 70 s to 175 s. In order to directly compare our maps with the maps of RITZWOLLER and LEVSHIN (1998) we sample the group-velocity dispersion curves for both Love and Rayleigh waves at 35 s, 40 s, 50 s, 60 s, 70 s, 80 s, 90 s, 100 s, 125 s, 150 s, and 175 s period.

Our data set is derived by applying the above method to digital seismograms from the Global Seismographic Network (GSN) of Incorporated Research Institutions for Seismology (IRIS), the Chinese Digital Seismograph Network (CDSN), the Global Telemetered Seismograph Network (GTSN), and the MEDNET and GEOSCOPE networks from the period 1989–1997. Earthquake source parameters are taken from the Harvard centroid-moment tensor catalog (DZIEWONSKI et al., 1981). To avoid

problems associated with short paths and with caustics, only stations in the range 30°
to 150° from the earthquake are used. For minor-arc paths, we determine phase-
dispersion curves of seismograms from earthquakes with $M_W > 5.5$ and depth less
than 50 km. We augment the data set of phase-velocity dispersion measurements
used in EKSTRÖM et al. (1997) by approximately 50%, and use a total of 198,998
minor-arc paths. In addition, we measure major-arc paths for events with $M_W > 6.5$,
resulting in 31,903 paths. We select the data based on a quality of fit parameter which
is determined in the measurement process. Depending on the period of measurement,
between 20% and 25% of the data satisfy the selection criteria and are retained in the
data set.

We discard measurements which are outliers and appear to be in error. For
minor-arc paths we remove measurements more than 2.5 standard deviations from
the mean, which eliminates fewer than 2.2% of the data. For major-arc paths there
are more outliers and the cut-off is reduced to 1.25 standard deviations, which
removes 8–13% of the data. The number of measurements which pass the quality
selection and this cut-off is shown in Table 1. The distribution of paths is typical for
a global study: better in the Northern Hemisphere, although the major-arc paths help
in the Southern Hemisphere. Figure 1 shows representative path coverages for
Rayleigh waves at 80 s and Love waves at 35 s.

The measurement uncertainty is estimated using a method of comparing pairwise-
similar paths. For all pairs of measurements recorded at a given station, we calculate
the distance between the earthquakes. The distance between the earthquakes, s
(separation), defines the "pairwise-similarity" of the pair. We also define the quality
of the pair, q, to be the worse of the two quality values of the two measurements. In
addition, we calculate the difference between the two group-velocity measurements,

Table 1

Number of measurements which pass quality check at each period

Period (s)	Minor-arc		Major-arc	
	Love	Rayleigh	Love	Rayleigh
35	24805	49477		
40	25264	50068		
45	25199	49886		
50	36895	64704		
60	36666	64556		
70	36648	64936	5030	7619
80	36564	64813	5122	7885
90	36636	65034	5003	8146
100	36479	64755	4956	8074
125	34983	63606	5018	7828
150	26345	56773	4790	7755
175	26393	56661	4586	7679

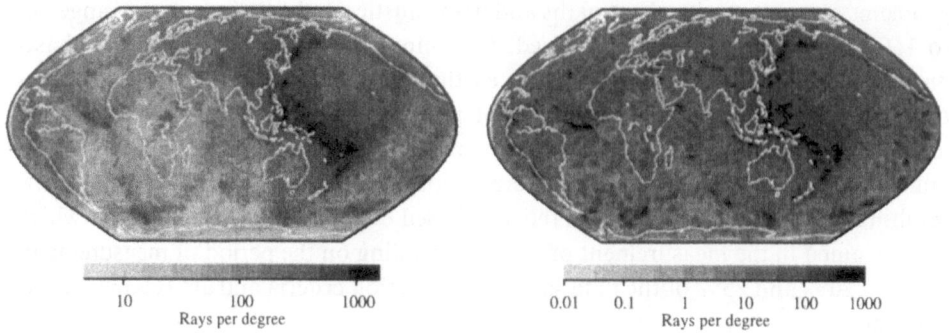

(a) Minor-arc Rayleigh wave 80 s   (b) Major-arc Rayleigh wave 80 s

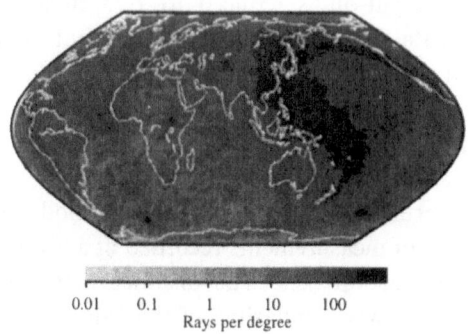

(c) Minor-arc Love wave 35 s

Figure 1

Raypath density for various models in ray degrees per $(\text{deg})^2$. The density is calculated by dividing the sum of the length of all raypaths within each cell by the area of the cell.

d. Note that d would be zero if there were no errors in the measurement or in the source parameters and the measurements were from earthquakes at the same location. We then divide the pairs into groups, for example one group is all pairs with s between $1°$ and $1.5°$ and q corresponding to 'B' quality paths. For each group we calculate one-half the root-mean-square of the differences,

$$e = \frac{1}{2}\sqrt{\frac{\sum_i^N d_i^2}{N}} \tag{5}$$

where N is the number of pairs in the group. If we have a normal distribution of errors, e is the empirical error for that group of pairs, that is if one calculates the difference between all values in a normal distribution, the root-mean-square is twice the standard deviation of the distribution.

However, the earthquakes for each pair of paths are at slightly different locations. Consequently the errors estimated by this method include the effect of the difference in group-velocity between the two paths. Since this is the signal we model, we attempt to determine and remove this effect. From Table 2 and Figure 2 we can see that the errors increase approximately linearly with separation s. The table also shows that the quality grade is quite useful – the lower quality paths have a larger error. We fit the dependence upon separation by assuming a relationship with distance for the

Table 2

Error estimates for average group-velocity perturbations $\overline{\delta U}/U_0$ (in percent) using the method of pairwise-similar paths for 80 s Rayleigh wave minor-arc data. Rows show different maximum earthquake separation distances. Columns are for different quality groups. The best fit gives a slope 0.0523%/(deg) for dependence upon distance and a "zero-distance" errors for quality A paths of 0.470%, quality B 0.715%, and quality C 0.898%. See Figure 2 for a graphical representation

Distance	Quality A	Quality B	Quality C
1.5°	0.549	0.790	0.975
1.75°	0.563	0.805	0.988
2°	0.576	0.817	1.00
2.25°	0.586	0.832	1.02
2.5°	0.597	0.846	1.03
2.75°	0.609	0.859	1.04

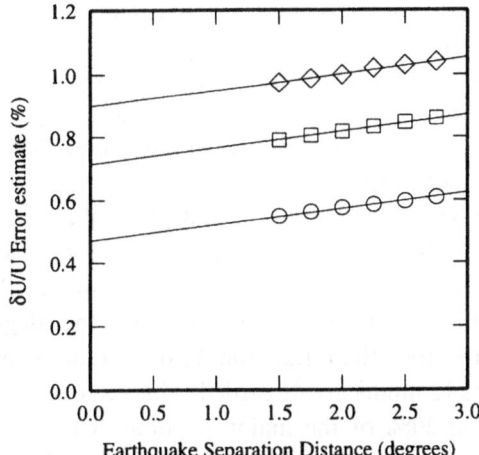

Figure 2

Example calculation of error estimates from pairwise-similar paths. Symbols are the error estimate within each category, with circles indicating best quality, squares intermediate, and diamonds are the lowest acceptable quality group. They are plotted at the maximum separation between earthquakes in their category. The straight lines are the fit to the estimate, and the y intercept is used as the error estimate for the given quality group.

errors, which is the same for all quality groups. For the error of group, e_{ij} from Equation (5), with separation in the range s_{j-1} to s_j in quality grouping i, we have

$$e_{ij} = E_i + Ss_j \; , \qquad (6)$$

where E_i is the error for group i at separation zero, and S is the slope of the increase of error with earthquake separation distance. This allows us to solve for the slope and error for each quality group at each period. This is graphically represented in Figure 2. To assign uncertainties for each measurement, we assign the value E_i for the quality group of the measurement. The uncertainties range from around 0.5%, for the best quality paths, to around 1.0%. Using the uncertainties derived from this method, the initial variance of the data/uncertainty for all paths varies between 2.3 and 5.0, that is, the average uncertainty is between 20% and 40% of the measurement, depending on the frequency and wave type.

3. Inversion

To determine global group-velocity maps from individual measurements of group velocity, we follow a two-step inversion procedure. We perform the same procedure for Love and Rayleigh waves at each frequency. For 70 s period and longer, we use major-arc measurements in addition to minor-arc measurements. We weigh each datum by the inverse of its uncertainty estimate.

In the first step of the inversion we invert for a long-wavelength model. We use all measurements which satisfy our quality criteria. The basis functions are surface spherical harmonics up to degree 12, described by 169 spherical harmonic coefficients. This results in a model with minimum wavelength of 3000 km. We use a least-squares inversion, and this parameterization requires no damping or regularization. The long-wavelength model explains most (70–95%, depending on frequency) of the variance in our data.

In the second step we invert for a higher degree spherical harmonic map. The maximum degree for which we invert (and which is used for our final model) is degree 40, which allows for description of features as small as 500 km. The data are selected using the quality criteria and the predictions of the degree 12 model. All measurements that are more than 1.25 standard deviations away from the value predicted by the degree 12 model are discarded. This removes approximately 25% of the minor-arc data and 30% of the major-arc data. This is a greater number of outliers than for phase-velocity data (EKSTRÖM et al., 1997), probably because the group velocity is derived from derivatives of the phase-velocity dispersion curve, which are more sensitive to errors. Of course some of these "outliers" may be correct data, but we think it is a better choice to exclude these points. For the inversion we again perform a least-squares fit to the data, but models expanded beyond degree 20 require regularization. We choose to damp the first spatial derivative of the output model. The strength of the damping is chosen conservatively (in our opinion) to produce models which are unlikely to have spurious features, but are probably more smooth than the actual Earth.

As is typical in studies of Earth structure, the smaller scale structure only explains a fraction of the variance that remains after the large-scale structure is accounted for. In this study we explain only about 10% of the remaining variance after the degree 12 inversion, even with a degree 40 spherical harmonic map. To determine whether these fits are statistically significant, we calculate $\sqrt{\chi^2/(N-M)}$, where χ is the root-mean-square misfit, N is the number of data points, and M is the number of model parameters. This statistic becomes smaller as the degree of the model is increased, which indicates that the use of more parameters is justified (PRESS *et al.*, 1992). The remaining variance indicates that we are underestimating the errors in our data. The additional errors could result from systematic effects, such as systematically mislocated earthquakes, which are not recovered by the pairwise-similar path estimate of errors.

In Figures 3 and 4, we illustrate the fits provided by the models. Over 90% of the initial variance in the Rayleigh-wave data is explained at all periods. For Love waves, the long-period (>100 s) fits are between 80% and 90%, the intermediate periods (60–90 s) between 90% and 95%, and the short-period fits are better than 90%. The short-period Rayleigh-wave models also fit extremely well. However, this higher fit for short periods is somewhat deceptive – the initial variance is substantially higher for short periods because there are larger long-wavelength group-velocity anomalies. The final residuals are actually larger at the short periods even though the fit is better. The improvement of the high-resolution model appears very small because so much of the signal is fit by the long-wavelength model. Since our data coverage is uneven, it is impossible to further refine the long-wavelength parameters without allowing smaller scale structures in our models. The residual misfit of the predictions of the final model to the data $(\overline{\delta U/U_0})$ is about 0.6% and somewhat smaller at longer periods.

4. Maps

To illustrate our final models, we show maps at two frequencies. All of the degree 40 maps of group velocity are available electronically, as described in the Acknowledgements.

We present two maps as examples, a short-period Love-wave model (for which we have the least data), and a long-period Rayleigh-wave model (for which we have the most data). The maps are shown in Figure 5. The apparent resolution of the 90 s Rayleigh wave map is high everywhere; structures as small as 1000 km appear to be modeled globally. However, the fewer data constraints on the 40 s Love-wave map cause less small-scale structure in some poorly sampled regions, such as Antarctica.

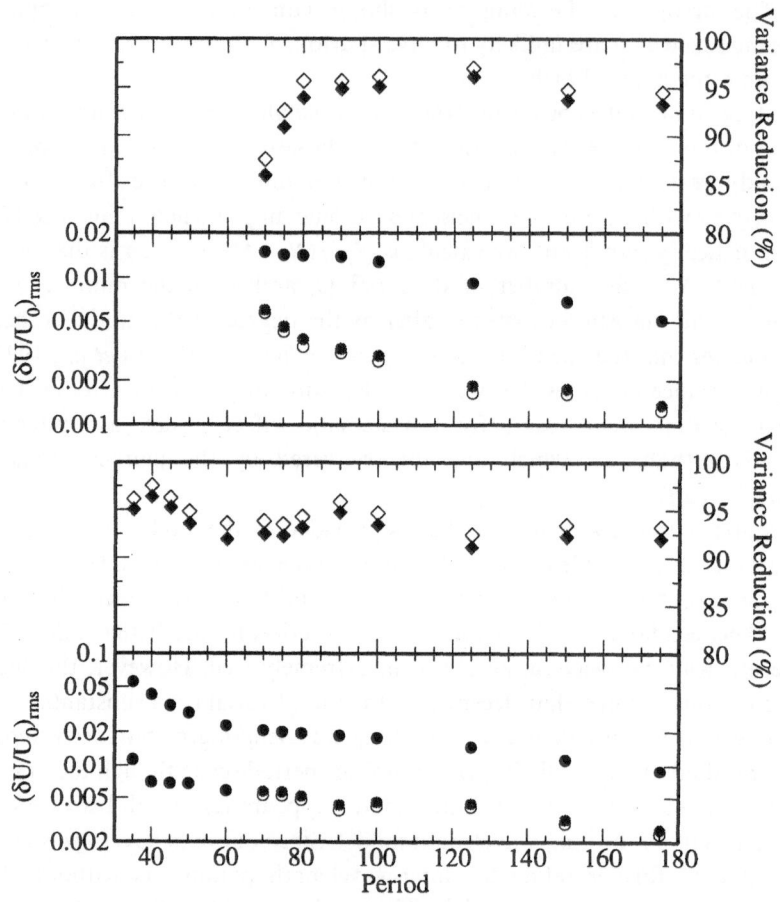

Figure 3

Fits of Rayleigh-wave models to data. Top panels are for major-arc data, bottom for minor-arc data. Solid circles are the initial root-mean-square (rms) of all measurements at each frequency, gray circles are the rms of the residual for the degree 12 models, and open circles are the rms of the residuals for the degree 40 models. The diamonds are the variance reduction to the weighted data, again, gray for the degree 12 models and open for the degree 40 models.

To understand the character of the maps, it is important to know how the group-velocity structure relates to the seismic velocities in the earth. The maps are primarily sensitive to shear-wave speeds in the crust and top few hundred kilometers of the upper mantle. The most shallow sensitivity is that of short-period Love waves, which are primarily sensitive at depths less than 30 km – crust in the continents, lithosphere in the oceans. The Rayleigh waves have a maximum sensitivity somewhat deeper (for the same frequency).

The sensitivity to differing depths leads to differences in the overall character of the models. As noted before, the short-period models have larger variations than the long-

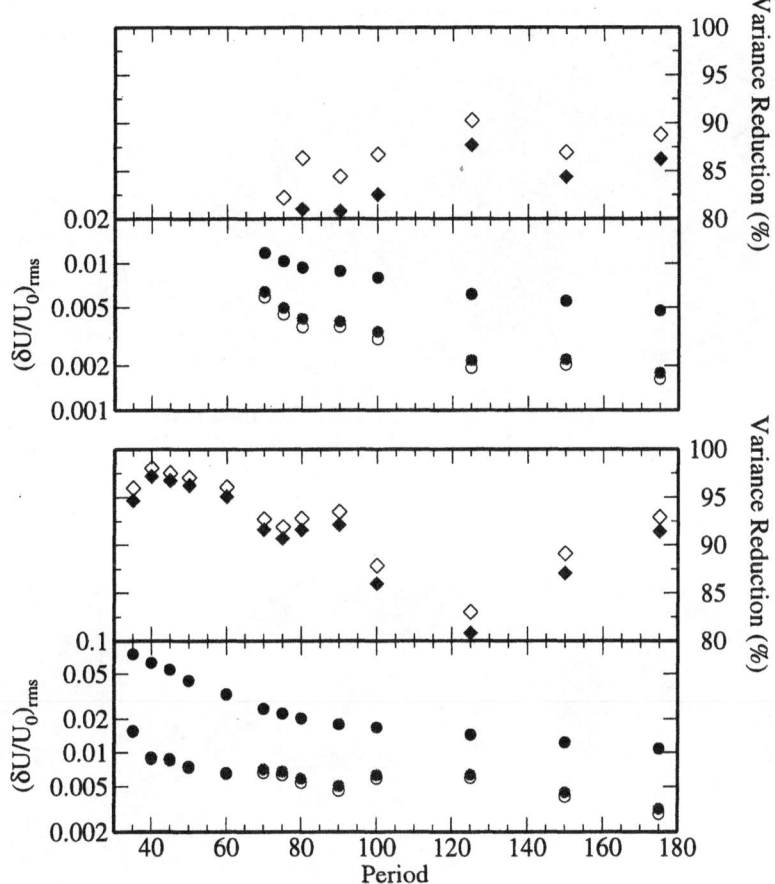

Figure 4
Fits of Love-wave models to data. See Figure 3 for explanation.

period models (as can be seen in the spectra of the models in Figs. 6 and 7). This arises because of the strong contrast at the Mohorovičić discontinuity. The spectra also show that the long-wavelength terms dominate the maps. The power in degrees greater than 12 is typically one or two orders of magnitude below the power of low-degree structures. These low-degree structures also contribute very strongly to the measurements since the variations with shorter wavelengths cancel out along the path.

The dominant features in our new maps are very similar to what has been seen previously in upper-mantle studies. The 40 s Love-wave map (Fig. 5a) matches the contrasts between oceans and continents very closely. The map images some relatively small structures such as New Zealand and the Falklands continental shelf. The thick crust of the Ontong Java plateau is also apparent. The deep crustal roots under mountain ranges appear clearly, for example under the Andes and the

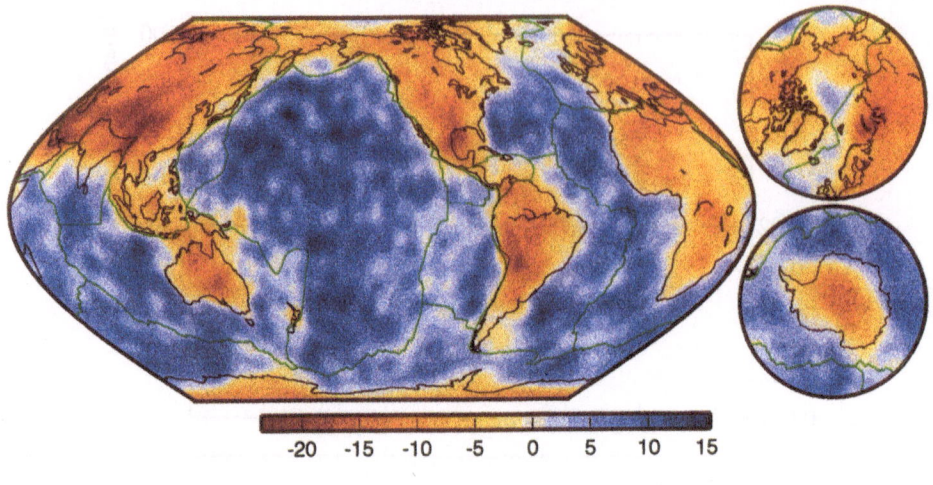

$$-20 \quad -15 \quad -10 \quad -5 \quad 0 \quad 5 \quad 10 \quad 15$$

(a) Love wave 40 s

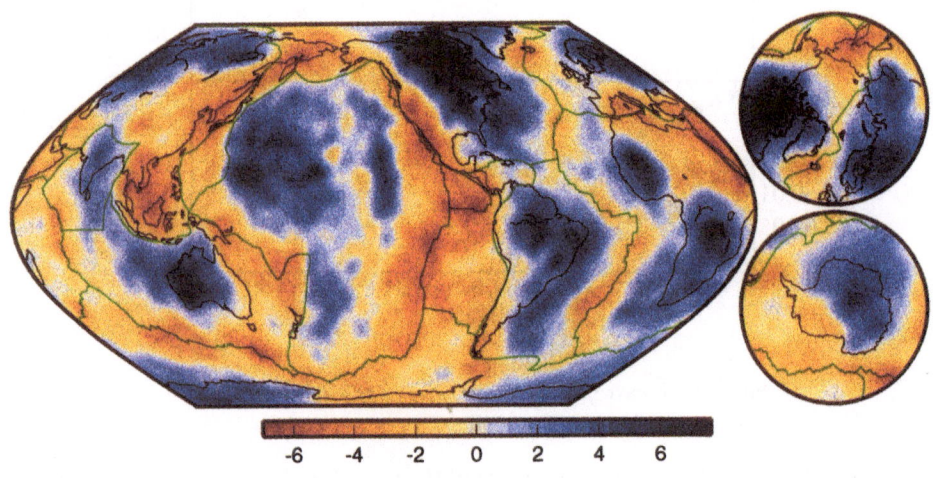

$$-6 \quad -4 \quad -2 \quad 0 \quad 2 \quad 4 \quad 6$$

(b) Rayleigh wave 90 s

Figure 5
Group velocity in percent variations of $\delta U/U_{\mathrm{PREM}}$.

Himalayas and Tibetan Plateau. Several regions of thick sediment can be seen, such as in the Kara Sea/Barents Sea and in the Gulf of Mexico. However, elsewhere in the oceans there is little coherent structure.

The 90 s Rayleigh-wave map (Fig. 5b) presents a very different picture. This map has a maximum sensitivity to seismic velocities at roundly 100 km depth. The dominant features are high-velocity lithospheric continental roots and low-velocity

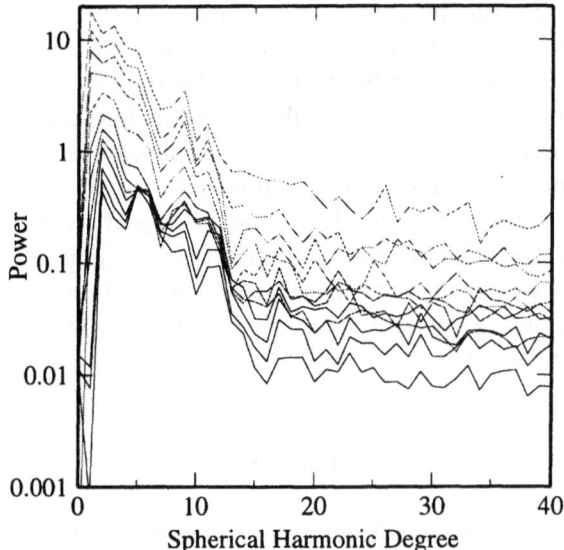

Figure 6
Spectra of Love-wave models. Maps for periods less than 60 s are lightest gray and maps 100 s and longer period are in black.

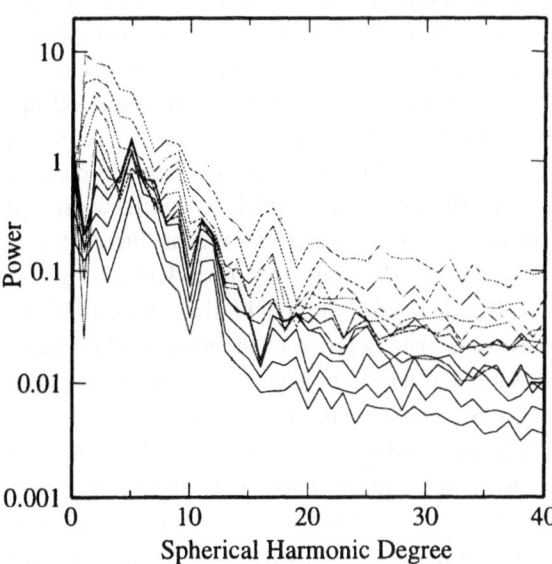

Figure 7
Spectra of Rayleigh-wave models. Maps for periods less than 60 s are lightest gray and maps 100 s and longer period are in black.

plate boundaries. One notable feature in this map is the clear division between east and west Antarctica. There are two places where the low velocities do not correspond with the plate boundaries from NUVEL-1 (DeMets *et al.*, 1990). The boundary between North America and Eurasia in Siberia appears further east in our maps, and the boundary between North and South America appears to be further south. We also note one surprising feature in the 90 s Rayleigh wave map: a north-south linear structure west-central in the Pacific. This structure appears to be a possible artefact of erroneous data, however we have confirmed that it is present in models derived from major- and minor-arc data separately. This feature may be a result of endeavoring to model azimuthally anisotropic material with an isotropic parameterization (see Larson, *et al.*, 1998, Figs. 8 and 13).

5. Comparison

To assess the agreement of global models with a regional model, we compare our maps with the high-resolution regional study of Eurasia of Ritzwoller and Levshin (1998). Figure 8a shows a visual comparison of the two studies for Love waves at 40 s, and Figure 8b is for Rayleigh waves at 100 s. It is apparent that the maps are very similar. The largest differences occur near the edge of the maps, particularly on the eastern edge of the region. This difference is due to the lack of coverage in the regional study. The global maps of this study agree with the long-established result that the lithosphere is fast under old basins (Dorman, 1969). In the central part of the maps, structures larger than about 700 km agree quite well. There are, of course, smaller scale features in the regional study, as small-scale features are not parameterized in the global modeling. Also, in the long-period map, the amplitudes of many small-scale anomalies are noticeably smaller in the global model.

In the 40 s Love-wave maps (Fig. 8a), the correlation is clear and the amplitudes agree well. As we survey the maps from north to south, we see striking agreement. Above 60°N, both models exhibit small anomalies in the east, a slow region with a maximum amplitude of about 17% in north central Russia, and various 5–10% slow anomalies in the west. Between 45°N and 60°N, there is less variation, although both models show slower regions near Belarus, at the north end of the Caspian Sea, north of the Kazakhstan-Russia western border, and northwest of Lake Baikal. The strongest anomalies are in the 30°N to 45°N range, with both models showing slow anomalies of 20–25% at the south end of the Caspian Sea and in eastern and western Tibet (but not as slow in the center). There is a sharp contrast between very slow velocities and only slightly slow anomalies west of 110°E. Between 15°N and 30°N, we again see matching strong anomalies, with both models showing regions 15–20% slow in the Persian Gulf and in Bangladesh.

In the 100 s Rayleigh-wave map (Fig. 8b), the correlation is still extremely high, but the amplitudes do not agree as well. Moving from east to west, we see a number

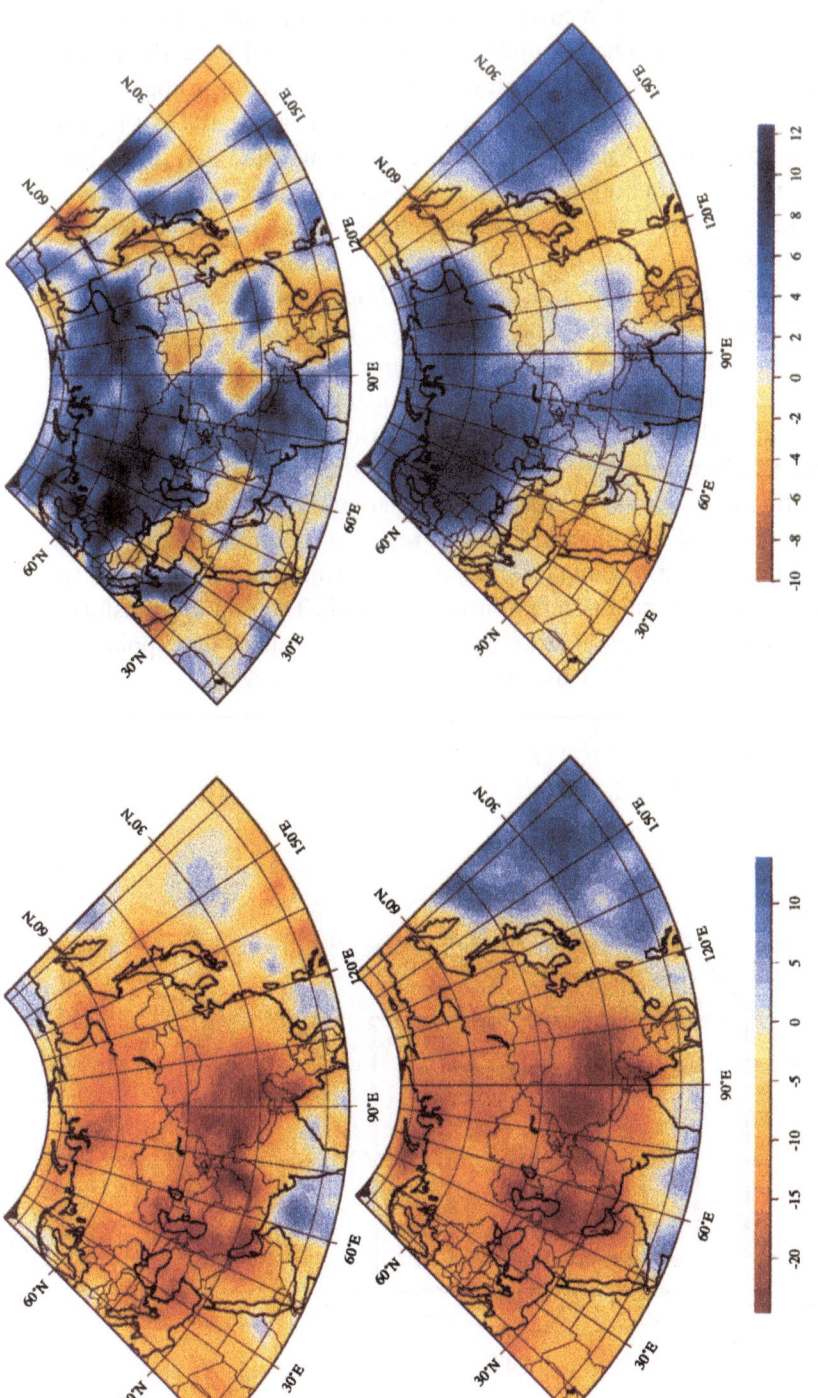

(a) Love wave 40 s

(b) Rayleigh wave 100 s

Figure 8

Comparison of group-velocity maps from RITZWOLLER and LEVSHIN (1998) study of Eurasia (top) with the global result of this study (bottom). Maps are plotted at the same scale, in percent variations of $\delta U / U_{\text{PREM}}$.

of matching features. Both models recover a line of slow regions along the coast of Asia, with slower spots in Kamchatka, under Sakhalin Island, off the coast of the Korean Peninsula, and off the coast of central China. A fast region is apparent in central Siberia. A belt of slow velocity connects Tibet to the coast, and a fast region in southeast China. Southeast Asia is slow, with a maximum anomaly of 3–5%. Central Russia is slightly fast in both maps and Kazakhstan has a complex structure. Tibet is characterized as a slow anomaly, whereas India is fast (although the location of this anomaly disagrees by a few hundred kilometers). There is disagreement in European Russia – both maps show a fast region, although the regional study has considerably larger amplitudes. Turkey and Iran are slow in both studies. The studies disagree in the Arabian Sea, however the Red Sea shows a 4–5% slow anomaly in both studies. In the rest of Africa there is scant data constraining the regional model.

To further determine the agreement between the two studies, we calculate several statistical parameters. Since the errors in the RITZWOLLER and LEVSHIN (1998) maps may be large near the edges, we have chosen two smaller interior regions to compare statistically: a large region of Eurasia and a smaller region of Central Asia, as shown in Figure 9. These regions are rectangular in equal-area projections, allowing us to easily calculate a two-dimensional discrete Fourier transform. The Fourier transform of a model $f(x_i, y_j)$ over a region R of size X by Y which is discretized in N_x points in the x direction and N_y points in the y direction is (PRESS et al., 1992)

$$F_{ij} = \frac{1}{\sqrt{XY}} \sum_{k=0}^{N_x-1} \sum_{l=0}^{N_y-1} e^{2\pi \iota x_k / \lambda_i^x} e^{2\pi \iota y_l / \lambda_j^y} f(x_k, y_l) ,$$

(7)

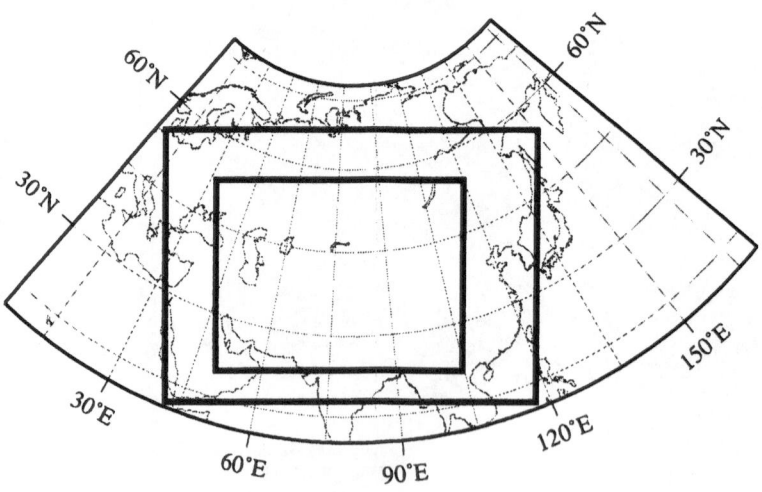

Figure 9
Boxes show regions used for statistical comparisons.

and $\lambda_m^x = X/x_m$ and $\lambda_m^y = Y/y_m$ are the wavelengths. From the Fourier transform coefficients, we calculate the amplitude spectra in the $A_x(\lambda_i)$ and $A_y(\lambda_j)$ directions by

$$A_x(\lambda_i^x) = \sqrt{\sum_j (F_{ij})^2} \qquad (8)$$

and

$$A_y(\lambda_j^y) = \sqrt{\sum_i (F_{ij})^2} \; . \qquad (9)$$

The amplitude spectra for maps at several frequencies are shown in Figures 10 and 11. These spectra allow comparison of the amplitude of the anomalies in the two

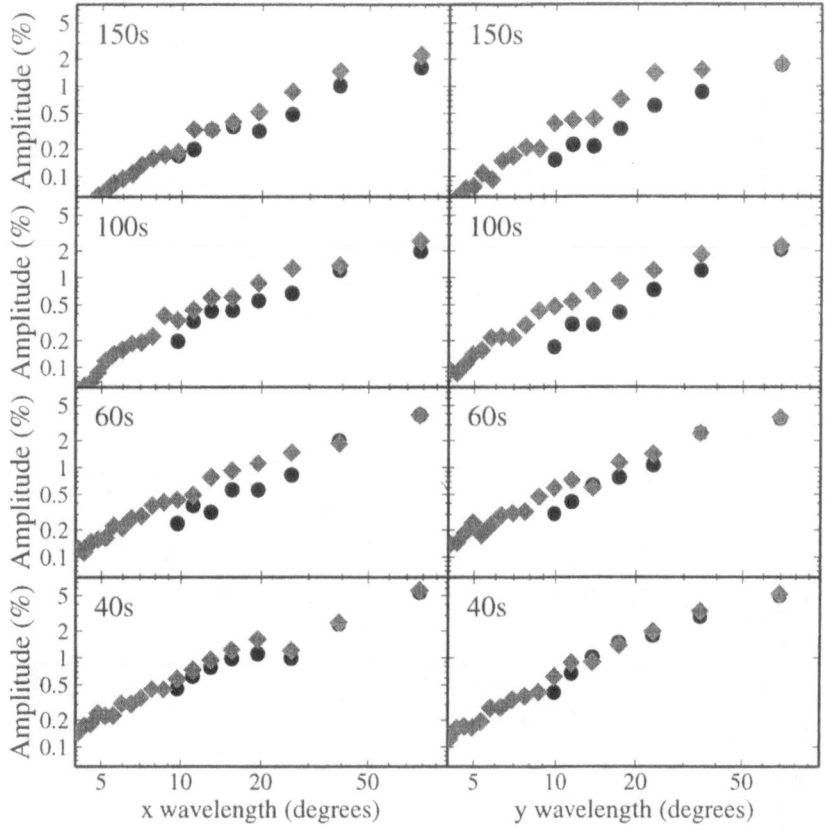

Figure 10

Spectra of Love-wave maps in Eurasia (computed in the large region in Fig. 9). Left panels are the amplitude spectra in the X direction (roughly east-west), and right panels are the amplitude spectra in the Y direction (roughly north-south). Circles are this study, diamonds are spectra of models from RITZWOLLER and LEVSHIN (1998). The period is in the upper left corner of each panel.

studies. We see excellent agreement between the two models, particularly for wavelengths longer than 30 degrees. At the shorter periods we see quite good agreement down to 15 degree wavelength. For shorter wavelengths, the global maps of this study consistently have lower amplitudes than the regional maps, especially for the long periods. The difference in amplitude for long-period Rayleigh waves is particularly large – the amplitude of the regional map is two to four times as large as the global map. To determine if this is due to our damping, we performed the inversions for Rayleigh waves without any damping. The effect on the amplitudes was insignificant for wavelengths longer than 1400 km. We return to discuss this discrepancy at the end of this section.

We calculate the correlation coefficient between the maps from the two studies at several different frequencies. The correlation coefficient, C, is defined in two dimensions by

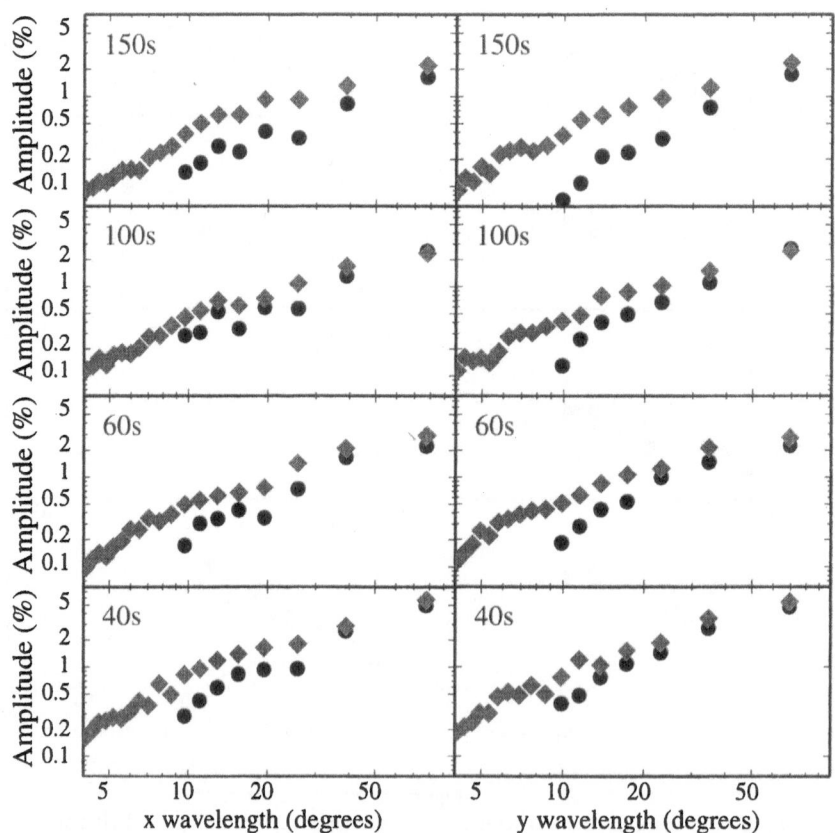

Figure 11
Spectra of Rayleigh-wave maps in Eurasia, see Figure 10 for an explanation.

$$C = \int_R \frac{f_1(\mathbf{r})f_2(\mathbf{r})}{\sqrt{[f_1(\mathbf{r})]^2 + [f_2(\mathbf{r})]^2}} \, d^2\mathbf{r} \ , \tag{10}$$

where f_1 and f_2 are the functions we are comparing (after the average value of each has been removed) and R is the region of comparison. We first compare the models over the large region in Figure 9. As shown in Figures 12 and 13, the overall correlation coefficients are extremely high – between 0.8 and 0.9 for Rayleigh waves, and between 0.7 and 0.9 for Love waves. However, this correlation may be a function of agreement at long wavelengths. To investigate this, we have removed the long-wavelength structure from the maps to test the agreement at shorter length scales. We filtered the models to allow only features with wavelengths between 1000 and 3100 km (equivalent to degrees 13–40 of a spherical harmonic expansion). We determine the correlation between these filtered maps, and the correlation remains very good – well above the 99% confidence level. Even though most of the signal can be explained by only long-wavelength structure, the smaller features in the models still agree. The agreement between these models, which are derived from completely different data-sets, reinforces our conclusion from statistical tests that we are justified in modeling small-scale structure. We note that the Rayleigh-wave

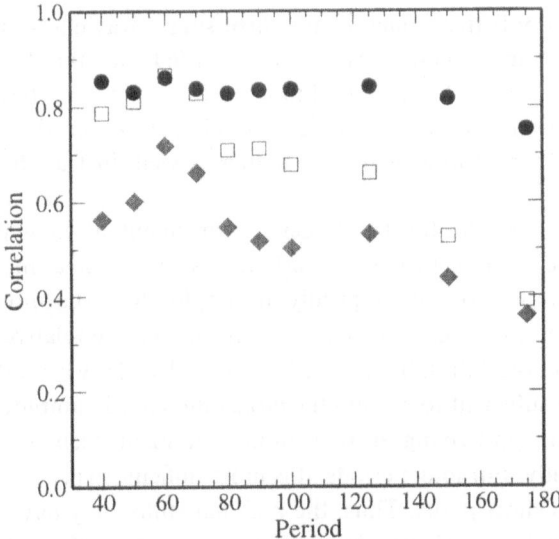

Figure 12

Correlation of Rayleigh-wave models in this study to models of RITZWOLLER and LEVSHIN (1998). Black circles are the total correlation in the large region of Figure 9, gray diamonds are the correlation for wavelengths 900–3000 km in the large region, and open squares are the correlation for wavelengths 900–3000 km in the small region. For the 145 parameters necessary to describe this resolution, the 99.99% confidence level is 0.32, and the 99% level is 0.25.

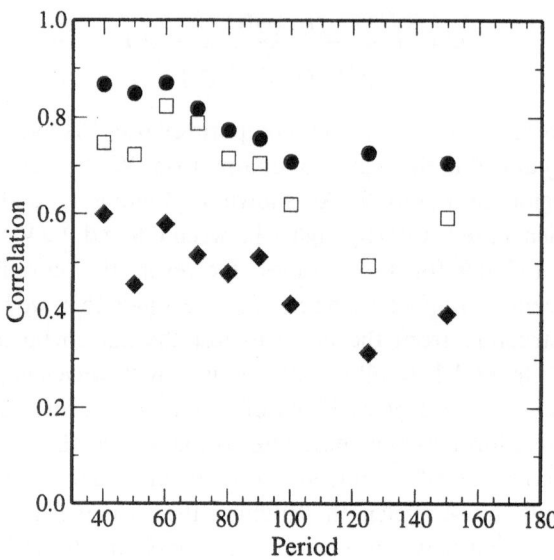

Figure 13
Correlation of Love-wave models in this study to models of RITZWOLLER and LEVSHIN (1998). See Figure 12 for explanation.

agreement is generally better, which is not surprising – Rayleigh-wave dispersion is easier to measure since these waves are recorded on the less-noisy vertical component. Also, because fundamental-mode Rayleigh waves have a low group velocity, they are easier to isolate from higher modes. However, the correlation plots show that the long-period models do not match as well, just as the comparison of spectra indicates.

Several possible reasons for the larger disagreement at long periods may be considered. As the period increases, surface waves become more sensitive to anomalies deeper in the earth, and typically the amplitude of the anomalies decreases with depth. Therefore, the signal is smaller, so the errors are relatively larger, which would lead to lower correlation between different models. However, the lower signal-to-error ratio is not sufficient to explain the large difference in amplitudes. There may be problems in the underlying dispersion-measurement technique for the long periods in either study. For long periods, it is more difficult to accurately measure the group velocity with short paths. Thus, the regional study may have larger errors at long periods and have insufficient damping to reduce the effect of errors. On the other hand, in this study, the smoothing constraints applied in the measurement of the phase-dispersion curve may limit the magnitude of the group velocity at long periods. Further research with different measurement methods will be necessary to determine the smaller-scale amplitudes at long periods.

6. Conclusions

We have produced global maps of surface-wave group velocity for both Love and Rayleigh waves in the period range 35 s to 175 s. The maps of group velocity can be useful for predicting the group arrival time of surface waves, which has applications in the location of earthquakes and explosions. These global models are particularly useful for monitoring purposes in regions where there are limited or no regional studies of group velocity, such as Africa and South America. The maps could also be used as starting models for regional studies to obtain even higher-resolution models where data coverage is sufficient.

At large scale-lengths, our new models agree well with previous work. In several regions, small features are shown which correlate well with variations that could be expected from surface geology. We have compared the maps to those of a regional study, and the correlation between the maps is extremely high. Global models are complementary to regional studies, as the quality of regional studies typically becomes poor near the edges of the model. Global models also provide information regarding less accessible regions of the Earth's surface.

Acknowledgements

We greatly appreciate the Generic Mapping Tools software (WESSEL and SMITH, 1998), which was used to produce all of the figures. This material is based upon work supported under a National Science Foundation Graduate Fellowship. We are grateful to M. Ritzwoller and A. Levshin for making digital versions of their maps easily available over the worldwide web. We thank J. Tromp and G. Laske for helpful discussions and for comments on this paper. We also extend appreciation to A. Curtis and A. Levshin for constructive reviews.

The degree 20 and degree 40 maps at all periods are available in several different formats from http://www.seismology.harvard.edu/projects/surfwave/LE00-Pageoph.

REFERENCES

CHEVROT, S., MONTAGNER, J. P., and SNIEDER, R. (1998), *The Spectrum of Tomographic Earth Models*, Geophys. J. Int. *133*(3), 783–788.

CURTIS, A., TRAMPERT, J., SNIEDER, R., and DOST, B. (1998), *Eurasian Fundamental Mode Surface Wave Phase Velocities and their Relationship with Tectonic Structures*, J. Geophys. Res. *103*(B11), 26,919–26,947.

DEMETS, C., GORDON, R. G., ARGUS, D. F., and STEIN, S. (1990), *Current Plate Motions*, Geophys. J. Int. *101*(2), 425–578.

DORMAN, J. *Seismic surface-wave data on the upper mantle*. In *The Earth's Crust and Upper Mantle*, Volume 13 of *Geophysical Monograph*, pp. 257–265 (Washington, DC, American Geophysical Union 1969).

DZIEWONSKI, A., CHOU, T., and WOODHOUSE, J. (1981), *Determination of Earthquake Source Parameters from Waveform Data for Studies of Global and Regional Seismicity*, Geophys. J. R. astr. Soc. *86*, 2825–2852.

DZIEWONSKI, A. M., and ANDERSON, D. L. (1981), *Preliminary Reference Earth Model*, Phys. Earth Planet. Int. *25*(4), 297–356.

DZIEWONSKI, A. M., BLOCH, S., and LANDISMAN, M. (1969), *A Technique for the Analysis of Transient Seismic Signals*, Bull. Seismol. Soc. Am. *59*(1), 427–444.

EKSTRÖM, G., TROMP, J., and LARSON, E. W. F. (1997), *Measurements and Global Models of Surface Wave Propagation*, J. Geophys. Res. *102*(B4), 8137–8157.

KEILIS-BOROK, V. I. (ed.) *Seismic Surface Waves in a Laterally Inhomogeneous Earth*. (Norwell, mass: Kluwer Academic Publishers 1989).

KNOPOFF, L., and SCHWAB, F. A. (1968), *Apparent Initial Phase of a Source of Rayleigh Waves*, J. Geophys. Res. *73*, 755–760.

LARSON, E. W. F., TROMP, J., and EKSTRÖM, G. (1998), *Effects of Slight Anisotropy on Surface-wave Propagation*, Geophys. J. Int. *132*(3), 654–666.

LASKE, G., and MASTERS, G. (1996), *Constraints on Global Phase Velocity Maps from Long-period Polarization Data*, J. Geophys. Res. *101*(B7), 16,059–16,075.

LEVSHIN, A., RATNIKOVA, L., and BERGER, J. (1992), *Peculiarities of Surface-wave Propagation across Central Eurasia*, Bull. Seismol. Soc. Am. *82*(6), 2464–2493.

LEVSHIN, A. L., RITZWOLLER, M. H., and RESOVSKY, J. S. (1999), *Source Effects on Surface Wave Group Travel Times and Group Velocity Maps*, Phys. Earth Planet. Int. *115*(3–4), 293–312.

LEVSHIN, A. L., YANOVSKAYA, T. B., LANDER, A. V., BUKCHIN, B. G., BARMIN, M. P., RATNIKOVA, L. I., and ITS, E. N. (1989), *Seismic Surface Waves in a Laterally Inhomogeneous Earth*, Chapter 5. In Keilis-Borok (1989).

MONTAGNER, J.-P., and TANIMOTO, T. (1990), *Global Anisotropy in the Upper Mantle Inferred from the Regionalization of Phase Velocities*, J. Geophys. Res. *95*, 4797–4819.

MUYZERT, E., and SNIEDER, R. (1996), *The Influence of Errors in Source Parameters on Phase Velocity Measurements of Surface Waves*, Bull. Seismol. Soc. Am. *86*(6), 1863–1872.

NISHIMURA, C. E., and FORSYTH, D. W. (1988), *Rayleigh Wave Phase Velocities in the Pacific with Implications for Azimuthal Anisotropy and Lateral Heterogeneities*, Geophys. J. R. astr. Soc. *94*, 479–501.

PASSIER, M. L., and SNIEDER, R. K. (1995), *On the Presence of Intermediate-scale Heterogeneity in the Mantle*, Geophys. J. Int. *123*(3), 817–837.

PRESS, W. H., TEUKOLSKY, S. A., VETTERLING, W. T., and FLANNERY, B. P., *Numerical Recipies in C: The Art of Scientific Computing* (2nd edn.) (Cambridge University Press 1992).

RITZWOLLER, M. H., and LEVSHIN, A. L. (1998), *Eurasian Surface Wave Tomography: Group Velocities*, J. Geophys. Res. *103*(B3), 4839–4878.

ROSA, J. W. C., (1987), *A Global Study on Phase Velocity, Group Velocity and Attenuation of Rayleigh Waves in the Period Range 20 to 100 Seconds*, Ph. D. Thesis, Massachusetts Institute of Technology, Cambridge, MA, USA.

TRAMPERT, J., and WOODHOUSE, J. H. (1995), *Global Phase Velocity Maps of Love and Rayleigh waves between 40 and 150 Seconds*, Geophys. J. Int. *122*, 675–690.

TRAMPERT, J., and WOODHOUSE, J. H. (1996), *High Resolution Global Phase Velocity Distributions*, Geophys. Res. Lett. *23*(1), 21–24.

VAN HEIJST, H. J., and WOODHOUSE, J. H. (1999), *Global High Resolution Phase Velocity Distributiuons of Overtone and Fundamental Mode Surface Waves Determined by Mode Branch Stripping*, Geophys. J. Int. *137*, 601–620.

WANG, Z., and DAHLEN, F. (1995), *Validity of Surface-Wave Ray Theory on a Laterally Heterogeneous Earth*, Geophys. J. Int. *123*, 757–773.

WANG, Z., TROMP, J., and EKSTRÖM, G. (1998), *Global and Regional Surface-wave Inversions; A Spherical-spline Parameterization*, Geophys. Res. Lett. *25*(2), 207–210.

WESSEL, P., and SMITH, W. H. F. (1998), *New, Improved Version of the Generic Mapping Tools Released*, EOS Trans. AGU *79*(47), 579.

ZHANG, Y.-S., and LAY, T. (1996), *Global Surface Wave Phase Velocity Variations*, J. Geophys. Res. *101*(B4), 8415–8436.

(Received June 30, 1999, revised February 5, 2000, accepted March 1, 2000)

 To access this journal online:
http://www.birkhauser.ch

Pure appl. geophys. 158 (2001) 1401–1423
0033–4553/01/081401–23 $ 1.50 + 0.20/0

| Pure and Applied Geophysics

Optimization of Cell Parameterizations for Tomographic Inverse Problems

Wim Spakman[1] and Harmen Bijwaard[1]

Abstract—We develop algorithms for the construction of irregular cell (block) models for parameterization of tomographic inverse problems. The forward problem is defined on a regular basic grid of non-overlapping cells. The basic cells are used as building blocks for construction of non-overlapping irregular cells. The construction algorithms are not computationally intensive and not particularly complex, and, in general, allow for grid optimization where cell size is determined from scalar functions, e.g., measures of model sampling or *a priori* estimates of model resolution. The link between a particular cell j in the regular basic grid and its host cell k in the irregular grid is provided by a pointer array which implicitly defines the irregular cell model. The complex geometrical aspects of irregular cell models are not needed in the forward or in the inverse problem. The matrix system of tomographic equations is computed once on the regular basic cell model. After grid construction, the basic matrix equation is mapped using the pointer array on a new matrix equation in which the model vector relates directly to cells in the irregular model. Next, the mapped system can be solved on the irregular grid. This approach avoids forward computation on the complex geometry of irregular grids. Generally, grid optimization can aim at reducing the number of model parameters in volumes poorly sampled by the data while elsewhere retaining the power to resolve the smallest scales warranted by the data. Unnecessary overparameterization of the model space can be avoided and grid construction can aim at improving the conditioning of the inverse problem. We present simple theory and optimization algorithms in the context of seismic tomography and apply the methods to Rayleigh-wave group velocity inversion and global travel-time tomography.

Key words: Model parameterization, tomography inverse problems.

1. Introduction

The subject of this paper is the design of highly irregular cell parameterizations and their use in tomographic problems. It originated from our aim to globally image structural detail of the order of 50–100 km in the mantle with travel-time tomography (Bijwaard *et al.*, 1998). Imaging such detail on the basis of a regular model parameterization (e.g., Fig. 1a) would require millions of cells for the upper mantle only. This number of unknowns can be reduced by about 1/4 to 1/3 if a cell model is used that conserves cell surface area (e.g., Fig. 1b). Although inversions with that many unknowns could be attempted, one is facing a severely

[1] Vening Meinesz Research School of Geodynamics Institute of Earth Sciences, Utrecht University, Budapestlaan 4, 3584 CD Utrecht, The Netherlands. E-mail: wims@geo.uu.nl

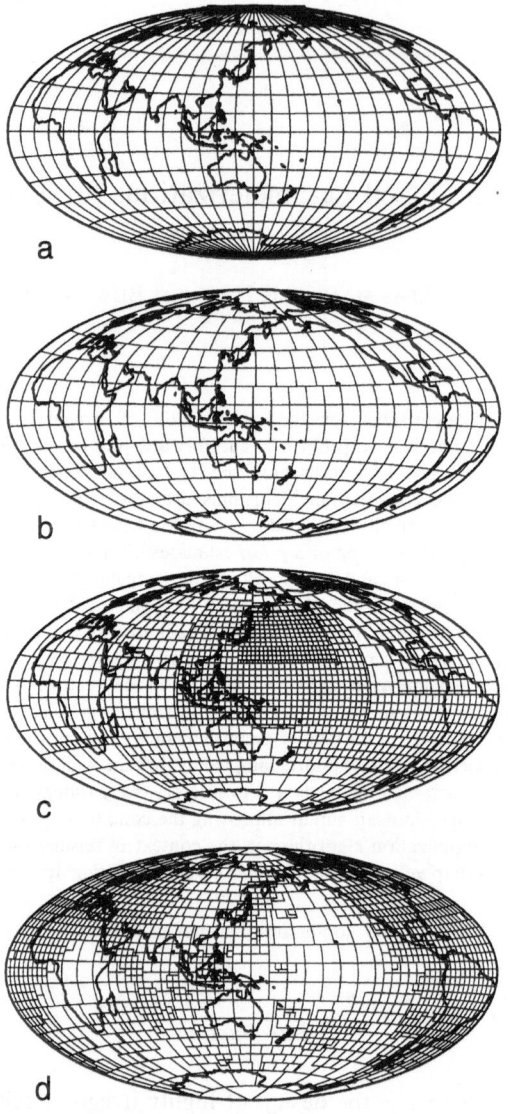

Figure 1

Four examples of cell models. a) A regular cell model based on a simple division of the earth's surface in equi-angular latitude and longitude intervals, in this case 12°. b) A cell model based on conservation of cell surface area. Latitudinal cell dimension is 12°, the longitudinal cell size adapts as a function of latitude. This type of model is called an *equal surface area cell model* or *ESA*-model. In the depth direction cell shapes are conical. c) A cell model based on overlaying regionally defined *ESA* cell models with different cell sizes. Also in the depth direction the cell division can be irregular. The cells are non-overlapping. d) A cell model in which cell dimensions are constrained by some scalar function defined or the mantle volume.

ill-conditioned inverse problem because large mantle volumes are abundantly overparameterized with respect to their poor ray sampling. Long computation times together with the fact that regularization is difficult to tune locally renders such huge inversions impractical at present. These problems motivated us to develop grid optimization algorithms leading to irregular cell models with the ultimate aim that cell dimensions should locally be in accord with the resolving power of the data. Such irregular models would necessarily lead to a large reduction in model parameters and generally yield a better conditioned inverse problem with the additional advantage that obtaining a solution is less demanding on explicit regularization.

Irregular parameterization has already been the topic of several studies offering a range of techniques for cell and node parameterizations. For relatively small size inverse problems (with hundreds to few thousands of unknowns) grid optimization can be achieved by simultaneous inversion for seismic velocities and node position (MICHELINI, 1995), interactively changing grid boundaries (VESNAVER, 1996), or by inversion for node position with genetic algorithms (CURTIS and SNIEDER, 1997). For larger size tomographic problems FUKAO et al. (1992) introduced regional grid densification where a grid of small cells is embedded in a global mantle model with larger cells. A similar approach has recently been implemented in spherical spline parameterizations for use in surface waveform inversion (WANG et al., 1998). ZHOU (1996) adopted a stack of regular global cell models of different cell sizes in which larger cells are overlapping with smaller ones. SAMBRIDGE et al. (1995) and SAMBRIDGE and GUDMUNDSSON (1998) provide a method of triangulation of irregularly spaced data as a tool for constructing irregular grids. None of these techniques provides a computationally efficient solution (for both small and large size inverse problems) for more general grid design problems. For instance: the construction of a cell grid in which cell dimensions lead to minimum hit count differences between adjacent cells, or a cell grid in which cell size adapts to *a priori* estimates of model resolution. In this paper we describe a method, based on cell parameterization, that solves these grid optimization problems.

We have adopted a strategy based on a starting point originally suggested by ABERS and ROECKER (1991): The mantle volume of interest is subdivided into an underlying or basic regular grid with small cells which are used (by some optimization algorithm) to construct larger cells in areas which are e.g., poorly sampled and poorly resolved. In Abers and Roecker's application grid construction is performed mainly by hand (G. Abers, personal communication) which puts a constraint on the size and the complexity of the parameterization problem one can still handle. Our extension is the development of automated and fast algorithms for the construction of a wide class of complex grids that can easily be built from millions of basic cells. We will demonstrate that the forward problem needs to be computed once only (on the basic grid) after which the grid optimization algorithms allow for easy experimenting with different (ir)regular grids at minimal computational costs. This adds considerable versatility to working with cell parameterizations.

In the following section we derive simple theory to demonstrate how to operate with irregular grids in seismic tomography without explicit computations involving their geometrical complexity. The construction will be the topic of later sections, followed by applications involving inversions of synthetic data, Rayleigh group-velocity inversion and global travel-time tomography.

2. Theory: Irregular Cell Models and Mantle Tomography

Define on the mantle volume V_M the usual inner product rule for any two piecewise continuous slowness fields $f(\mathbf{r})$ and $g(\mathbf{r})$:

$$(f, g) = \int_{V_M} f(\mathbf{r}) g(\mathbf{r}) \mathrm{d}V \tag{1}$$

Assume a *basic grid* of non-overlapping cells on which constant amplitude functions $b_j(\mathbf{r})$, $j = 1, 2, \ldots, N_b$ are defined:

$$b_j(\mathbf{r}) = \begin{cases} B_j^{-\frac{1}{2}} & \mathbf{r} \text{ inside cell } j \text{ with volume } B_j \\ 0 & \text{elsewhere} \end{cases} \tag{2}$$

Note that $(b_i, b_j) = \delta_{ij}$ (with δ_{ij} the Kronecker symbol). The constant amplitude of b_j functions is not restrictive for the analysis. Higher order functions, e.g., quadratic, can equally well be used on the cell domain although these lead to normalization factors different from $B_j^{-1/2}$. The b_j cells become the building blocks in the construction of irregular cell models. A computationally efficient choice will prove to be a regular cell model based on an equi-angular longitudinal and latitudinal division of the earth's surface.

From the basic grid cell models $c_k(\mathbf{r}), k = 1, \ldots, N_c$ are constructed with in practice $N_c \ll N_b$. We require that the c_k cells (with volumes C_k) are also non-overlapping and normalized (similar to (2)). The relation between the c_k functions and the b_j functions is given by:

$$c_k(\mathbf{r}) = \sum_{j=1}^{N_b} P_{kj} b_j(\mathbf{r}), \quad P_{kj} = (c_k, b_j) = \frac{B_j^{1/2}}{C_k^{1/2}} \Delta_{kj}. \tag{3}$$

$\Delta_{kj} = 1$ when b_j is a building block of cell c_k and is 0 otherwise. The Δ_{kj} are the only unknown coefficients in (3) because the cell volume is obtained as $C_k = \sum \Delta_{kj} B_j$.

The implementation of irregular cell parameterization certainly depends on the type of inverse problem. Here we give a simple derivation for use in travel-time tomography. The basic steps involved will be similar for other tomographic problems. The travel-time delay Δt_i acquired along the i-th seismic raypath L_i is:

$$\Delta t_i = \int_{L_i} \Delta s(\mathbf{r}) dl_i + \delta t_i \ . \tag{4}$$

$\Delta s(\mathbf{r})$ represents the unknown seismic slowness anomaly field of the earth and δt_i absorbs linearization and observation errors, and source and receiver terms. δt_i can be ignored here. Projecting $\Delta s(\mathbf{r})$ on the basis c_k gives

$$\Delta s(\mathbf{r}) = \sum_{k=1}^{N_c} s_k c_k(\mathbf{r}) + \delta s(\mathbf{r}) \tag{5}$$

in which $s_k = (\Delta s, c_k)$ are the unknown projection coefficients. $\delta s(\mathbf{r})$ is the projection residue which can be small for a sufficiently detailed basis c_k. Substituting (5) into (4) leads to (neglecting δs)

$$\Delta t_i = \sum_{k=1}^{N_c} s_k \frac{A_{ik}^c}{C_k^{1/2}} \ . \tag{6}$$

The coefficients A_{ik}^c are the arc lengths of the i-th ray in cells c_k. The computation of these coefficients can be time-consuming and troublesome if performed explicitly on the geometry of an irregular grid. With (3) the A_{ik}^c can be related to arc lengths A_{ij}^b of raypaths in the basic grid:

$$A_{ik}^c = \sum_{j=1}^{N_b} \Delta_{kj} A_{ij}^b \ . \tag{7a}$$

The A_{ik}^c are obtained by merely summing all arc lengths in the basic cells b_j that are contained by cell c_k. For M delay times (and corresponding rays) equation (7a) yields in matrix notation

$$\mathbf{A}_c = \mathbf{A}_b \mathbf{\Delta}^T \ . \tag{7b}$$

Using (6) and (7) the matrix equation to construct and to invert on an irregular grid becomes

$$\Delta \mathbf{t} = \mathbf{A}_b \mathbf{\Delta}^T \mathbf{C}^{-1/2} \mathbf{s}_c \ . \tag{8}$$

$\Delta \mathbf{t}$ is the delay time vector, $\mathbf{C}^{-1/2}$ is a diagonal matrix with the coefficients $C_k^{-1/2}$, and $\mathbf{s}_c$ is the model vector. The s_k in (5) assembled in $\mathbf{s}_c$ are not the physical slowness anomalies. The latter are defined as the volume average of $\Delta s(\mathbf{r})$ over the spatial domain of $c_k(\mathbf{r})$. The physical cell slowness is $\hat{\mathbf{s}}_c = \mathbf{C}^{-1/2} \mathbf{s}_c$ which after substitution in (8) and with (7b) transforms (8) into the usual tomographic equation but now defined on an irregular grid

$$\Delta \mathbf{t} = \mathbf{A}_c \hat{\mathbf{s}}_c \ . \tag{9}$$

Equation (8) conveniently factorizes the standard forward equation of travel-time tomography for the implementation of irregular grid construction and inverse problem weighting (if needed). Assuming that Δ can be constructed with moderate effort, equation (8) avoids the numerical complexity of obtaining A_c from explicit computation on the irregular grid. A particular b_j cell can only be a building block of exactly one c_k cell, therefore each column j of Δ contains only one non-zero element, a "1", at a particular row k (Fig. 2a) which allows collapsing of Δ into a pointer $k = p(j)$ of length N_b relating each b_j to its host cell c_k. Consequently grid construction concentrates only on creating the grid pointer $k = p(j)$.

The strategy for using irregular cell models can be summarized as follows (Fig. 2b). First, we define a regular basic grid for which A_b is computed. This is the major, but quite common, computational task in the forward problem. A_b must be computed only once. The next step is the construction of an irregular cell model consisting of cells c_k. This task effectively results in the pointer array $k = p(j)$ providing all information needed to compute A_c. Experimenting with different c_k models for inversion only requires the (cheap) multiplication of equation (7). During this multiplication the hit count (or other sampling measures) of the irregular grid

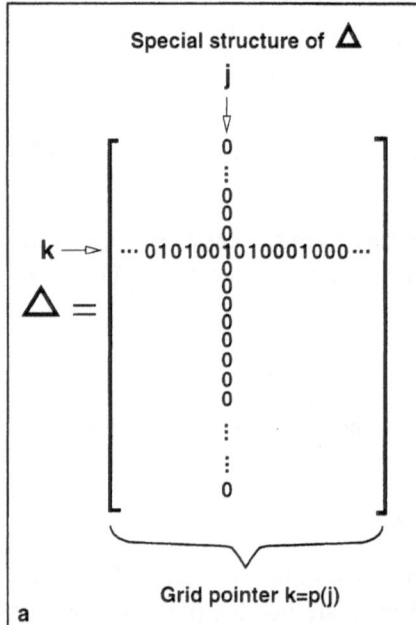
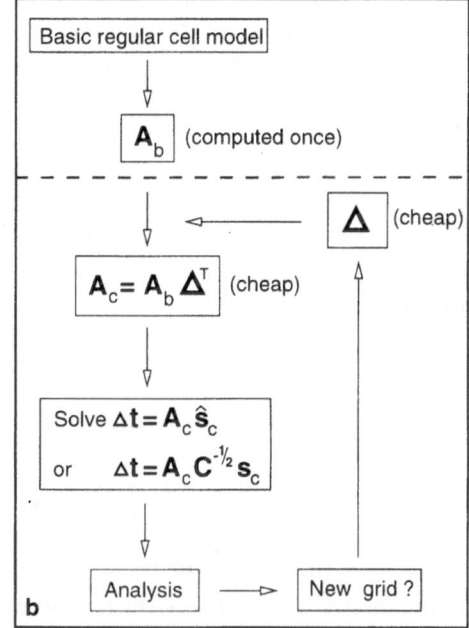

Figure 2

The grid pointer and its role in the forward and inverse problems of tomography. The left panel a) illustrates the structure of Δ and how this matrix can be collapsed into the grid pointer $k = p(j)$. For each basic cell j there is only one constructed cell k. The right panel b) shows that the use of Δ efficiently facilitates tomographic experiments with different cell grids at negligible computational cost in the forward problem.

also can be computed. The problem left is to ascertain efficient methods for finding $k = p(j)$.

3. Construction of Irregular Cell Models

We concentrate on global applications in spherical coordinates. Local problems or problems in Cartesian geometry follow the same grid construction rules. A basic cell grid is simply defined as a regular division of the earth's surface in cells with equal angular lateral dimensions, e.g., of 0.5°. The depth discretization is in layers of constant cell thickness. Such grids allow cheap computation of cell numbers from spatial coordinates. Constructing a grid pointer basically requires looping over the basic cell index j and assigning cell numbers k. We start with simple examples to illustrate basic steps for grid construction.

Regional densifications in regular cell models are obtained by first initializing the pointer with cell numbers k of a model with the largest cells to be used. Next a regional model with smaller cells is defined with cell numbering that is non-overlapping with the global model. The grid pointer is updated by looping over j and by assigning for j cells in the regional grid new cell numbers k. This process can be repeated for other regions and for different levels of cell size. At the end the grid pointer is sorted and renumbered in closed sequence to remove gaps in k-cell numbering.

Equal Surface Area (ESA) cell grids (Fig. 1b) are more complex in geometry than regular cell models but can still be systematically numbered. ESA grids (and overlays of ESA grids, Fig. 1c) are easily constructed following the same steps as for overlaying regular grids. In particular, the lateral cell dimensions of ESA grids can be increased with depth to approximate an Equal Volume (EV) cell model of the mantle. EV grids greatly reduce the number of model parameters needed for mantle parameterization as compared to simple regular or ESA models.

The grid pointers are obtained at negligible computational cost. After pointer construction A_c can also be computed cheaply and (8) can be solved for s_c. Note that all geometrical aspects of an irregular cell c_k can always be retrieved from the b_j cells for which $p(j) = k$. Appendix A gives a simple algorithm for determining neighbor-cell information which can be helpful for more complex grid construction and inverse problem regularization. Finally, after inversion the solution s_c can be expanded on the basic grid by assigning its values to a vector s_b of length $N_b (do\ j = 1, N_b; k = p(j); s_b(j) = s_c(k); enddo)$. Hence the cell numbering of a c_k model is unimportant for both the forward and inverse problems.

3.1 Grid Optimization: Constrained Cell Models

ESA and EV grids are particularly practical because we can exploit their surface (volume) conserving aspects in the construction of so-called constrained cell models

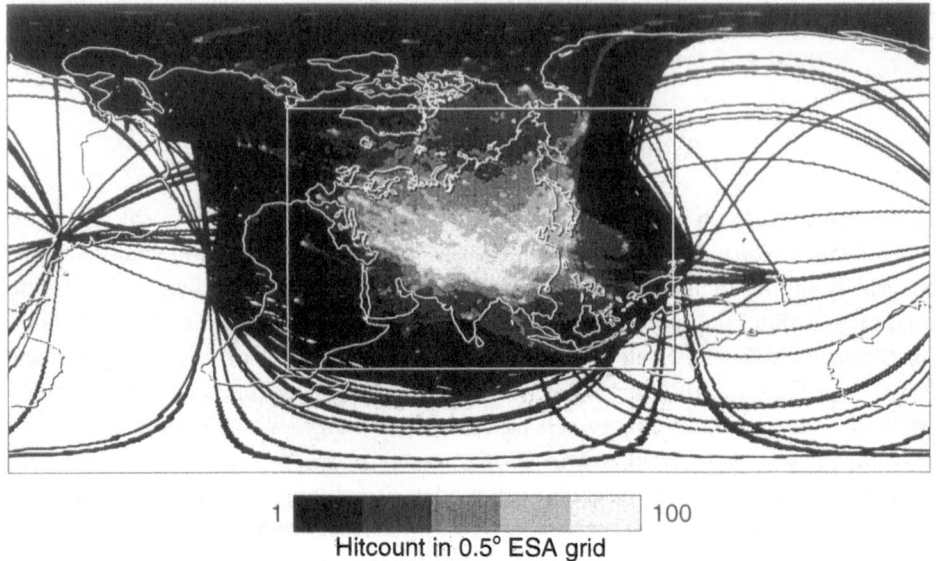

1 ▮▮▮▮▮▮▮▮▮▮▮▮ 100

Hitcount in 0.5° ESA grid

Figure 3
Hit count in a 0.5° ESA grid produced by 13,040 great circle paths associated with a Rayleigh-group velocity data set of RITZWOLLER and LEVSHIN (1998). The white box encompasses the region for which regional model and true fit are computed.

(Fig. 1d). The cell division of a constrained cell model satisfies a particular scalar measure defined on the study region. We will use a model sampling measure but other measures can be thought off. The grid algorithms lead to grid optimization subject to constraints such as: cells are at least sampled by N seismic rays and hit count differences between adjacent cells are minimal.

It is convenient to design the constrained grids in a practical example which, in later sections, also involves inversion of synthetic and real data on these grids. For the construction we use 13,040 great circle paths associated with Rayleigh-wave group-velocity measurements at period $T = 50$ determined by RITZWOLLER and LEVSHIN (1998). The basic matrix A_b is computed once on a 2-D regular basic grid with 0.5° equi-angular cells and consists of $720 \times 360 = 259,200$ cells. The hit count in a 0.5° ESA grid is depicted in Figure 3 and shows great global variability. We decide to work with minimum cell size of 2° which considerably reduces computation time for our synthetic inversion experiments. Four different grids are constructed.

GRID-I is just a simple global 2° ESA grid. It consists of 10,498 cells and is needed for comparison of inversion results with those obtained on constrained grids. The hit count is shown in Figure 4-I.

▶

Figure 4
Hit count and cell boundaries for grid-I to grid-IV. For grid-I only the hit count is shown.

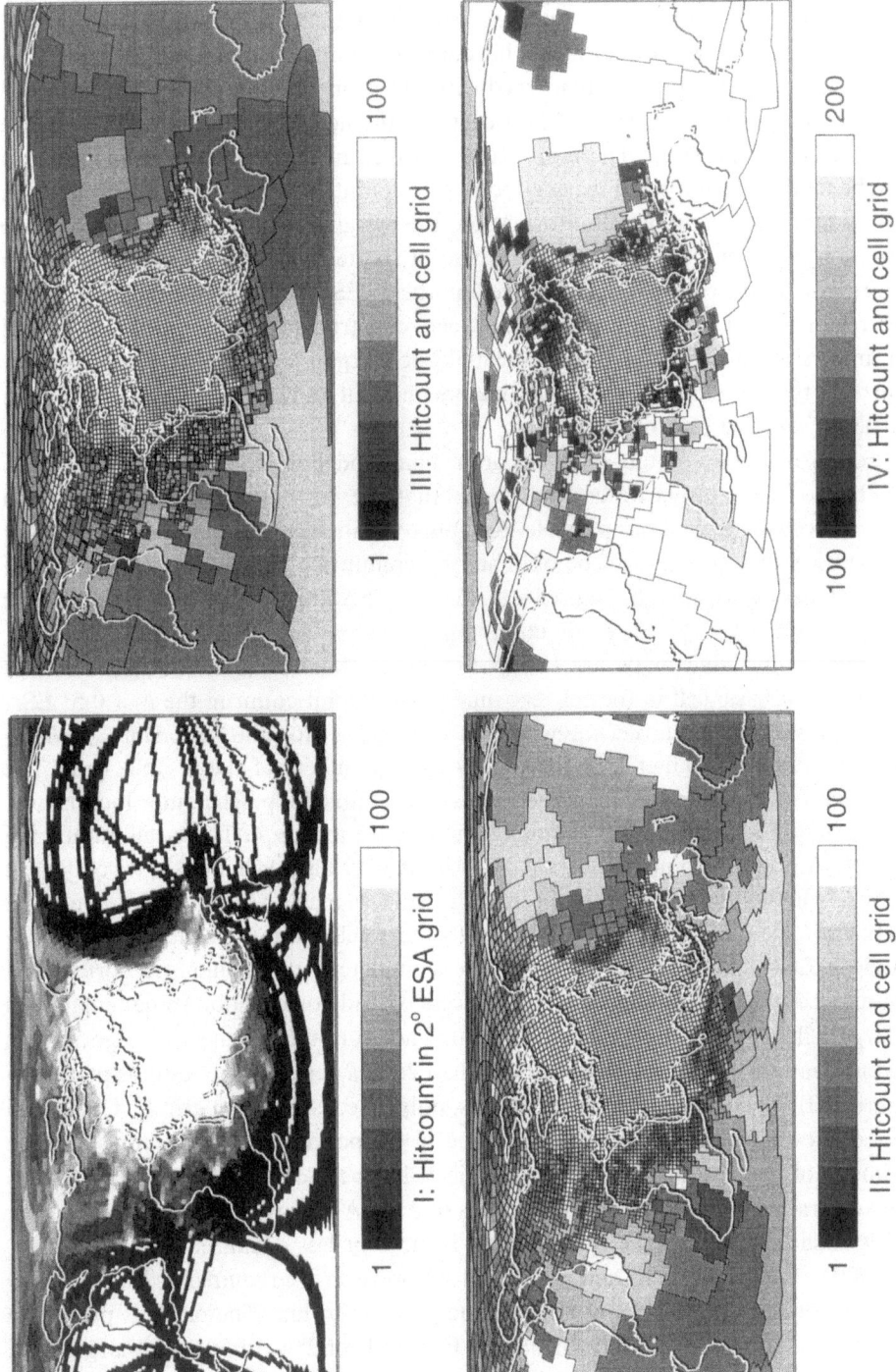

For *GRID-II* we require that cells in the final model should at least be sampled by a minimum number of rays (in our case 20). This requires a *cell merging algorithm*. First, neighbor-cell information and hit count is computed on a 2° ESA grid. ESA cells with hit count below 20 are merged with one or more neighbors by changing cell numbers in the grid pointer $p(j)$. This destroys the neighbor cell information of the grid one is manipulating. Therefore, cells involved in the renumbering process can only be used once in a renumbering loop. After pointer sorting and redetermination of hit count and neighbor cell information, the merging process is repeated. For the current application convergence is obtained in six iterations. The result is shown in Figure 4-II. In the central part of the region the 2° ESA cells are unaffected. Outside this region larger irregular cells have been constructed. The number of model parameters is reduced from 10,498 to 3805. The merging operation has considerably improved the sampling of model parameters and still allows for imaging 2° detail in the better sampled part of the model.

For *GRID-III* we require that cells in the final model obtain hit count as closely as possible to some target value and hence that the hit count differences between adjacent cells are as small as possible. An approximation of hit count as a function of cell position and volume can be obtained by computing hit count on a number of ESA cell models with increasing cell size. We use ESA grids with $N \times 0.5°$ cells where N goes from 4 to 12. Each hit count map is expanded on the basic grid. For convenience the nine maps are collapsed into a *cell-size map*: The number N is assigned to a basic cell in the cell-size map when the hit count in the $N \times 0.5°$ ESA grid is closest to the target value (50 in our case). For the *hit count equalizing algorithm* the grid pointer is initialized by the k numbers of the 6° ESA grid. The algorithm for determining $k = p(j)$ consists of a double loop. The outer loop is over the $N \times 0.5°$ ESA grids and is running from $N = 12$ to $N = 4$. The k-cell numbering of the ESA grid is recomputed. In the inner loop, over basic cell index j, we only need to check whether the value of the j cell in the cell-size map equals N. If this condition is satisfied, the pointer value k is replaced by the cell number in the current (more detailed) ESA grid. After the loops are completed, the pointer is sorted and renumbered in closed sequence. In the process of grid construction some remainders of large cells may have no or low hit count. Therefore the cell merging algorithm is used to clean the grid from cells with less than 25 hits per cell. The result is plotted in Figure 4-III. The grid consists of 3703 cells, only 98 cells less than grid II. Compared to grid II even larger cells have been placed in the poorest sampled regions and the perimeter of the core of 2° ESA cells in grid II has been changed to larger cells. The more central part of the region still consists of 2° ESA cells with hit count larger than 50. Outside this central part the hit count is more or less equalized.

GRID-IV is constructed similar to grid-III however we required at least 100 hits per cells for both the hit count and cell merging algorithm. The result is plotted in Figure 4-IV. It consists of only 2074 cells. ESA cells of 2° with hit count larger than 100 are only present in the core of Eurasia. High hit counts of very large cells result from a merging with smaller cells with hit counts of 100 or more. For grid-III we merged cells with a lower target value (25) than used for the hit count construction (50) which partly avoids this effect.

Grid optimization with the cell-merging algorithm or with the hit count equalizing algorithm aims at obtaining a better conditioned inverse problem by improving the sampling of model parameters in poorly sampled regions. Sampling constraints are just examples and can be replaced by other measures, e.g., cell models adapting to solution geometry, e.g., strong gradients/discontinuities, to estimates of spatially varying resolution (e.g., NOLET et al., 1999), to seismicity patterns, to topography or tectonic regionalization; the possibilities are many and depend on the particular application.

4. Synthetic Inversion Experiments

First we must establish how tomographic solutions are affected by our irregular cell parameterizations, how sensitive inversions on irregular grids are to regularization, and how underparameterization may influence the solution. For this purpose we perform a synthetic inversion experiment in which solutions are computed on all four cell models.

4.1 Experiment Design

The tomographic model obtained by RITZWOLLER and LEVSHIN (1998) from the inversion of the 13,040 group-velocity measurements is shown in Figure 5. The spatial characteristics of this solution show some correlation with the earth's topography, which lead us to devise a test aimed at imaging topography. This problem is cast as a group-velocity inversion experiment in which the earth's topography represents the group velocities. By numerical integration the following synthetic delay times d are determined:

$$d = \int\limits_{\text{ray}} \frac{1}{V_0[1 + t(\theta, \phi)]} \, ds - \frac{L}{V_0} \; ,$$

where L represents the arc length of surface-wave great circle path, and $t(\theta, \phi)$ are anomalous seismic velocities with respect to the reference velocity of $V_0 = 1$ km/s. The values of t are taken as the average of ETOPO5-topography over $0.5°$ ESA cells and are scaled to range between $+0.1$ and -0.1 (corresponding to $+10$ km and -10 km).

As a consequence of the projection residual in (5), even in case of ideal resolution, solutions determined on the four grids will never be able to reproduce the detail of true topography t. The underparameterization causes inconsistency between data and model predictions. This can also exist in real data experiments whenever the data are sensitive to smaller-scale earth structure than modeled by the cell (or other)

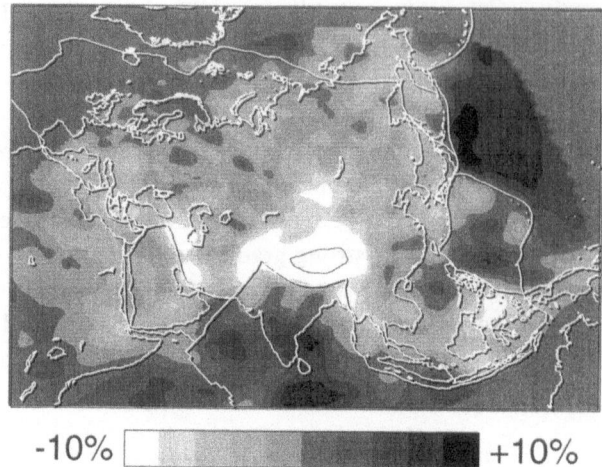

-10% ▢▨▨▨▨▨▨▨▨▨▨▨▨ +10%

Figure 5

Rayleigh-wave group velocities at period 50 s obtained by RITZWOLLER and LEVSHIN (1998) from an inversion of 13,040 data. In the Himalaya region an extra line contour (−20%) is added. Note that the solution is truncated in e.g., the Pacific and Atlantic.

parameterization used. To quantify effects of underparameterization we compute delay time values d_p from the projections of $t(\theta, \phi)$ on each grid. The differences $d - d_p$ result from underparameterization and lead to RMS data inconsistencies RMS $(d - d_p)/$RMS (d_p) of 3.84%, 7.23%, 8.14%, and 15.48% for grid-I to grid-IV, respectively. Hence, the inversion experiments are not noise-free and include effects of underparameterization.

The projected topography maps effectively yield for each grid the *exact solution* to the inverse problem. The comparison with inversion results leads to a measure of *model fit*. Inversion results are also compared to the true topography $t(\theta, \phi)$ which yields *true fit*. We define model (or true) fit as: RMS $(M - M_e)/$RMS (M_e) where M is a solution and M_e is either the exact model or the true topography. In real data inverse problems one is obviously interested in how well the model fits real earth structure, measured by the "true fit". However, the general resolution analyses based on a generalized inverse of $\mathbf{A}_c$ always lead to a measure of "model fit", because it is linked with the parameterization used. The results will reveal a difference between the two.

4.2 Inversion Results

In each grid-experiment solutions are computed for 25 different combinations of amplitude and second derivative damping. Values of 1, 500, 1000, 2000, and 3000 are used for both types of damping. We did not apply cell volume or hit count weighting of the model parameters; damping parameters relate directly to the matrix elements. The inversion algorithm is *LSQR* of PAIGE and SAUNDERS (1982) which has gained wide application in tomography since NOLET (1985). The zero-damped solutions were stopped after 200 *LSQR* iterations. Due to the regularization, all other inversions required less than 200 iterations for the normalized data misfit to become stable within one part per million.

Figure 6 summarizes the results of the 4×25 inversions of the same synthetic data in terms of model fit, true fit, and data fit. Each column of panels in Figure 6 gives the results for grid-I to grid-IV, respectively. We computed model and true fit on a global and regional scale. The regional fit is computed for the solution inside the white box in Figure 3. Figure 7 displays examples of inversion results which gave a good (model and true) fit to topography in the well sampled part of the model.

Data fit (Fig. 6E): Except for the zero-damped models, the data misfits are similar to or larger than the noise level in the data, hence the errors due to underparameterization may not have affected the solutions to a large extent.

Regularization: Solutions computed on the irregular grids (II–IV) are much less sensitive to damping than those computed on the regular grid (I). We attribute this effect to a better conditioning of the inverse problem as a result of improved sampling of the model parameters. It is well know that regularization changes the eigenvalue spectrum of the original inverse problem. The relatively small amount of damping we apply only regularizes the lower part of the spectrum which is often associated with poorly resolved and high variance solution components. The fact that the solutions on the irregular grids are less sensitive to regularization indicates that the reparameterization itself has modified the lower end of the eigenvalue spectrum associated with the grid-I inversion and has improved the conditioning of the inverse problem. The reduction in model parameters with respect to grid-I does not cause the reduced sensitivity to damping: a 4° ESA grid would encompass about 2600 cells with many cells unsampled or very poorly sampled and the inverse problem on this grid is therefore as ill-conditioned as that on grid-I.

Model fit (Fig. 6A, C): The solutions plotted in Figure 7 have comparably high regional model fit and are nearly equivalent in the well sampled part of the model (Eurasia) (irrespective of the different noise levels for each grid solution). We noticed, however, that in the well sampled region solutions on the irregular grids benefit from slight amplitude damping to suppress small-scale amplitude variations. Hence, underparameterization of poorly sampled regions may have an effect on the quality of the solution in the well sampled parts, although the effects can be controlled. We observe an increase in global model fit from left to right in row C. Apparently, reparameterization for improved parameter sampling leads to better convergence to the exact global solution. This improvement can basically also be achieved for regular grids by spatially variable regularization with first and second derivative damping which have an interpolating and smoothing effect on the solution. However, these regularization operators are known for their slow convergence (VANDECAR and SNIEDER, 1994) and would require more computation time to lead to a completely interpolated solution. We note that inversions on grid-I already require a factor between 2 and 3 more computation time compared to solutions on the irregular

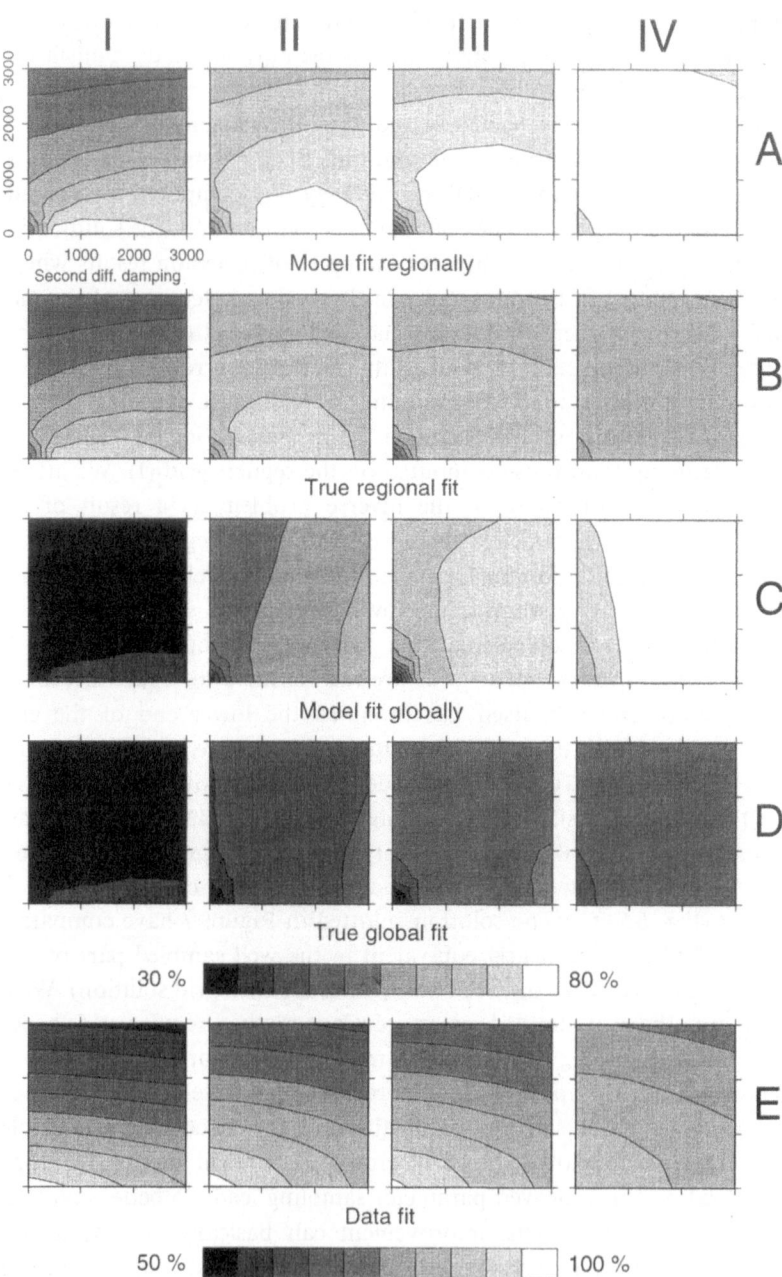

Figure 6

Summary of results obtained in 4 ∗ 25 synthetic inversions. Each column refers to results obtained on a particular grid (I to IV). The rows summarize: regional model fit (row A), true regional fit (B), global model fit (C), true global fit (D), and data fit (E). Each panel shows results for combinations of second derivative regulation (horizontal axis) and amplitude damping. See upper left panel for units. The contouring scale below row D pertains to results plotted in A to D.

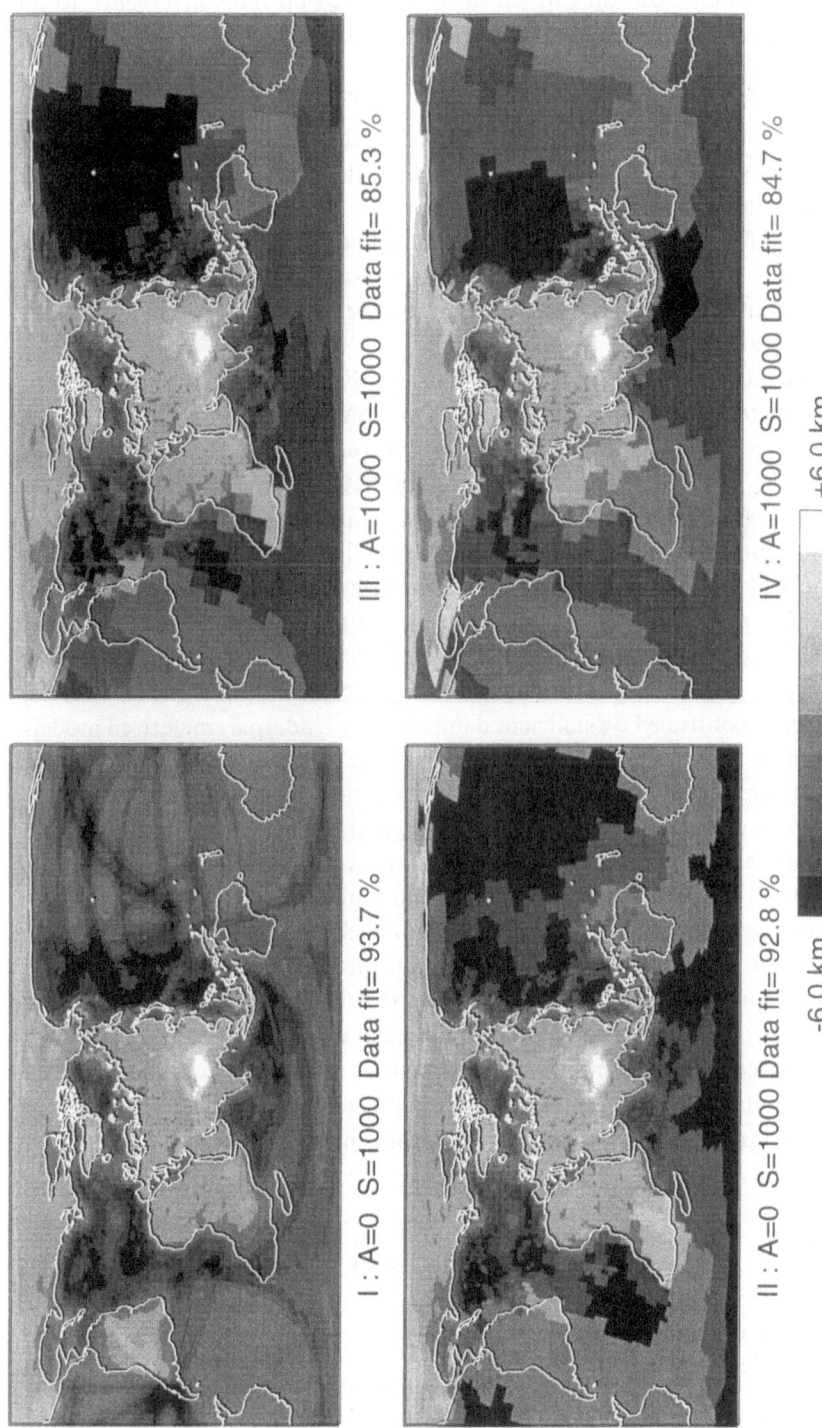

Figure 7

Synthetic inversion results with high regional model fit and true fit for all four grids. The amount of amplitude damping (A) and second derivative damping (S) is indicated.

grids. Hence, reparameterization can considerably speed-up the convergence to an acceptable least-squares solution and can replace the strong regularization otherwise needed in poorer sampled regions on a regular grid.

True fit (Fig. 6B, D): Good regional fit is obtained for grids I, II, and III, however it is somewhat less for grid-IV. Apparently the resolution at 2° is sufficient for most of this region and grid-IV, with many larger cells in the central region, is an overly restrictive parameterization for the current problem. Globally, we again infer an increase of true fit for irregular grids, although the difference between grid-I and grid-III is only 8%. From inspection of Figure 7 we can also infer that the irregular grids to a better job in fitting the topography outside the well sampled region, albeit they yield topography averages over huge cells.

The experiment was aimed at studying the sensitivity of solutions for different (ir)regular parameterizations, in particular: how the solution in a well resolved part is affected by changing the parameterization elsewhere. This is a general problem independent of the particular experiment geometry, but which needs to be addressed for our type of irregular parameterization. The main conclusion of this test is positive: in the well resolved part of the model (Eurasia) we obtained quite similar solutions on all grids although for different damping parameters. In more detail we infer the following: (i) Our grid optimization based on hit count constraints leads to better conditioned inverse problems which are therefore less demanding on tuning of regularization. (ii) Undersampling of structure in regular and irregular grids affects the solution but can be controlled by sufficient damping. (iii) Underparameterized models may be well resolved formally (good model fit) but can still perform less well in fitting true earth structure. Hence, slight overparameterization of actual earth structure would lead to more useful resolution estimates. Irregular parameterization allows for a compromise between overparameterization in regions well sampled and underparameterization elsewhere, while effects of undersampling can be controlled. In this respect the main advantage of irregular grids is computational efficiency in the inversion stage (in our experiment a factor 2-3) without trading in on solution quality. Inversions based on the computation of generalized inverses (for instance with Singular Value Decomposition) are decidedly more computer intensive than inversions with *LSQR*. For such methods considerably more computation time reduction can be achieved by inversion on irregular cell grids. Recall that our implementation of grid optimization also leads to substantial savings in the forward problem (cf. Fig. 2).

5. Real Data Experiments

5.1 Rayleigh Wave Group Velocity Inversion

Actual group velocities have a different sensitivity for earth structure than synthetic data and may have complex noise statistics (including effects of forward model linearization). We inverted the data of RITZWOLLER and LEVSHIN (1998) on all four grids used in the synthetic experiment. The aim here is to closely approximate the result of Ritzwoller and Levshin (Fig. 5). They used a Backus-Gilbert inversion approach which does not require *a priori* model parameterization. They also invoke an *a posteriori* smoothing of the inversion result using a spatially dependent filter. After experimentation with damping parameters we found acceptable solutions for zero amplitude damping and a value of 3700 for second derivative damping. The need of increased damping relative to the synthetic experiments is probably related to larger data errors and/or the sensitivity of Rayleigh waves for the detail of crust/ lithosphere structure. Figure 8 shows the inversion results for the four grids which must be compared to Figure 5. The data fit for grids I to IV are: 64.9%, 64.0%, 64.1%, and 63.6%, respectively. The data fit obtained by RITZWOLLER and LEVSHIN (1998) for the model displayed in Figure 5 is 63.2%. All solutions compare very well with the result of Ritzwoller and Levshin. The differences are mainly at the level of local smoothness determined by our 2° cells and their *a posteriori* smoothing. The computational speed up with respect to inversions on grid-I varies again by factors ranging from 2 to 3. The synthetic test and this real data application demonstrate that our irregular cell grid approach is a useful and efficient tool for cell parameterization of tomographic problems.

5.2 Global Travel-time Tomography

Additional value of our implementation of irregular cell grids is best exemplified by the global travel-time application of BIJWAARD *et al.* (1998). They used about 7.6 million *P*- and *pP*-delay times, selected from the data set of ENGDAHL *et al.* (1998), for the construction of A_b on a 0.6° basic grid with 26 cell layers covering the entire mantle. Next they applied the hit count equalizing algorithm for construction of a global mantle parameterization. This 3-D application does not complicate the algorithm. A target value was taken of 500 hits per cell in the upper mantle and of 1000 in the lower mantle, which avoids placing too large cells in the upper mantle and too small cells in the lower mantle. In the upper mantle the smallest cells allowed have dimensions of 0.6°. A minimum cell size of 1.2° is taken for the upper part of the lower mantle, and below 1100 km the smallest cells are 1.8°. The final irregular cell model encompasses $N_c = 277,565$ cells constructed from the $N_b = 4,680,000$ basic cells; a reduction in model parameters of about 94%. Compared to the computation of A_b (about 10 CPU hours on an SGI-R8000 processor, raytracing excluded) the grid is constructed at low computational costs. Apart from the determination of hit count on several ESA grids prior to grid construction, which is a one-time computation, the construction of the final grid takes only about 10 minutes which allowed for quick experimenting to obtain suitable target values for hit count. In those mantle volumes where data sampling is sufficiently high, the model parameterization allows for imaging detail on the order of 60 km in the upper mantle, of 100 km between depths

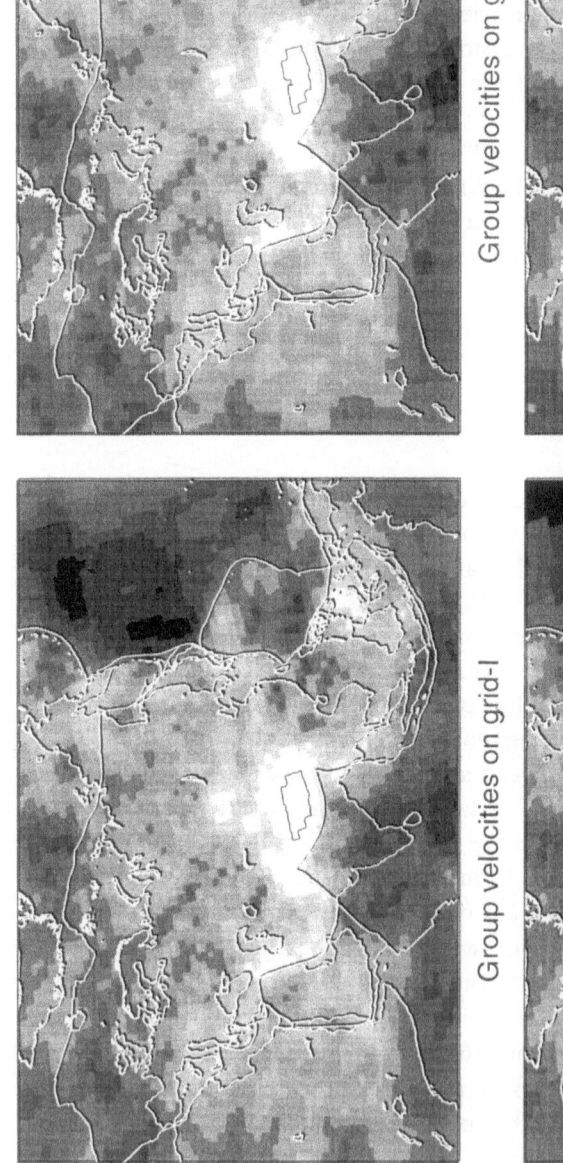

Group velocities on grid-I

Group velocities on grid-III

Group velocities on grid-II

Group velocities on grid-IV

-10% +10%

◄

Figure 8
The results of inverting Rayleigh-wave group velocity measurements at 50 s on all four grids. The damping is the same for all four solutions: no amplitude damping and a value of 3700 for second derivative damping. An extra contour line (−20%) is added for the Himalaya region. Compare to Figure 5.

of 660 to 1100 km, and of 150 to 200 km below 1100 km. Hence the large numerical reduction in the cells does not impair the potential for imaging small details in well sampled regions. Inversions on this cell grid take about 3 days (200 iterations with *LSQR* and I/O included). Regularization was performed by minimizing the second derivative of the solution. Local tuning of regularization proved most effective by making the regularization cell size dependent with larger damping for larger cells.

Figure 9 depicts a regional slice through the tomographic model with the irregular grid superposed and demonstrates how the cell model allows for the detailed imaging

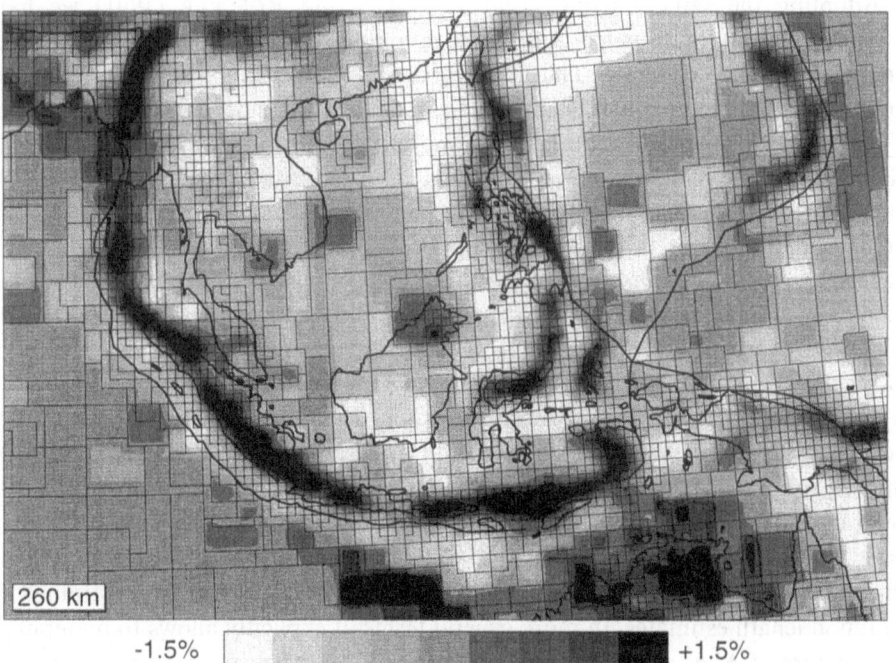

260 km

-1.5%   +1.5%

Figure 9
Part of the *P*-wave velocity anomaly model of BIJWAARD *et al.* (1998) with superposed cell division of the model. The cut displays the southeast Asian region at a depth of 260 km. Dark, relatively thin, and elongated zones represent the image of subduction below the Sunda arc (to the west and south of Indonesia), and the complex pattern of subduction below the eastern Sunda margin (Banda, Sulawesi, Philippines, Taiwan), and below the Mariana's (upper left). At this depth the subduction zones are sampled by the smallest cells (0.6° × 50 km).

of slab structure on global mantle scale. Structures on the order of 100 to 250 km are nicely sampled by the smallest cells. The use of an irregular grid in this global application has been fundamental for obtaining a mantle model that focusses on fine details of upper mantle (e.g., slab subduction) and lower mantle structure (e.g., BIJWAARD and SPAKMAN, 1999a; VAN DER VOO et al., 1999). A coarser parameterization employing the same number of model parameters in a regular grid does not yield such an approximation of these slab structures (see BIJWAARD et al., 1998). We note that an inversion on an ESA grid of 0.6° and with the same depth division requires about 3.2 million model parameters. We estimate that an inversion on this regular grid would require 5 to 10 times longer computation time (15–30 days). Further, it would be very demanding on local tuning of spatially dependent regularization. Hence this inversion would be rather impractical at present.

6. Discussion

Adopting the same starting point as ABERS and ROECKER (1991) we have developed new and automated grid optimization algorithms to construct arbitrarily complex irregular cell grids. The algorithms are simple and fast, do not involve complex computations, are basically scale and application independent, avoid *ad hoc* solutions to irregular cell parameterization, and require only minor adaptation of existing tomography code based on cells. Our approach leads to a practical generalization of cell-based tomography because it allows inversion on both regular and irregular grids at negligible cost in the forward problem (cf. Fig. 2b). This efficiency creates versatility in experimenting with many different grids of arbitrary complexity and is not offered as inexpensively by other parameterization techniques. Effects of underparameterization (a problem in any type of parameterization) can be analyzed by experimentation with different grids. With respect to solutions obtained on regular cell models our experiments indicate that solutions on constrained irregular grids are of comparable quality in well sampled model regions and may improve solutions in poorly sampled regions, all at considerably lower computational cost. This is particularly advantageous for very large size inverse problems where working with dense regular grids would become prohibitively expensive and difficult to tune. Further, our approach allows for grid optimization constrained by resolution length estimates. In more general terms: it explicitly allows to manipulate the structure of $A_c^T A_c$, i.e., implicitly control the eigenvalue spectrum of the normal matrix.

The only restriction on the design of the final grid is the basic cell size. However, the basic cell size can be chosen arbitrarily small and is only limited by available computer memory to store the pointer in the grid algorithm programs and for the matrix mapping. Physical constraints only apply to cells in the constructed model and not to the basic building blocks.

Cell parameterizations with constant amplitudes may seem restrictive on the approximation of structure or on the quality of 3-D ray tracing applications. For instance, WANG *et al.* (1998) advocate the use of higher order parameterizations from which slowness derivatives for raytracing can be readily obtained. These derivatives can, however, also be well approximated from cell parameterizations. By projecting with (5) a local linear slowness $s(\mathbf{r}) = s(\mathbf{r}_0) + \mathbf{\Gamma} \cdot (\mathbf{r} - \mathbf{r}_0)$ centered about cell midpoint $\mathbf{r}_0$ in a 3-D cell model, $\mathbf{\Gamma}$ is obtained exactly by finite differencing of cell amplitudes. From a similar but 1-D derivation involving local quadratic behavior it can be shown that the second derivative is overestimated by only 8%. Hence local smoothness up to second order can also be provided by a sufficiently detailed cell parameterization. Cell boundaries can be ignored where 3-D raytracing is concerned. In good approximation velocities and gradients can be determined from interpolation of cell values (BIJWAARD and SPAKMAN, 1999b). We remark that also higher order functions (e.g., quadratic) can be defined on the cell domains.

At present, the only other method that allows for the construction of arbitrarily complex irregular grids for both small- and large-scale tomographic inverse problems is the Tetrahedra/Voronoi cell method of SAMBRIDGE and GUDMUNDSSON (1998). As differences we mention that (i) their $\mathbf{A}_c$ must be computed explicitly on the geometrical complexity of the grid of tetrahedra/Voronoi cells (for which they provide fast routines), (ii) their method does not offer grid optimization in the sense of constrained grid construction. A quick and approximate solution to the latter would be to use our cell algorithms as preprocessor for the determination of node positions.

7. Conclusions

We started with a simple demonstration, condensed in Figure 2b, that cheap construction of grid pointers $k = p(j)$ would enhance the versatility of using regular and irregular cell parameterizations for tomographic studies. The grid pointer approach circumvents the recomputation of the forward problem and avoids computations on the geometrical complexity of irregular cell grids. The advancement we make is that automated design of irregular cell parameterizations is indeed inexpensive. This allows for easy experimentation with different grids to solve the inverse problem. Further, our algorithms facilitate grid optimization in which cell model construction is constrained by some scalar function defined on the model volume. This proves powerful because it allows quantified and controllable design of irregular parameterizations. We took hit count to constrain grid construction although many other possibilities exists (e.g., model resolution lengths). Our experiments demonstrate that reparameterization can improve the conditioning of the inverse problem. Demands on regularization are reduced and local tuning is made effective by making regularization weights cell-size dependent. Generally a large reduction in model parameters is achieved. This leads to considerable computational savings for the inverse problem (with respect to solving on a regular grid) without sacrificing solution quality and makes huge inversions for detailed mantle structure possible.

Acknowledgements

We are grateful to Mike Ritzwoller and Anatoli Levshin for providing us with the group-velocity data set. Jeroen Tromp, Alberto Michelini and two anonymous reviewers offered constructive comments. Support is received from the Netherlands Geosciences Foundation (GOA) with financial aid from the Netherlands Organisation for Scientific Research (NWO) through the Pionier project PGS 76-144, through hardware grant NWO-750.396.02, and for support of H.B. through grant NWO-750.195.13. This work was conducted under the program of the Vening Meinesz Research School for Geodynamics (VMSG).

Appendix
A Neighbor-cell Algorithm

With two loops over the pointer array $k = p(j)$ it is possible to assemble for each constructed cell k the numbers j of basic cells hosted by each k cell and store this information in a help array of length N_b. Next we exploit the regularity of the basic grid to find neighbors in the *basic* grid of all j cells which are the building blocks of a particular k cell. For each of these j neighbors we determine if it is located in a constructed cell different from cell k (again using $k = p(j)$). If so, we have found a neighbor of k and its cell number is stored in a list. Once the j loop (for one k cell) is finished, sort the list of k numbers, remove the multiples, and store the result.

REFERENCES

ABERS, G. G., and ROECKER, S. W. (1991), *Deep Structure of an Arc-continent Collision: Earthquake Relocation and Inversion for Upper Mantle P- and S-wave Velocities Beneath Papua New Guinea*, J. Geophys. Res. *96*, 6379–6401.

BIJWAARD, H., and SPAKMAN, W. (1999a), *Tomographic Evidence for a Narrow Whole Mantle Plume below Iceland*, Earth Plan. Sci. Lett. *166*, 121–126.

BIJWAARD, H., and SPAKMAN, W. (1999b), *Fast Kinematic Raytracing of First and Later Arriving Seismic Phases*, Geophys. J. Int. *139*, 359–369.

BIJWAARD, H., SPAKMAN, W., and ENGDAHL, E. R. (1998), *Closing the Gap between Regional and Global Travel-time Tomography*, J. Geophys. Res. *103*, 30,055–30,078.

CURTIS, A., and SNIEDER, R. (1997), *Reconditioning Inverse Problems Using the Genetic Algorithm and Revised Parameterization*, Geophys. *62*, 1524–1532.

ENGDAHL, E. R., VAN DER HILST, R. D., and BULAND, R. P. (1998), *Global Teleseismic Earthquake Relocation with Improved Travel Times and Procedures for Depth Determination*, Bull. Seismol. Soc. Am. *88*, 722–743.

FUKAO, Y., OBAYASHI, M., INOUE, H., and NENBAI, M. (1992), *Subducting Slabs Stagnant in the Mantle Transition Zone*, J. Geophys. Res. *97*, 4809–4822.

MICHELINI, A. (1995), *An Adaptive-grid Formalism for Travel-time Tomography*, Geophys. J. Int. *121*, 489–510.

NOLET, G. (1985), *Solving or Resolving Inadequate and Noisy Tomographic Systems*, J. Comp. Phys. *61*, 463–482.

NOLET, G., MONTELLI, R., and VIRIEUX, J. (1999), *Explicit, Approximate Expressions for the Resolution and a posteriori Covariance of Massive Tomographic Systems*, Geophys. J. Int. *138*, 36–44.

PAIGE, C. C. and SAUNDERS, M. A. (1982), LSQR: An algorithm for sparse linear equations and sparse least squares, ACM Trans. Math. Soft. *8*, 43–71.

RITZWOLLER, M. H., and LEVSHIN, A. (1998), *Eurasian Surface Wave Tomography: Group Velocities*, J. Geophys. Res. *103*, 4839–4878.

SAMBRIDGE, M., BRAUN, J., and MCQUEEN, H. (1995), *Geophysical Parameterization and Interpolation of Irregular Data Using Natural Neighbours*, Geophys. J. Int. *122*, 323–342.

SAMBRIDGE, M., and GUDMUNDSSON, O. (1998), *Tomographic Systems of Equations with Irregular Grids*, J. Geophys. Res. *103*, 773–781.

VANDECAR, J. C., and SNIEDER, R. (1994), *Obtaining Smooth Solutions to Large, Linear, Inverse Problems*, Geophys. *59*, 818–829.

VAN DER VOO, R., SPAKMAN, W., and BIJWAARD, H. (1999), *Tethyan Subducted Slabs under India*, Earth Planet. Sci. Lett. *171*, 7–20.

VESNAVER, A. L. (1996), *Irregular Grids in Seismic Tomography and Minimum-time Raytracing*, Geophys. J. Int. *126*, 147–165.

WANG, Z., TROMP, J., and EKSTRÖM, G. (1998), *Global and Regional Surface-wave Inversions: A Spherical Spline Parameterization*, Geophys. Res. Lett. *25*(2), 207–210.

ZHOU, H.-W. (1996), *A High-resolution P-wave Model for the Top 1200 km of the Mantle*, J. Geophys. Res. *101*, 27,791–27,810.

(Received September 2, 1999, accepted March 15, 2000)

 To access this journal online:
http://www.birkhauser.ch

Pure appl. geophys. 158 (2001) 1425–1444
0033–4553/01/081425–20 $ 1.50 + 0.20/0

▌Pure and Applied Geophysics

Surface Wave Velocities Across Arabia

T. A. Mokhtar,[1] C. J. Ammon, R. B. Herrmann[2] and H. A. A. Ghalib[3]

Abstract — The group-velocity distribution beneath the Arabian Plate is investigated using Love and Rayleigh waves. We obtained a balanced path coverage using seismograms generated by earthquakes located along the plate boundaries. We measured Love- and Rayleigh-wave group-velocity dispersion using multiple filter analysis and then performed a tomographic inversion using these observations to estimate lateral group velocity variations in the period range of 5–60 s. The Love- and Rayleigh-wave results are consistent and show that the average group velocity across Arabia increases with increasing period. The tomographic results also delineate first-order regional structure heterogeneity as well as the sharp transition between the Arabian shield and the Arabian platform. Systematic differences are observed in the distribution of the short-period group velocities across the two provinces, which are consistent with surface geology. The slower velocities in the platform reveal the imprint of its thick sedimentary section, while faster velocities correlate well with the exposed volcanic flows in the shield. Shear-wave velocity models for the two regions, obtained from the inversion of the group velocities, confirm results from previous studies of higher *S*-wave velocity in the upper crust beneath the shield. This may be due to the present remnants of the oceanic crust (ophiolite belts) associated with the island arcs evolutionary model of the Arabian shield.

The mapping of the surface-wave group velocity using a large data can be used in constraining the regional structure at existing and planned broadband stations deployed in this tectonically complex region as part of the seismic monitoring under CTBT.

Key words: Arabia, crust, surface waves, seismic tomography.

Introduction

The deployment of broadband seismic stations within the Arabian shield (VERNON and BERGER, 1998) provided an excellent opportunity to study the seismic structure of the Arabian plate using high quality seismic signals previously unavailable for this part of the world. A number of recent studies have made use of the recorded broadband data (e.g., SANDVOL *et al.*, 1998; MCNAMARA *et al.*,

[1] Department of Geophysics, Faculty of Earth Sciences, King Abdulaziz University, P.O. Box 80206, Jeddah, 21589, Saudi Arabia.

[2] Department of Earth & Atmos. Sci., Saint Louis Univ., 3507 Laclede Ave., St. Louis, MO, 63103, USA. E-mail: rbh@eas.slu.edu, E-mail: ammon@mantle.slu.edu

[3] Multimax Inc., 1090 N. Highway A1A, Suite D, Indialantic, Florida 32903, USA. E-mail: hghalib@multimax.com

1997; RODGERS *et al.*, 1999; and MOKHTAR, 2001). In this paper we present maps showing the group-velocity variations across the Arabian plate. One goal of our effort is to increase the lateral resolution by incorporating as many as possible of the available wave propagation paths across the Arabian plate and particularly incorporate observations in which the records were made within the plate. The analysis performed is a tomographic inversion of broadband surface wave group-velocity dispersion measurements. The results presented provide images for both Love- and Rayleigh- wave group velocities for 5–60 s periods. While this methodology has been well developed for many years, the application of the method to a large data set in this tectonically complex region makes this study a unique contribution. Short-period group velocities are of particular importance to the CTBT monitoring since this type of data is missing in many other studies related to this program. Improved measurements of surface-wave dispersion properties in Arabia will indeed assist in defining the crustal structure beneath existing and planned broadband stations deployed in this region as part of seismic monitoring under CTBT.

Tectonic Setting

The Arabian Plate consists of two major tectonic provinces: the Arabian shield and the Arabian platform (Fig. 1). The Arabian shield covers about one third of Arabia and consists of Precambrian gneiss and metamorphosed sedimentary and volcanic rocks that have been intruded by granites (POWERS *et al.*, 1966; BROWN, 1972). The shield consists of five micro-plates (STOESER and CAMP, 1985) which are separated by four ophiolite-bearing suture zones. These micro-plates are considered the remnants of Precambrian island arcs (SCHMIDT *et al.*, 1979) that accreted to form an Arabian neo-craton around 630 Ma and which in turn were subjected to subsequent intracratonic deformation and magmatism producing the present day shield (STOESER and CAMP, 1985). Widespread Tertiary and Quaternary volcanic rocks related to initial stages of the Red Sea formation are predominant along western Arabia in the shield (BROWN, 1972; COLEMAN, 1977).

The Arabian platform is a large sedimentary basin that comprises about two thirds of the plate and consists of Paleozoic and Mesozoic sedimentary layers that uncomformably overlap the basement rocks and gently dip to the east (POWERS *et al.*, 1966). The platform sediments increase in thickness to the east and reach a thickness of 10 km or more beneath the Mesopotamian foredeep (BROWN, 1972).

Five tectonic boundaries surround the Arabian plate (MAAMOUN, 1976). In the north and northwest lies the continental collision boundary between the Arabian, the Persian and Turkish plates along the Zagros and Taurus Mountains. A subduction boundary exists in the Gulf of Oman region in southern Iran (Makran subduction

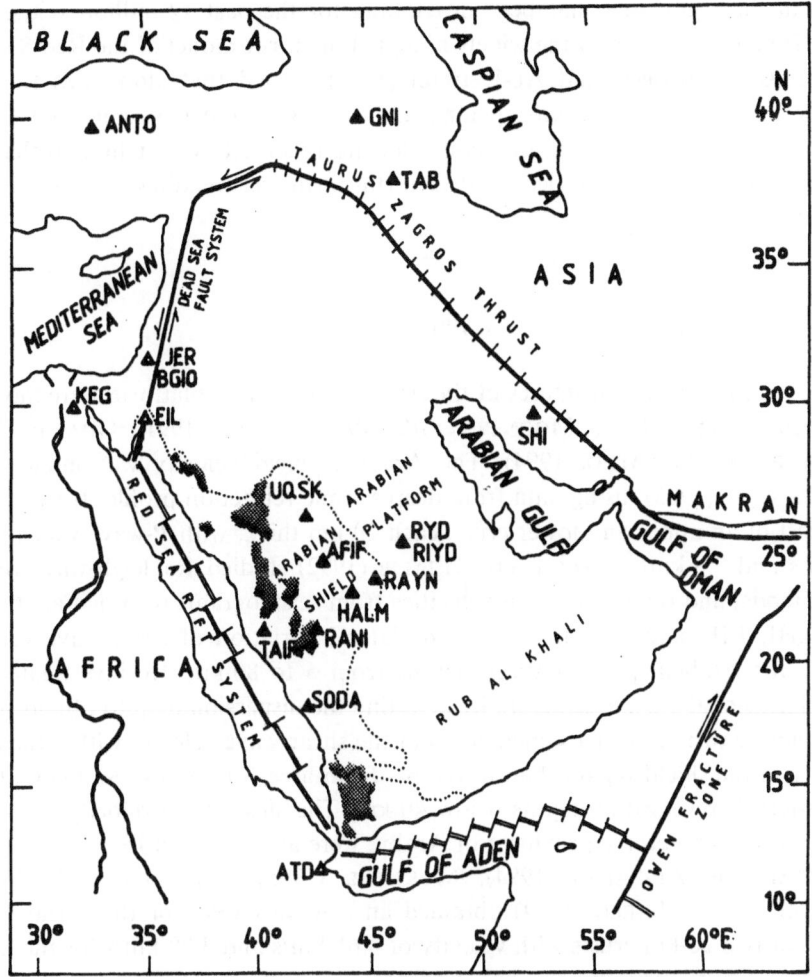

Figure 1
Tectonic provinces of the Arabian plate and the different plate boundaries. Seismic stations used in this study are shown. The shaded regions indicate Tertiary and Quaternary volcanic flows.

zone) and a transform fault boundary along the Owen fracture zone in the southeast. The Dead Sea fault system represents another transform fault in the northwest, and spreading is taking place along the Red Sea and Gulf of Aden axial troughs in the west and southwest. All of these boundaries are tectonically active and most of them produce an appreciable number of earthquakes, especially along the boundary associated with the Zagros Mountains belt and the active spreading axis along the Red Sea and the Gulf of Aden.

ROOBOL and AL-REHAILI (1997) suggested that the presence of a more or less linear distribution of volcanic fields in western Arabia is evidence for a new rift system in the region that has been developing for the past 12 million years. This system is independent from the widely accepted main rift model of the Red Sea and Gulf of Aden. ROOBOL and AL-REHAILI (1997) argued that along this trend of volcanic features, fractures are opening and closing and that seismic and micro-seismic activities are produced that coincide with this fracture system, and that this new rift system is related to deep seated north-northwest trending faults parallel to the Red Sea.

Previous Geophysical Investigations

The seismic velocity structures of the Arabian shield and platform were modeled by several studies (e.g., MOONEY *et al.*, 1985; BADRI, 1991; GHALIB, 1992; MOKHTAR and AL-SAEED, 1994). The Arabian shield velocity was modeled by MOONEY *et al.* (1985), using data from deep seismic refraction profile. It was found to consist of an upper and lower crust, each 20-km thick, with P-wave velocities of 6.3 km/s and 7.0 km/s, respectively. GHALIB (1992) studied Rayleigh-wave funda-mental mode, and obtained group velocities from seismograms recorded at stations TAB, SHI, EIL, and JER to outline the lateral variation of shear-wave velocity beneath the Arabian plate between depths from 5 to 80 km. He investigated the three-dimensional seismic structure by inverting the dispersion data to obtain shear-wave velocity variations. He concluded that the shear-wave velocity within the crust is higher in the shield region than in the platform area for depths less than 40 km. The pattern is reversed at depths below 40 km. The shear-wave velocity structures of the two major tectonic provinces of Arabia were also modeled by several studies (MOKHTAR and AL-SAEED (1994), MOKHTAR (1995), RODGERS *et al.* (1999)). MOKHTAR and AL-SAEED (1994) obtained an S-wave model for the shield which consists of two 20-km crusts with velocity of 3.61 km/s and 3.88 km/s for the upper and lower crust, respectively. The underlying mantle material has an S-wave velocity of 4.61 km/s, and the eastern and southern parts of the Arabian platform were modeled by upper and lower crusts of comparable thickness, but with S-wave velocity of 3.4 km/s and 4 km/s, respectively. The upper mantle velocity beneath the platform has a shear-wave velocity of 4.4 km/s. MOKHTAR (1995) used waveform modeling of surface waves to verify the Arabian platform models of MOKHTAR and AL-SAEED (1994).

The depth to the mantle is about 43–45 km beneath the platform and it decreases to the southwest and reaches about 38–40 km in the southwestern part of the shield (MOONEY *et al.*, 1985; MOKHTAR and AL-SAEED, 1994). SANDVOL *et al.* (1998) estimated the lithospheric mantle and crustal velocity structures beneath the Arabian shield through the modeling of teleseismic P waves recorded by the temporary broadband array used in this study. Application of the receiver function techniques showed that the crustal thickness of the shield area varies from 35 to

40 km in the west adjacent to the Red Sea, to 45 km in central Arabia. These results are consistent with previous results from refraction data and surface wave inversion. RODGERS *et al.* (1999) used waveform modeling of surface waves to model the lithosphere structure of the Arabian shield and platform along two independent paths from the Zagros and the Gulf of Aqaba to stations in the Arabian shield. They reported fast crustal velocities for the shield and slower crustal velocities for the platform, and that the Arabian shield has a thinner crust with lower mantle velocities and higher Poisson's ratio beneath it compared to the Arabian platform.

RITZWOLLER and LEVSHIN (1998) and RITZWOLLER *et al.* (1998) produced tomographic maps from surface-wave group velocities across all of Eurasia, central Asia, western China, and parts of the Middle East. These maps were at a length scale intermediate between regional and global surface-wave studies and extend for periods in the range of 20–200 s. Group-velocity maps from these studies for 20–30 s period display low velocity anomalies associated with most of the known sedimentary basins across Eurasia, especially those associated with Love waves. This was clearly demonstrated for the Arabian platform sedimentary basin. On the other hand, 20 s Rayleigh- and Love-waves data correlate well with the high velocity regions associated with the massive basaltic flows in northern Ethiopia near the southwestern boundary of the Arabian plate. These regions were, however, among the regions that were poorly sampled at short periods by these two studies.

Observations

We measured surface-wave group velocities generated by earthquakes located along the boundaries of the Arabian plate in the Red Sea, Gulf of Aqaba, Gulf of Aden, western Iran, Turkey, and the Dead Sea fault system. Our observations were compiled from four different sources: 1) Digital broadband seismograms from the Saudi Arabian 1995/1996 temporary seismic network which include about 494 seismograms from earthquakes that occurred during the period December 31, 1995 to September 15, 1996. These data represent about 50% of the Rayleigh-wave and about 60% of the Love-wave data; 2) digital seismograms recorded by the permanent broadband stations in the region for the period between 1990 and 1996, which represents 26% of the Rayleigh-wave and 36% of the Love-wave data; 3) analog observations of Rayleigh waves from the regional WWSSN stations recorded between 1970 and 1979, which represent 19% of the total Rayleigh-wave data; and 4) analog observations from RYD station recorded between 1981 and 1987 and represents 5% of both Rayleigh- and Love-wave data.

RAYLEIGH WAVES RAY PATHS

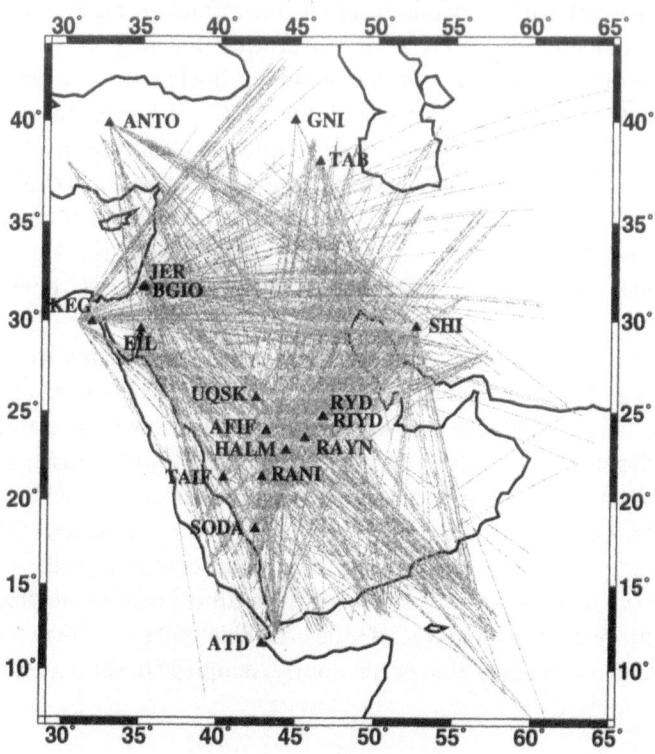

Figure 2
Rayleigh waves raypaths coverage for periods of 11–13 s.

In Figures 2 and 3 we present the great circle path coverage along which dispersion measurements were made for both Rayleigh and Love waves. Surface-wave data in the period range 5 to 60 s were used. Mainly seismic stations inside the plate or very close to its tectonic boundaries recorded these data, avoiding long paths that may traverse a number of complicated tectonic provinces and geological features. A total of 987 Rayleigh-wave and 682 Love-wave paths were available. A maximum of 916 rays for Rayleigh and 682 rays for Love were used in the period range of 11–13 s (this is the average of the total number of raypaths over the three periods). The number of great circle paths decreases for longer periods and reaches 376 for Rayleigh and 178 for Love at 56–60 s periods. The seismic stations used include the digital broadband stations of AFIF, HALM, RAYN, RIYD, RANI, TAIF, UQSK, and SODA; regional broadband GSN seismic stations of ANTO, GNI, BGIO, KEG; and a Geoscope station ATD; and the WWSSN stations from which analog data were digitized (TAB, SHI, EIL, JER, and RYD). These stations provided significant coverage of both the Arabian shield and platform. There is also

LOVE WAVES RAY PATHS

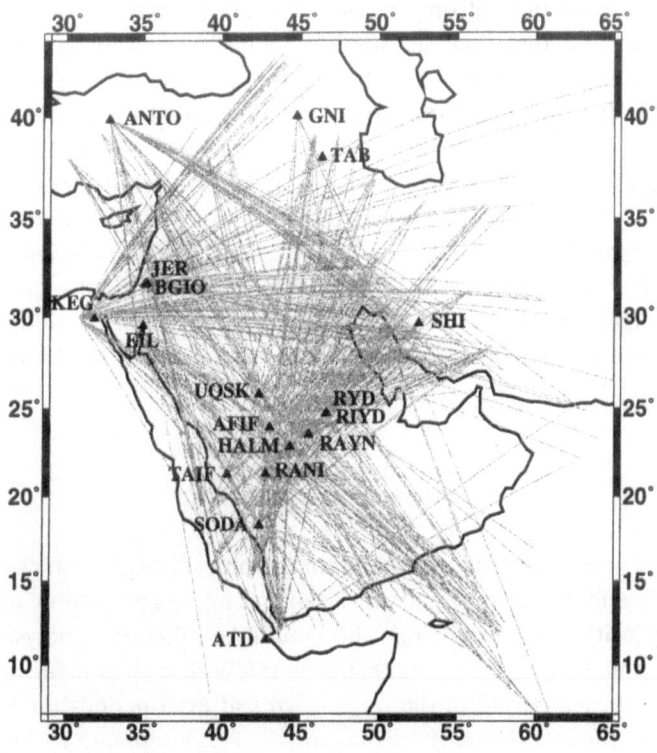

Figure 3
Love waves raypaths coverage for periods of 11–13 s.

significant coverage of the western part of Iran and all of Iraq. We succeeded in presenting excellent path coverage across most of Arabia, except for its southeastern part east of Al-Rub Al-Kahli. The low seismic activity associated with Makran subduction zone and Owen fracture zone and a lack of a seismic station near or in Al-Rub Al-Kahli contributed mainly to the poor coverage of this region. Earthquakes epicentral data were obtained from USGS Earthquake Data Reports (EDR).

The regional scale tomographic maps presented below provide significant constraints on the shear velocity and crustal thickness of the Arabian plate. Maps showing the distribution of surface-waves group velocities propagating at regional distances in the period range of interest in this study are very useful in constraining the regional structure of the crust and upper mantle. There are, however, potential sources of errors that may affect the estimated group velocity maps. These include the uncertainties in the origin times, epicentral locations, depth of earthquakes used, and spectral notches. Although it is important to point out these potential pitfalls of group-velocity measurements and interpretation, especially because of their strong impact at short periods, we assume that their effects are small and that their accumulated impact does not greatly change the estimated measurements. This is because of the large number of samples used at each period for the different paths, and the high signal-to-noise ratio of the data used. We also assume that the measured waves have propagated along the great circle path connecting the source and the receiver. RITZWOLLER and LEVSHIN (1998) discussed the problems expected to result from several factors such as off-great circle propagation, azimuthal anisotropy, and systematic event mislocations, and argued that these effects should not alter the tomographic maps of group velocities strongly beyond the resolution estimates they had. In addition, our use of short distance paths helps minimize the likely deflection of the path from the great circle arc.

Methods of Analysis

We present the results of regional tomographic inversion of dispersion data from Rayleigh and Love waves across Arabia using single station measurements of group velocity and applying the multiple filtering analysis technique (DZIEWONSKI and HALES, 1972; HERRMANN, 1987). The dispersion measurements were obtained at each period in the range 5–20 s at even periods only in the range 22–60 s. Observations from each period are inverted separately and the images from adjacent periods were averaged. We parameterized the regional slowness variations using a uniform, one-degree by one-degree, grid of constant-slowness cells. Group-velocity maps were produced using a conjugate-gradient least-squares algorithm (PAIGE and SANDERS, 1982). Laplacian smoothness constraints were incorporated in the inversion and thus we minimized a combination of group travel-time misfit and a Laplacian measure of a two-dimensional model roughness. The balance between group delay misfit and minimal roughness is selected empirically by running inversions with a range of smoothness importance weights and selecting the value that produces the simplest model and still satisfactorily matches the observed group delays. The resulting velocity variations were obtained as percent velocity perturbations from the average velocity of all the measurements, which was the initial model. These variations were subsequently incorporated in the starting model and are presented as contour representation of the variations of the group velocities of Love and Rayleigh waves across Arabia.

Figure 4 presents the average group velocities of Rayleigh waves estimated from all the paths used. A comparison is made with the group-velocity measurements of MOKHTAR and AL-SAEED (1994) for the Arabian shield and the Arabian platform. The group-velocity values for all of Arabia are higher than the corresponding

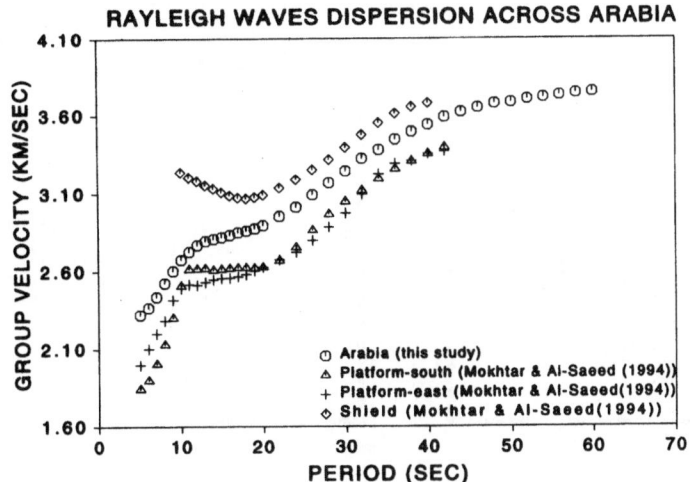

Figure 4

Average group velocities for Rayleigh waves in Arabia compared to those of the Arabian shield and the Arabian platform as obtained by MOKHTAR and AL-SAEED (1994).

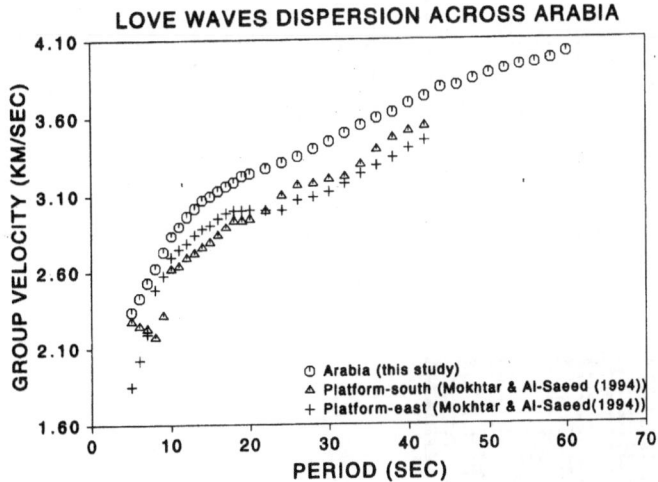

Figure 5

Average group velocities for Love waves in Arabia compared to those of the Arabian platform.

values for the Arabian platform and are lower than those representing the Arabian shield. Similar conclusions can be drawn for Love waves (Fig. 5), however, MOKHTAR and AL-SAEED (1984) were not able to obtain reliable Love-wave data for the Arabian shield from events located in the Red Sea. VERNON and BERGER (1998) pointed out the difficulty in distinguishing all phases along the propagation

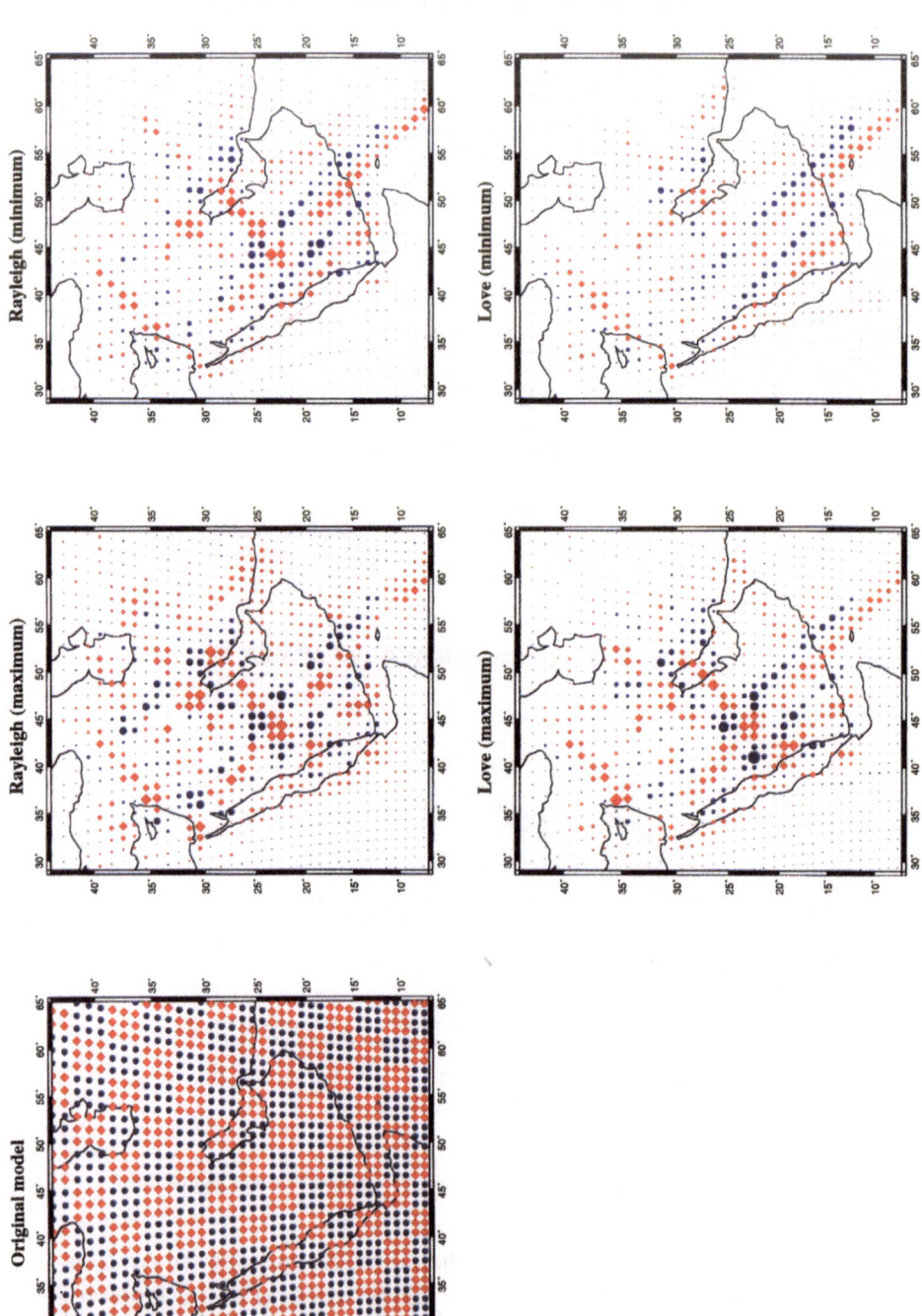

Figure 6

Checker-board test results showing the original velocity perturbation model using a feature of 3° × 3° in dimension, and the tomographic inversion results for both Rayleigh and Love waves using the maximum and minimum number of rays. Diamonds identify cells with relatively low velocity, circles identify cells with relatively high velocity.

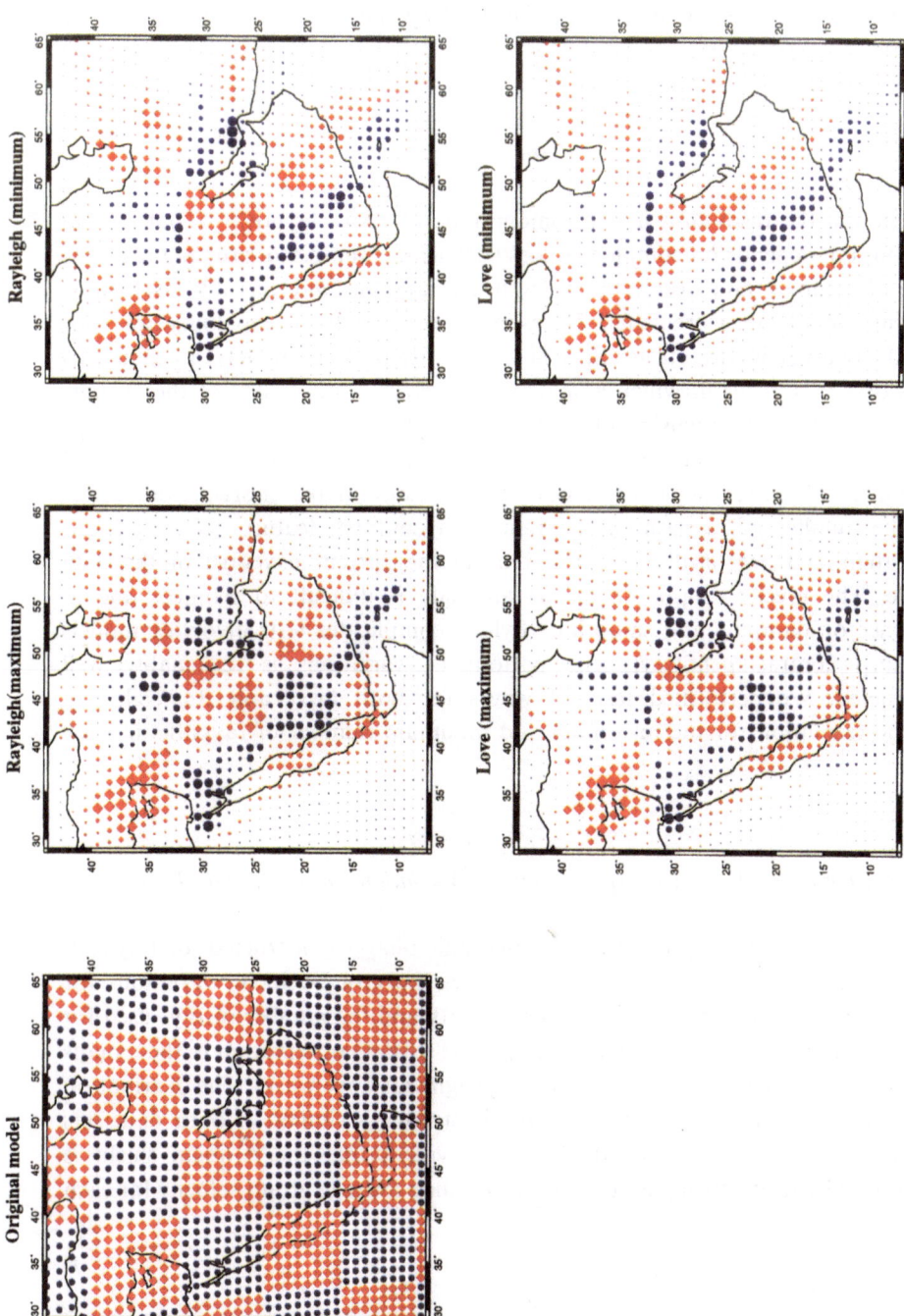

Figure 7

Checker-board test results using a 8° × 8° velocity perturbation model. Diamonds represent regions of relatively low velocity, circles represent regions with relatively high velocity.

path extending from the Red Sea to stations at the middle of the plate. They reported weak to non-existent *S* phases followed by emergent *Lg* phases along this path.

Resolution

Estimating the resolution in the tomographic inversion is not a trivial task in that resolution depends on complex factors such as the number of crossing rays, the density of sources and receivers, as well as random and possible systematic uncertainty in the measurements. To estimate our resolving capability we used standard "checker-board" tests. Specifically we tested two models. Each model consists of $1° \times 1°$ cells in which a velocity perturbation of $\pm 10\%$ of the average was chosen to have a dimension of $3° \times 3°$ for the first model (Fig. 6), and $8° \times 8°$ for the second model (Fig. 7). Checker-board test results show that features of dimension $8° \times 8°$ are reasonably recovered especially in the short-period ranges where the number of raypaths is maximum. In contrast, features of dimensions smaller than that are hardly recognized using the current coverage of the data available. The precise amplitude of the anomaly is difficult to estimate due to dependence on damping and smoothing (which complicates tests involving models with sharp contrasts) however the pattern of variations is reasonably well reconstructed. We performed tests of waves for both Love and Rayleigh and tested the resolution for the maximum and minimum number of raypaths in each case.

Rayleigh- and Love-wave Group-velocity Variations in the Middle East

In Figures 8 and 9 we present the tomographic images constructed for Rayleigh and Love waves, respectively. Each map shows the group-velocity variations for three adjacent periods. The average group velocity and the number of rays used for that period range are indicated on each map. One striking observation in these maps is the consistency of the results from both Rayleigh and Love waves. The distribution of faster and slower than average regions is strikingly similar in both data sets especially in the short periods. It is clearly evident that the Arabian shield is characterized by relatively higher group velocity than average group velocity for each

▶

Figure 8a
Tomography map of the Arabian plate Rayleigh-wave group velocity for periods 5 to 24 s. The average group velocity and the average number of rays used are given for each period group. Contour interval is 0.1 km/s.

RAYLEIGH WAVES

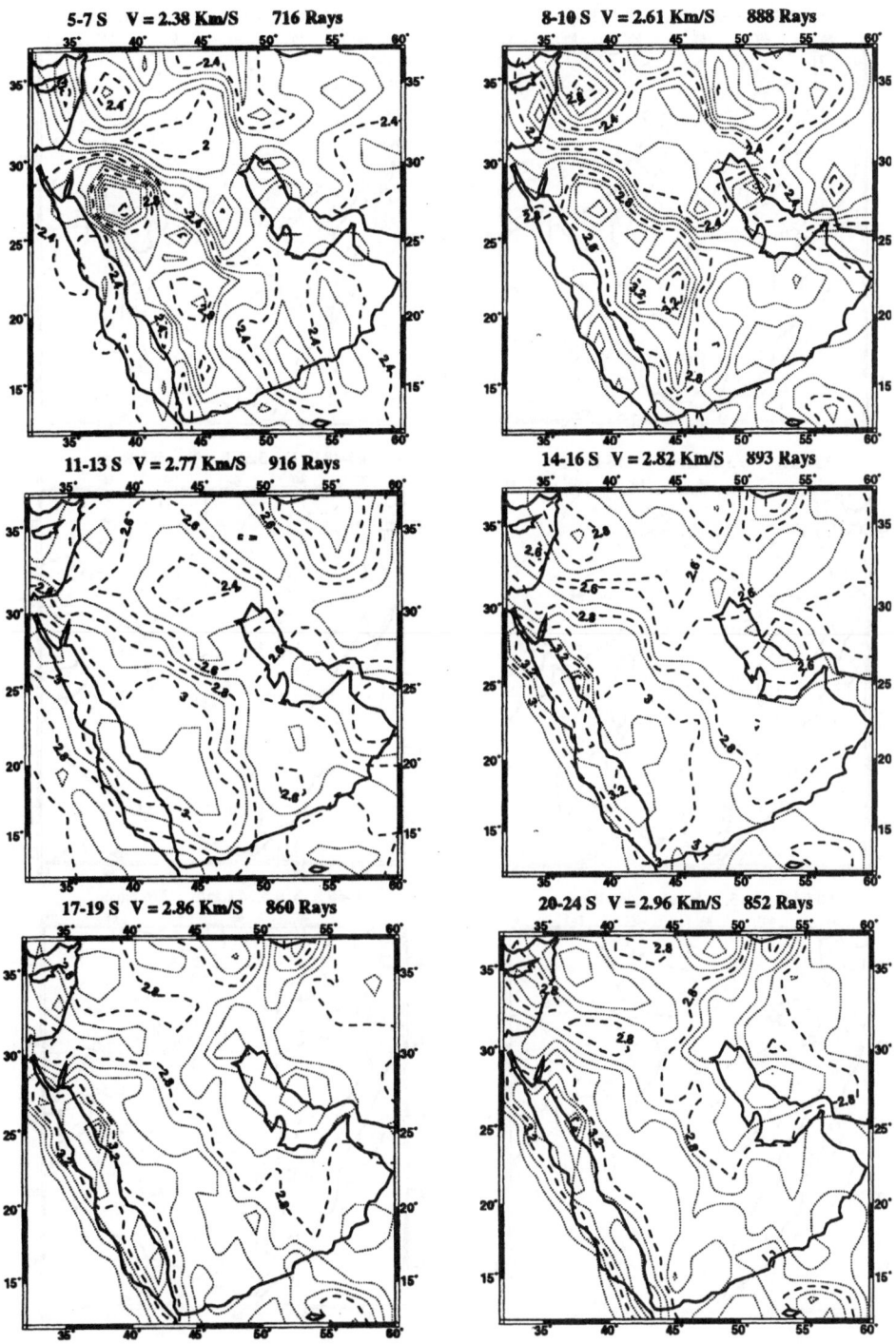

5-7 S   V = 2.38 Km/S   716 Rays

8-10 S   V = 2.61 Km/S   888 Rays

11-13 S   V = 2.77 Km/S   916 Rays

14-16 S   V = 2.82 Km/S   893 Rays

17-19 S   V = 2.86 Km/S   860 Rays

20-24 S   V = 2.96 Km/S   852 Rays

RAYLEIGH WAVES

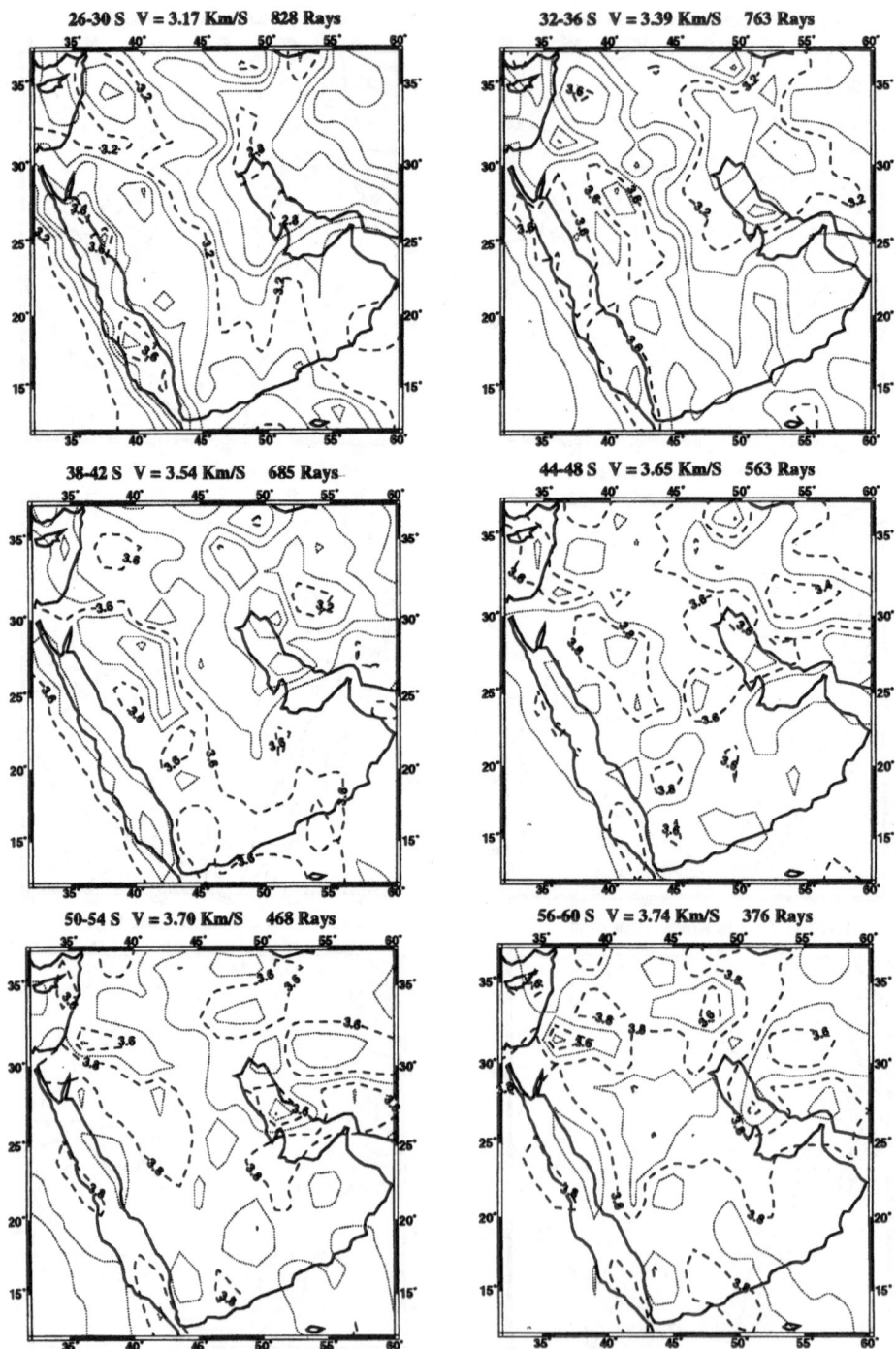

Figure 8b
Same as Figure (8a), except for periods 26 to 60 s.

LOVE WAVES

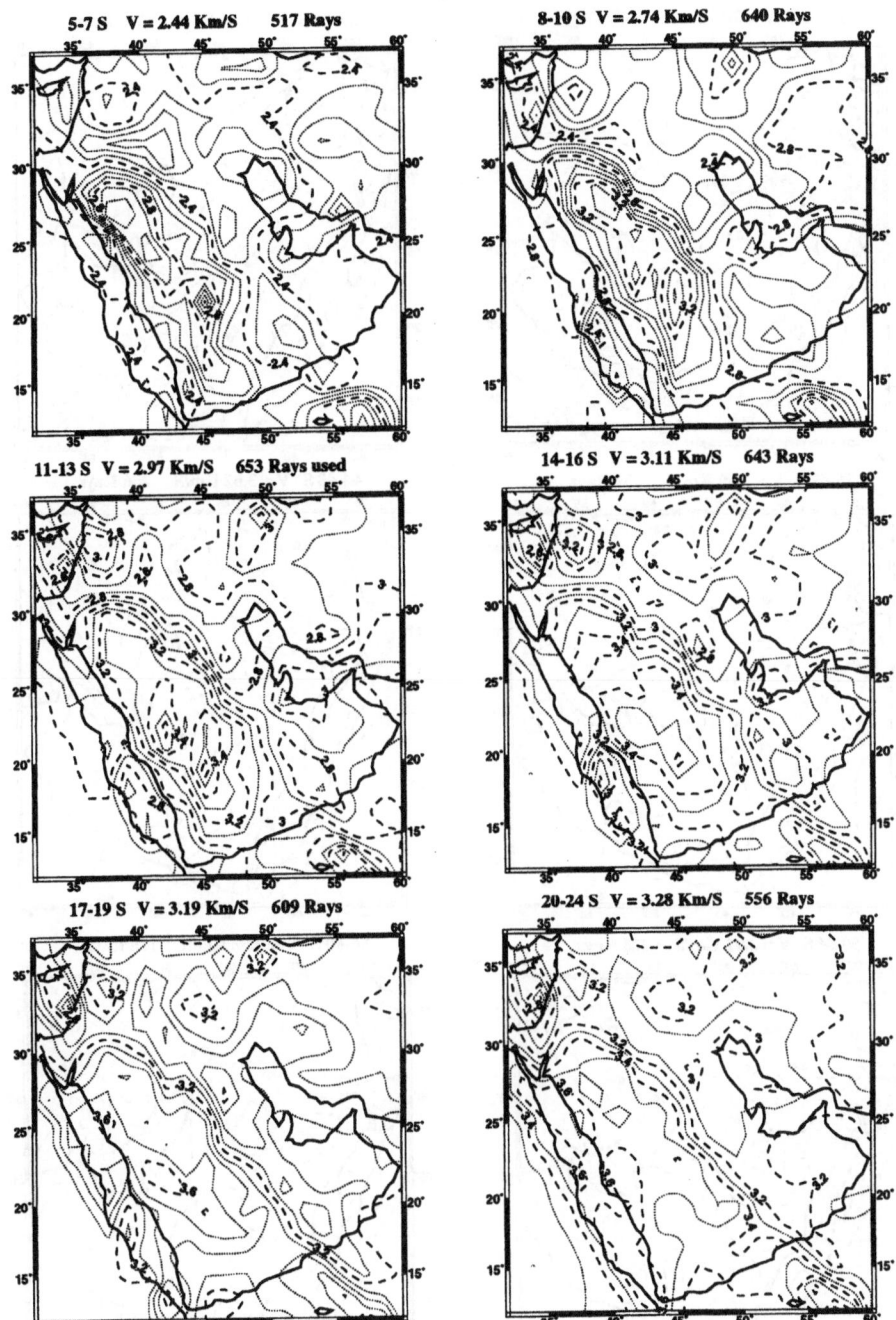

Figure 9a

Tomography map of the Arabian plate Love-wave group velocity for periods 5 to 24 s. The average group velocity and the average number of rays used are given for each period group. Contour interval is 0.1 km/s.

LOVE WAVES

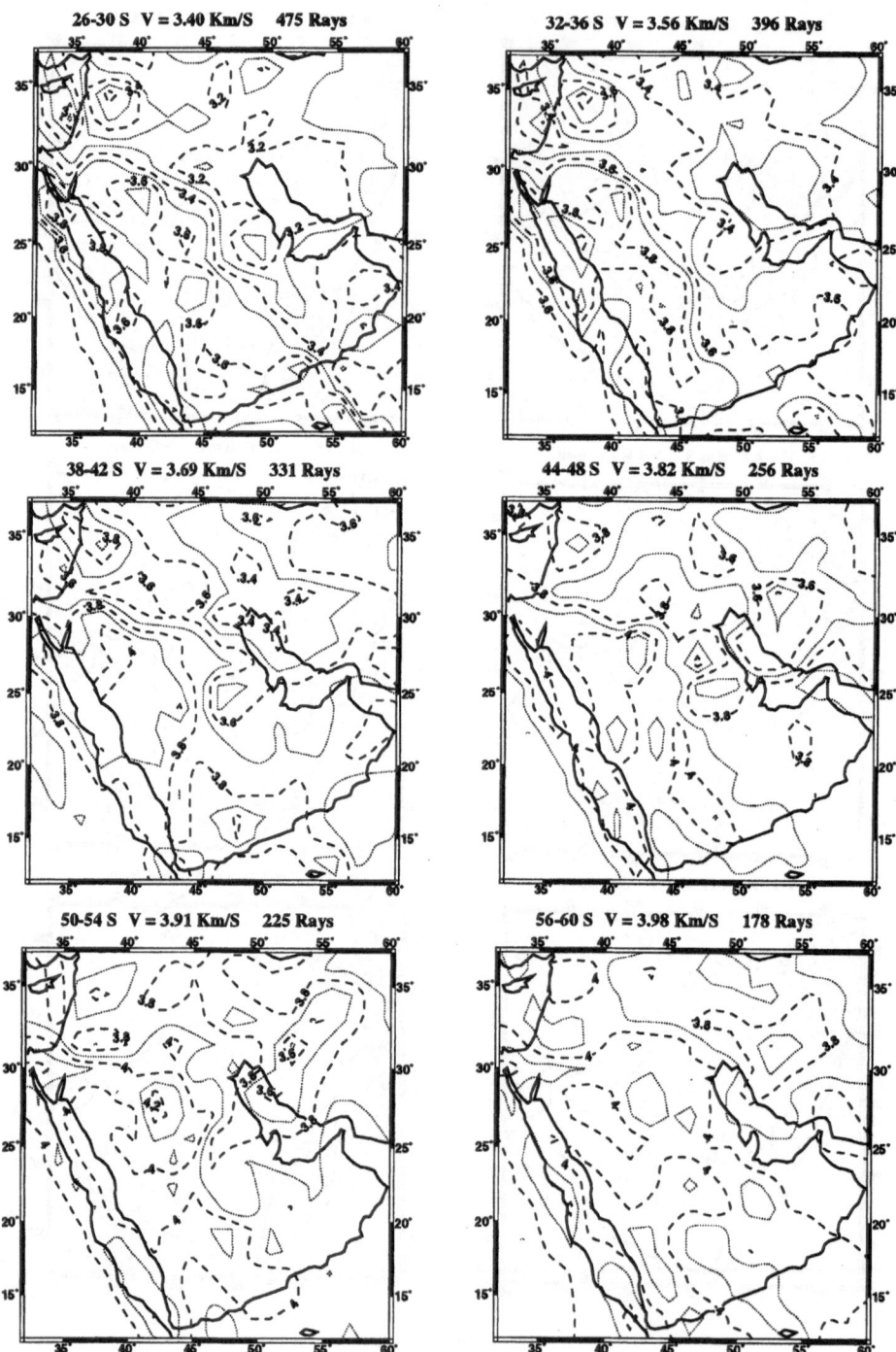

Figure 9b
Same as Figure (9a), except for periods 26 to 60 s.

group of periods, while the Arabian platform, which is covered by an eastward-thickening sedimentary section, has in general slower than average group velocities. This result is consistent with those of GHALIB (1992), MOKHTAR and AL-SAEED (1994), MOKHTAR (1995), RITZWOLLER and LEVSHIN (1998), RITZWOLLER et al. (1998), and RODGERS et al. (1999).

The boundary between the fast and slow regions exhibits sharp transition, and correlates well with the surface geology, especially at periods shorter than 16 s. The resolution diminishes for longer periods where the number of rays becomes smaller. The correlation with surface geology continues to longer periods, however, the difference between the group velocities of both regions becomes smaller with increasing period, and the correlation between the boundary of low and high velocities and surface geology becomes less obvious at periods higher than 40 s. The average group velocity of Love waves increases from about 2.44 km/s at 5–7 s to 3.98 km/s at 56–60 s, while that of Rayleigh waves increases from 2.38 km/s at 5–7 s to 3.74 km/s at 56–60 s.

We do not have adequate coverage to image the southeastern part of the Arabian plate where it is covered mainly by Al-Rub Al-Khali region. Also, the localized fast anomaly located east of the Gulf of Aden and the slow feature located in the eastern Mediterranean region should be interpreted as a result of bias in the data since there are not enough traverses crossing these two regions.

Discussions and Conclusion

In spite of the low resolution power of our data for features of dimensions less than $8° \times 8°$, it is interesting to observe the distribution of small size high local anomalies within the shield at periods shorter than 16 s. Their locations correlate well with the distribution of the Tertiary and Quaternary volcanic flows in the shield. ROOBOL and AL-REHAILI (1997) argued that along this trend of volcanic features, seismic and micro-seismic activities are produced that coincide with this fracture system which marks the location of a new rift system in the region. They did not, however, present the locations of earthquake epicenters that show the presence of recent seismicity that correlates with the suggested new rift system. In spite of this, and even if there were some manner of seismic activities associated with these fractures and volcanic fields, our results cannot support the argument that these surface features have any relationship to a deep-seated structure. One would expect the seismic velocity distribution to reflect the heterogeneity of the material beneath such a rift. In other words, high density and high seismic velocity from the upper mantle material would reveal itself in a somewhat linear trend parallel and in line with the proposed rift system and well distinguished from the surrounding crust that is being rifted. Due to the limited resolution of the data at hand with depth, we are not able to trace the extension of these features with depth. Thus, it would be

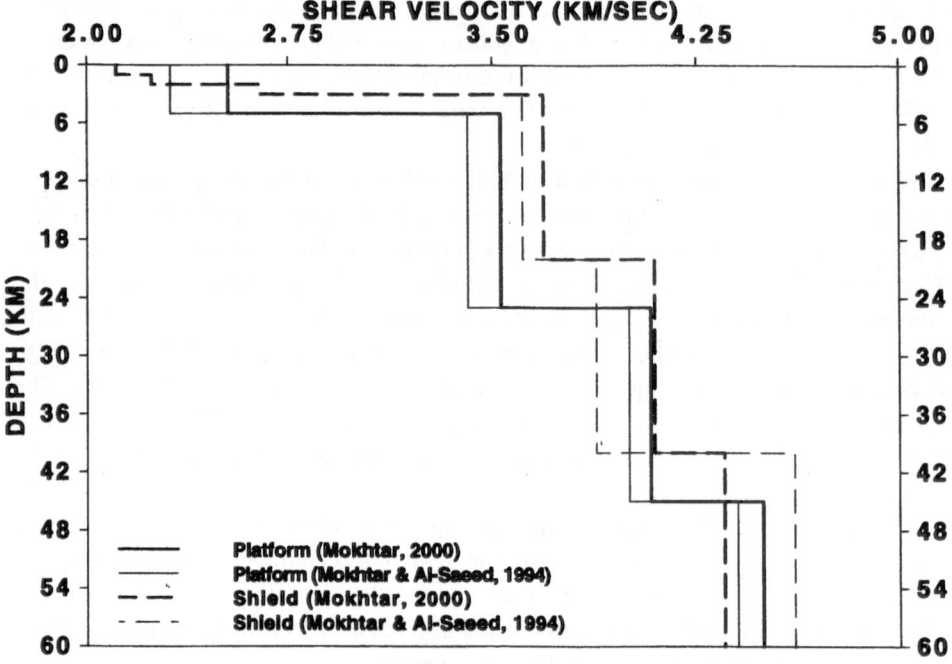

Figure 10
Comparison of the shear-wave velocity models of the Arabian shield and the Arabian platform.

premature to try to relate the possible heterogeneity within the shield to the suggested rifting system. Additional investigation is clearly required to address this issue further.

MOKHTAR (2001) presented models of the shear-wave velocity structures along eight different paths across Arabia, based on the group velocity dispersion data computed in this study. The models represent velocity structures along five different paths that are representative of the Arabian platform, and three additional paths that are representative of models of the Arabian shield. In Figure 10, we present the shear models of MOKHTAR (2001) for both the shield and the platform and compare them to the models presented by MOKHTAR and AL-SAEED (1994). The new models indicate that the crustal shear velocity of both regions is higher than previously estimated, and that the velocity of the upper crust is higher in the shield than in the platform. Similar velocities for the lower crust of both regions were obtained. Unreasonably high velocity of 3.87 km/s resulted at a depth of 4 km when a model of single crust of 36 km thickness is used for the platform as suggested by RODGERS *et al.* (1999).

The Arabian shield is believed to have evolved from island arcs formed during a series of subduction episodes during 950–715 Ma. These island arcs were subsequently juxtaposed by compressional orogeneis (SCHMIDT *et al.*, 1979; AL-SHANTI,

1992). Numerous ophiolitic belts of Pan-African age occur in Arabia and northeastern Africa and are interpreted as tectonically emplaced masses, which represents the remnants of the oceanic crust that were associated with the island-arc evolutionary model (GASS, 1979). The surface and shear waves higher velocity of the upper crust of the shield may be due to the presence of the remnants of oceanic crust in the upper crust as indicated by MOKHTAR and AL-SAEED (1994).

In conclusion, we have presented maps showing the variations of surface-wave group velocity in Arabia for the period range of 5–60 s. Group velocities of surface waves are higher in the Arabian shield than in the Arabian platform. Inversion for shear-wave models indicates that the higher group velocity of the shield is due to higher seismic waves velocity in the upper crust, which is consistent with the evolutionary model of the Arabian shield.

REFERENCES

AL-SHANTI, A. M. S. *The Geology of the Arabian Shield* (King Abdulaziz University Press, Jeddah (1992)), 219 pp. (in Arabic).

BADRI, M. (1991), *Crustal Structure of Central Saudi Arabia Determined from Seismic Refraction Profiling*, Tectonophysics *185*, 357–374.

BROWN, G. F. (1972), *Tectonic Map of the Arabian Peninsula*, Saudi Arabian Peninsula Map Ap-2. Saudi Arabian Dir. Gen. Miner. Resour.

COLEMAN, R. G., *Ophiolites. Ancient Oceanic Lithosphere?* (Springer-Verlag, Berlin (1977)), 229 pp.

DZIEWONSKI, A. M., and HALES, A. L., *Numerical Analysis of dispersed seismic waves*. In (Alder, B., Frenbach, S., and Rotenberg, M., *Methods in Computational Physics* (eds). (Academic Press, New York, NY, 1972) pp. 39–85.

GASS, I. G. (1979), *Evolutionary Model for the Pan-African Crystalline Basement, Evolution and Mineralization of the Arabian-Nubian Shield*, I. A. G. Bull. *3(2)*, 11–29.

GHALIB, H. A. A. (1992), *Seismic Velocity Structure and Attenuation of the Arabian Plate*, Thesis, Saint Louis University, Saint Louis, Missouri, USA, 350 pp.

HERRMANN, R. B. ed., *Computer Programs in Seismology, VOL I–VOL VII* (Saint Louis University, Saint Louis, Missouri, USA, 1987).

MAAMOUN, M. (1976), *La seismicite du Moyenet du Proche-Orient dans le cadre de la seismotectonique mondiale*, Thesis, Doc. Sci., Univ. Louis Pasteur, Strasbourg.

MCNAMARA, D. E., HAZLER, S. E., and WALTER, W. R. (1997), *Velocity Structures across Northern Africa, Southern Europe, the Middle East and the Arabian Peninsula from Surface Waves Dispersion*, EOS *78*, F499.

MOKHTAR, T. A., and AL-SAEED, M. M. (1994), *Shear Wave Velocity Structures of the Arabian Peninsula*, Tectonophysics *230*, 105–125.

MOKHTAR, T. A. (1995), *Phase Velocity of the Arabian Platform and the Surface Waves Attenuation Characteristics by Waveform Modeling*, Jour. King Abdulaziz Univ., Earth Sci. *8*, 23–45.

MOKHTAR, T. A. (2001), *Variations of the Crustal Structure of Arabia*, in preparation.

MOONEY, W. D., GETTINGS, M. E., BLANK, H. R., and HEALY, J. H. (1985), *Saudi Arabian Seismic Deep Refraction Profile: A Travel-time Interpretation of Crustal and Upper Mantle Structure*, Tectonophysics *111*, 173–246.

PAIGE, C. C., and SANDERS, M. A. (1982), *LSQR: An Algorithm for Sparse Linear Equations and Sparse Least Squares*, ACM Trans. Math. Softw. *8*, 43–71.

POWERS, R. W., RAMIRES, L. F., REDMOND, C. P., and ELBERG, E. L. (1996), *Geology of the Arabian Peninsula-sedimentary Geology of Saudi Arabia*. U. S. Geol. Surv. Prof. Pap. *560-D*, 147 pp.

RITZWOLLER, M. H., and LEVSHIN, A. L. (1998), *Eurasian Surface Wave Tomography: Group Velocities,* J. Geophys. Res. *103,* 4839–4878.

RITZWOLLER, M. H., LEVSHIN, A. L., RATNIKOVA, L. I., and EGORKIN, A. A. (1998), *Intermediate Period Group Velocity Maps across Central Asia, Western China, and Parts of the Middle East,* Geophys. J. Inter. *134,* 315–328.

RODGERS, A., WALTER, W., MELLORS, R. J., AL-AMRI, A. M. S., and ZHANG, Y.-S. (1999), *Lithospheric Structure of the Arabian Shield and Platform from Complete Regional Waveform Modeling and Surface Wave Group Velocities,* Geophys. J. Inter. *138,* 871–878.

ROOBOL, M. J., and AL-REHAILI, M. (1997), *Geohazards along the Makkah-Madinan-Nafud (MMN) Volcanic Line,* Saudi Arabian Deputy Ministry for Mineral Resources Technical Report *DMMR-TR-97-1,* 125–140.

SANDVOL, E., SEBER, D., BARAZANGI, M., VERNON, F., MELLORS, R., and AL-AMRI, A. (1998), *Lithospheric Velocity Discontinuities beneath the Arabian Shield,* Geophys. Res. Lett. *25,* 2873–2876.

SCHMIDT, D. L., HADLEY, D. G., and STOESER, D. B. (1979), *Late Proterozoic Crustal History of the Arabian Shield, Southern Najd Province, Kingdom of Saudi Arabia, Evolution and Mineralization of the Arabian-Nubian Shield.* I. A. G. Bull. *3(2),* 41–58.

STOESER, D. B., and CAMP, V. E. (1985), *Pan-African Microplate Accretion of the Arabian Shield,* Bull. Geol. Soc. Am. *96,* 817–826.

VERNON, F., and BERGER, J. (1998), *Broadband Seismic Characterization of the Arabian Shield. Final Scientific and Technical Report,* Philips Laboratory, 36 pp.

(Received July 24, 1999, revised December 20, 1999, accepted January 15, 2000)

 To access this journal online:
http://www.birkhauser.ch

Pure appl. geophys. 158 (2001) 1445–1474
0033–4553/01/081445–30 $ 1.50 + 0.20/0

❙ Pure and Applied Geophysics

A Surface Wave Dispersion Study of the Middle East and North Africa for Monitoring the Comprehensive Nuclear-Test-Ban Treaty

MICHAEL E. PASYANOS,[1] WILLIAM R. WALTER[1] and SHANNON E. HAZLER[2]

Abstract—We present results from a large-scale study of surface-wave group velocity dispersion across the Middle East, North Africa, southern Eurasia and the Mediterranean. Our database for the region is populated with seismic data from regional events recorded at permanent and portable broadband, three-component digital stations. We have measured the group velocity using a multiple narrow-band filter on deconvolved displacement data. Overall, we have examined more than 13,500 seismograms and made good quality dispersion measurements for 6817 Rayleigh- and 3806 Love-wave paths. We use a conjugate gradient method to perform a group-velocity tomography. Our current results include both Love- and Rayleigh-wave inversions across the region for periods from 10 to 60 seconds. Our findings indicate that short-period structure is sensitive to slow velocities associated with large sedimentary features such as the Mediterranean Sea and Persian Gulf. We find our long-period Rayleigh-wave inversion is sensitive to crustal thickness, such as fast velocities under the oceans and slow along the relatively thick Zagros Mts. and Turkish-Iranian Plateau. We also find slow upper mantle velocities along known rift systems. Accurate group velocity maps can be used to construct phase-matched filters along any given path. The filters can improve weak surface wave signals by compressing the dispersed signal. The signals can then be used to calculate regionally determined M_S measurements, which we hope can be used to extend the threshold of $m_b{:}M_S$ discriminants down to lower magnitude levels. Other applications include using the group velocities in the creation of a suitable background model for forming station calibration maps, and using the group velocities to model the velocity structure of the crust and upper mantle.

Key words: Surface waves, group velocity, dispersion, tomography, Middle East, North Africa.

Introduction

The Middle East and North Africa is a tectonically complex region resulting from the closure of the Tethys Sea and the convergence of the relatively stable African continent with Eurasia. A topographic map of the area, which indicates the location of many of the tectonic features, is shown in Figure 1. A thorough review of the geologic history of much of the area is given in HAZLER (1998), with a short synopsis provided here. Much of African geologic history has taken place in the pre-

[1] Geophysics and Global Security Division, Lawrence Livermore National Laboratory, P.O. Box 808, L-205, Livermore, CA 94551, USA. E-mail: pasyanos1@llnl.gov
[2] Department of Geology and Geophysics, University of Colorado, Boulder, CO 80309, USA.

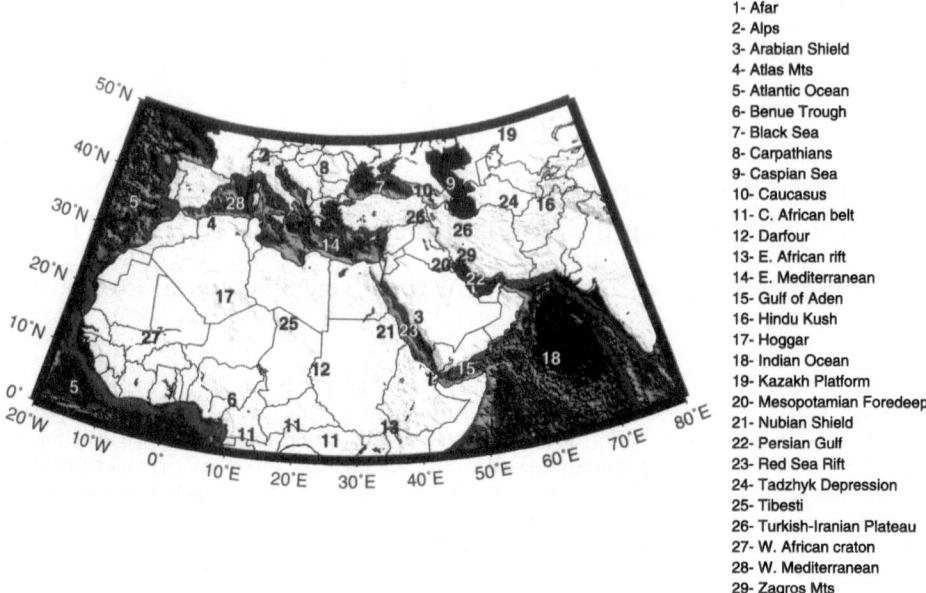

1- Afar
2- Alps
3- Arabian Shield
4- Atlas Mts
5- Atlantic Ocean
6- Benue Trough
7- Black Sea
8- Carpathians
9- Caspian Sea
10- Caucasus
11- C. African belt
12- Darfour
13- E. African rift
14- E. Mediterranean
15- Gulf of Aden
16- Hindu Kush
17- Hoggar
18- Indian Ocean
19- Kazakh Platform
20- Mesopotamian Foredeep
21- Nubian Shield
22- Persian Gulf
23- Red Sea Rift
24- Tadzhyk Depression
25- Tibesti
26- Turkish-Iranian Plateau
27- W. African craton
28- W. Mediterranean
29- Zagros Mts

Figure 1

Map of the Middle East / North Africa region indicating the location of geologic, tectonic and geographic features discussed in the text.

Cambrian. The African shields formed during the Archaen, while the Lower Proterozoic saw the assemblage of the West African craton. The northern African core was assembled in the Pan-African orogeny – an event which included the suturing of the West African craton to the East African craton; the formation of the Central African belt from the collision of the North African cratons with the Congo craton; the formation of the Mozambique belt in East Africa; and the collison of West Africa with North America. By the Paleozoic, Africa was at the center of Gondwana. The Hercynide orogeny, the result of the collision of Gondwana and Laurasia to create Pangea, was the first of two orogenic events to form the Atlas Mts.

The breakup of Pangea in the Jurassic saw the opening of the Atlantic (first with North America) and the closing of the Tethys Sea. The Benue trough represents a failed rift of this system. The rotation of Africa occurred first as convergence through subduction and increasingly as convergence through continental collision in the formation of the Tethys belt, producing the Alps, Atlas, Carpathians, Caucasus, and other ranges along the Alpine-Himalayan belt. Remnant oceanic regions are still found in the Mediterranean, Black, and Caspian Seas. In the Eocene, the Indian subcontinent started to collide with Eurasia, forming the Himalayas. Most recently, we have seen the opening of the Red Sea, Gulf of Aden, and East African rifts, separating the Arabian and Nubian shields and creating a triple junction at Afar. As a result of spreading along the Red Sea rift, there is active convergence between the Arabian and Eurasian plates which is forming the Zagros Mts., and the Turkish-Iranian Plateau. Large sediment accumulations (up to 10 km thick) are found along the foot of the collision zone in the Persian Gulf and Mesopotamian foredeep. Other large sediment accumulations occur in the Caspian, Black Sea, and Eastern Mediterranean.

A number of previous studies have attempted to characterize the earth structure in the Middle East and North Africa using surface waves. Earlier studies usually relied on one or several surface-wave paths. Using primarily surface-wave phase velocity but also group velocity and body wave velocities for several East African paths, GUMPER and POMEROY (1970) developed the one-dimensional AFRIC model. KNOPOFF and SCHLUE (1972) and KNOPOFF and FOUDA (1975) measured phase velocities along paths along the East African rift and in the Arabian Peninsula, respectively. DORBATH and MONTAGNER (1983) followed a regionalization of surface-wave measurements across Africa, based on cratonic and non-cratonic regions. ASUDEH (1982) combined two-station phase-velocity measurements for four events with body wave data to study Iran. Using a similar technique, TAHA (1991) measured phase velocities in Egypt. Single station group-velocity measurements across Turkey were made by both EZEN (1991) and MINDEVALLI and MITCHELL (1989).

HADIOUCHE et al. (1986) looked at both phase and group velocities along a single path. This study was followed by HADIOUCHE and ZÜRN (1992) who studied structure in the Afro-Arabian region by simultaneously inverting both phase and group velocities along a limited number of paths. MOKHTAR and AL-SAEED (1994) performed a similar study in the Arabian Peninsula using just group velocities. KNOX et al. (1998) employed two-station phase velocities to study the crust and upper mantle beneath Afar and western Saudi Arabia. HAZLER (1998) and HAZLER et al. (2001) studied group velocities in northern Africa using regionalizations based on geology. RODGERS et al. (1999) combined waveform modelling and group-velocity measurements to study the Arabian Peninsula. Recently, more involved studies have used many measurements and tomographic techniques to solve for isotropic (HADIOUCHE and JOBERT, 1988) and anisotropic models (HADIOUCHE et al., 1989) of the African continent. RITZWOLLER and LEVSHIN (1998) performed a group-velocity tomography of Eurasia that includes portions of Africa and the Middle East.

The purpose of this research is to improve surface-wave group-velocity maps and lithospheric shear-wave velocity models for northern Africa and the Middle East. Group-velocity maps allow phase-match filtering, which has the potential to lower M_s thresholds. M_s is an important discriminant measure and could help identify smaller magnitude events. Improved shear-wave velocity models can improve location and event assessment capabilities throughout the region. Both improved identification and location capabilities are important to monitoring the Comprehensive Nuclear-Test-Ban Treaty (CTBT). We wish to estimate the lateral variation of Rayleigh- and Love-wave group and phase velocity using several tomography techniques, and to invert the surface-wave measurements at grid points for shear-velocity structure. To date we have concentrated on the group-velocity component of this study since, unlike phase-velocity measurements, these measurements can be made without knowledge of the source mechanism.

Many of the changes in this study from previous years (MCNAMARA *et al.*, 1997) are a direct result of the "Workshop on the U.S. Use of Surface Waves in Monitoring the CTBT," which was held during the 1998 SSA meeting in Boulder, Colorado (WALTER and RITZWOLLER, 1998). As a consequence of the workshop, we have changed the way that we are making our measurements and the method we are using to invert the observations. Concurrently, we are continuing to increase the number of paths and expanding our coverage of the Middle East-North Africa region. Our database is actively being loaded with seismic data for all events that can yield quality surface-wave measurements. The increased density of future data will improve our overall resolution and allow us to interpret even finer scale structure, however, the present coverage is such that we do not expect significant large-scale changes in the results presented here.

Data and Measurements

Data Selection

For the surface-wave study, vertical and transverse component teleseismic and regional seismograms were selected for 1990–1999 from our research database (RUPPERT *et al.*, 1998). The waveform data was gathered from broadband, three-component, digital IMS, MEDNET (BOSCHI *et al.*, 1988), GEOSCOPE (ROMA-NOWICZ *et al.*, 1984) and IRIS stations plus the portable 1995–1997 PASSCAL deployment in Saudi Arabia (VERNON *et al.*, 1996). Figure 2 shows the distribution of earthquakes (circles) and broadband digital seismic stations (triangles) throughout the Middle East/North Africa region that are used in this study. We chose events and stations within this box in an effort to use regional paths and minimize the contributions from outside anomalies. Seismicity in the region is concentrated in southern Europe and the Middle East, with almost no seismicity within the African continent, leading to an uneven distribution of surface-wave measurements. To date, over 13,500 seismograms have been analyzed to determine the individual group velocities of 7–150 second Rayleigh and Love waves. Of these, quality group-velocity measurements have been made for 6817 Rayleigh-wave and 3806 Love-wave paths. Because of the difficulty in making short-period measurements at long epicentral distances, we are able to make significantly more measurements at longer periods. For example, while we made 5300 Rayleigh-wave measurements at 50 seconds, we have only been able to make 2100 measurements at 15 seconds and only 1000 at 10 seconds.

We have applied strict selection criteria to the earthquakes used in this study in order to insure that only high-quality surface wave travel times are used in the inversion. All of the measurements were made using a single analyst to ensure picking consistency. To eliminate potential errors in the group-velocity measurement process,

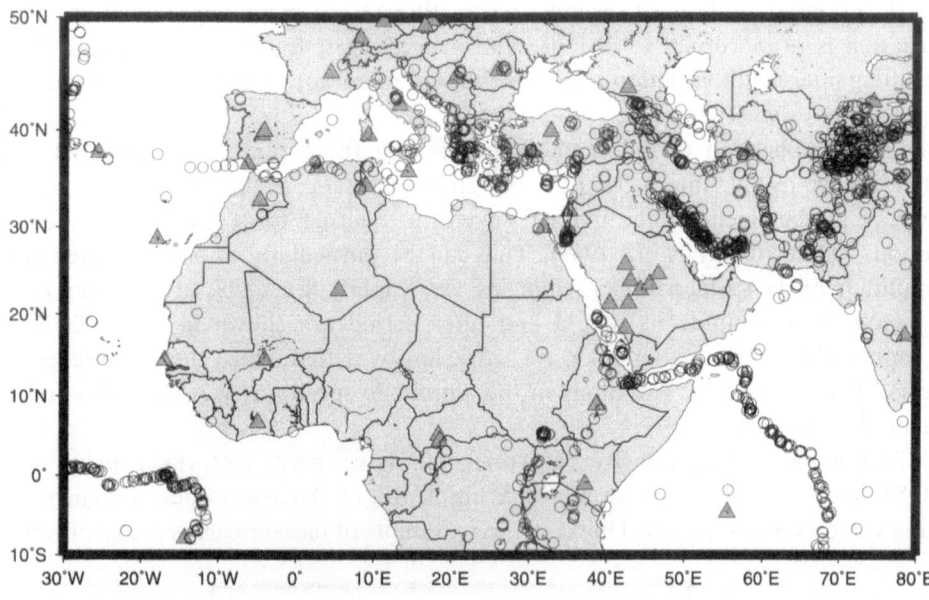

Figure 2

Distribution of earthquakes (circles) and broadband digital seismic stations (triangles) used in this study.

only travel times from high-quality relatively continuous dispersion curves were used. Qualitative assessments were made at the time of calculation and were used to eliminate spurious travel times. Furthermore, in order to eliminate measurements from stations with large timing problems, the surface waves are compared to the data set of *P*-wave picks that are more sensitive to timing problems. Finally, travel-time residuals with a velocity deviation greater than 25% from the data set mean were eliminated, so that oceanic paths, which have group velocities significantly faster than the mean, are included but that significant outliers which may be caused by other factors are excluded.

Group-velocity Measurements

To obtain the Rayleigh-wave dispersion curve, a narrow-band Gaussian filter is applied to the broadband vertical component, displacement seismogram over many different periods (e.g., HERRMANN, 1973). Similarly, to obtain Love-wave measurements, we can apply the same filter to data rotated into the transverse direction. The maximum amplitude at each period is picked on the envelope function and the arrival time corresponding to this maximum amplitude is used to compute the group velocity. We use an optimal Gaussian width parameter that minimizes the area of the dispersed wavetrain on the group velocity – period curve. The Gaussian filter width is a function of the source-receiver distance and the period is computed assuming PREM group velocities (DZIEWONSKI and ANDERSON, 1981), but held constant at short periods (less than 30 seconds) where PREM is a poor approximation. The variation is proportional to distance and hence largest for the longest paths. The resulting filter width variation is similar (but not identical) to that shown in Figure 5 of LEVSHIN *et al.* (1992).

One complication to this methodology is that the period of the narrow-band filter, known as the "filter period," is not always the same as the period of the peak-to-peak measurement of the filtered waveform, referred to as the "instantaneous period" (DZIEWONSKI *et al.*, 1969). This can be particularly important when the amplitude of the surface wave is changing very rapidly. Examples of the differences between the instantaneous period and filter period are shown in Figure 3. The recommendation of the surface-wave workshop was that, since the instantaneous period is not biased by variations in the amplitude spectra, it was the appropriate quantity to measure.

All of our group-velocity measurements have been performed using the PGSWMFA (PGplot Surface Wave Multiple Filter Analysis) code designed by Chuck Ammon of St. Louis University. An example of measurements made with the

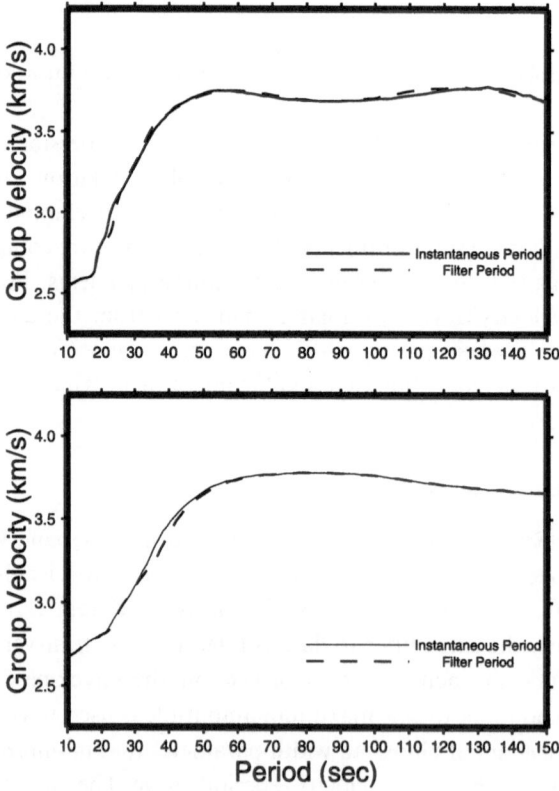

Figure 3
Group velocity measurements made using filter period (solid) and instantaneous period (dashed).

code is shown in Figure 4. The left panel shows the contours of the velocity-period spectrum that are used to make the dispersion measurement along with uncertainty estimates. The center panel shows the waveform under study, while the right panel shows the spectral amplitude as a function of period. Using PGSWMFA, we have remeasured waveforms to determine the group velocities as a function of instantaneous period instead of filter period.

Inversion

Inversion of Travel Times for Lateral Group-velocity Variation

The surface-wave travel time, for a given period, is expressed simply by $t = d\,s$ where t is the total travel time, d is source to receiver distance, s is slowness (inverse velocity). For estimating lateral group-velocity variations, the sampling region is gridded and the slowness for each grid cell is determined. The travel-time equation then becomes:

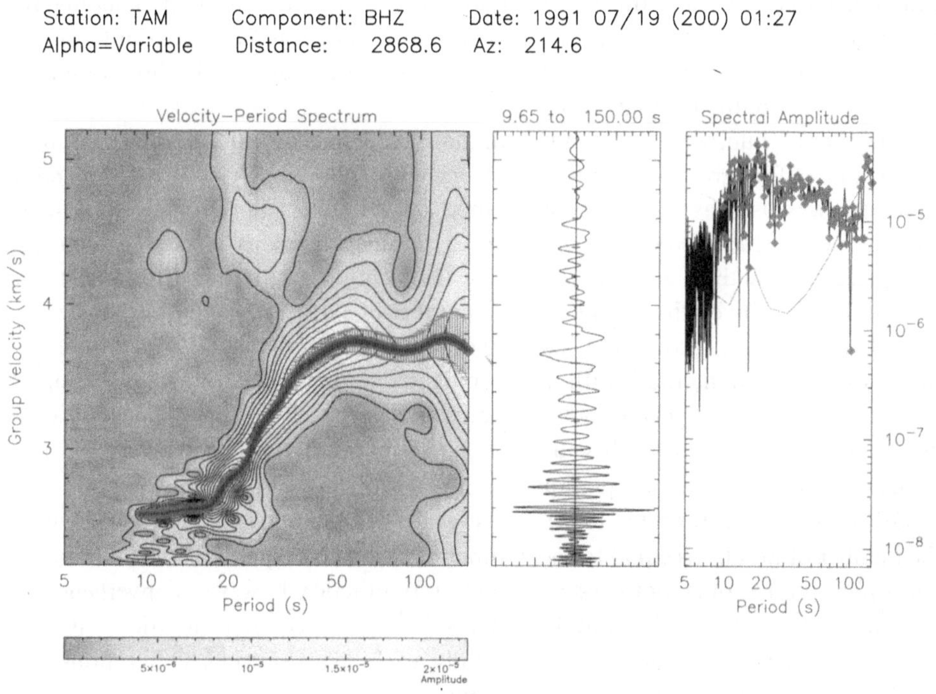

Station: TAM   Component: BHZ   Date: 1991 07/19 (200) 01:27
Alpha=Variable   Distance: 2868.6   Az: 214.6

Figure 4

An example of group velocity measurements made using the PGSWMFA program. The left figure shows the contours of the velocity-period spectrum which are used to make the dispersion measurement, along with uncertainty measurements. The center figure shows the Rayleigh wave waveform, while the right figure shows the spectral amplitude as a function of period.

$$t = \sum d_i s_i \ , \tag{1}$$

where d_i is the distance the ray travels in cell i and s_i is the slowness in cell i. For a number of paths, a series of these equations can be represented in matrix form as:

$$\boldsymbol{T} = \boldsymbol{DS} \ . \tag{2}$$

Additionally, the travel-time measurements can be relatively weighted by any number of factors such as measurement quality, path distance, event magnitude, etc. We choose to weight by both quality and distance. In practice, we find that these relative weighting factors have no significant effect on the overall inversion results.

We also choose to impose a smoothness constraint on the data by constructing the Laplacian of the slowness and requiring it to be zero. A series of these equations can be represented in matrix form as:

$$\lambda \Delta \boldsymbol{S} = \boldsymbol{0} \ , \tag{3}$$

where $\Delta \boldsymbol{S}$ is the Laplacian of the slowness. The weighting factor λ controls the tradeoff between fitting the travel times and smoothing the model. While this equation imposes a smoothness constraint, it also has the effect of damping the travel-time inversion. In order to avoid imposing any *a priori* constraints on the inversion, no effort was made to relax this constraint at "expected" discontinuities, such as ocean/continent boundaries.

There are a number of different methodologies available for inverting measured travel times for group slowness (and velocity). Previously, a backprojection technique was used as the inversion method on a smaller subset of group-velocity measurements (McNamara *et al.*, 1997). In most cases, it appeared that the backprojection method was able to resolve the location and pattern of the fast and slow anomalies. Given the relatively small number of surface-wave paths, however, it was often unable to recover the full amplitude of the anomalies. We have replaced the backprojection method with the conjugate gradient method. The conjugate gradient technique is a search method that works very well on sparse linear systems like the travel-time problem. Because there is no matrix inversion involved, it is well suited for large systems of equations. Like other conjugate gradient methods (i.e., LSQR, Paige and Saunders, 1982), convergence will theoretically be reached within the number of iterations equal to the number of constraint equations (i.e., number of paths plus number of smoothness constraints). In practice, however, convergence, as determined by both residuals and distances between successive iterations, is very rapid and achieved much sooner. In our study, each inversion runs 30 iterations. We chose a 2 degree by 2 degree grid for the inversion.

Uncertainty, Resolution and Damping

Uncertainties in the estimated group-velocity maps can be related to theoretical errors, such as off-great circle propagation, event mislocation, azimuthal anisotropy and source group time shifts, or non-theoretical errors such as measurement errors and path distribution. Bias from all of the theoretical sources is considered by RITZWOLLER and LEVSHIN (1998). They conclude that the largest source of bias from theoretical errors is probably from azimuthal anisotropy. A first-order, qualitative, measure of data set resolution can be obtained by inspecting the raypath distribution throughout the sampling region (Fig. 5). Though raypath density is important, azimuthal sampling is as significant. For a more quantitative assessment, resolution can be investigated with synthetic travel times computed through laterally varying "checkerboard" test velocity models. Using the 50 second period as an example, we compute the Rayleigh-wave travel time for each path (Fig. 5) through a model with $8° \times 8°$ checkers that vary in velocity by 5% about a mean of 3.61 km/s. The synthetic travel times are then inverted using the inversion methods described above. While LEVEQUE et al. (1993) has discussed some of the limitations of the checkerboard test, it is still a useful tool in assessing our data resolution. In particular, our ability to reproduce the input model is directly indicative of the density and azimuthal sampling of our data set. For example, in this case the location and amplitude of the checkerboard pattern is well-resolved throughout the Mediterranean, northeast Africa, the Arabian Peninsula, and the Middle East, but is poor in northern Eurasia, central and west Africa, and the Indian Ocean. This is due to the lack of crossing paths in these regions.

Our results depend, in part, on the value that we choose for the weighting factor λ. If the weighting factor is set too low, then the inversion is underdamped and the map exhibits streaking. If the weighting factor is set too high, then the inversion is overdamped and only very broad features will be resolved. When this number approaches the distances that the paths travel in each cell, then the travel time and smoothness have about equal weights. There is also some implicit damping due to the fewer number of iterations in the conjugate gradient method than the total number of constraint equations. Figure 6 shows the results at 20 seconds using factors of 0.1, 1.0, and 5.0, where 1.0 is equal overall weighting between travel time and smoothness constraints. For $\lambda = 0.1$, there is effectively no smoothing, the image is rather oscillatory, and the anomalies are on the order of $\pm 16\%$. The image does not change significantly from this model with damping parameters smaller than 0.1. By the time we reach $\lambda = 5.0$, the image is extremely smooth and the anomalies are on the order of $\pm 6\%$.

Obviously, as damping is decreased, we will increase our variance reduction. At the same time, however, we will have increased the effective number of free parameters in the model. For example, at $\lambda = 0$, the number of free parameters in the model is simply equal to the total number of blocks; at $\lambda = 1$, we have halved the effective number of free parameters; and as λ approaches infinity, the effective number of free parameters is reduced to 1. It is useful, therefore, to compare the residual variances with regard to the degrees of freedom. By relating variances and degrees of freedom for a wide range of damping parameters, we find an optimal

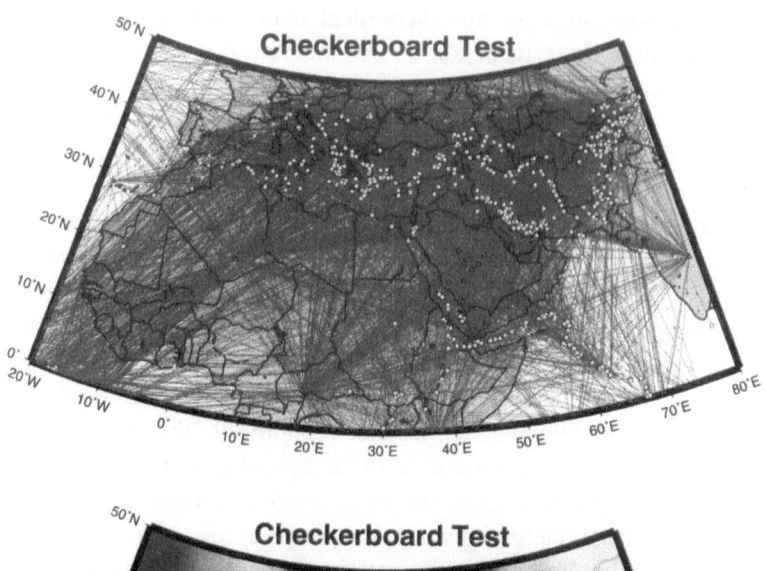

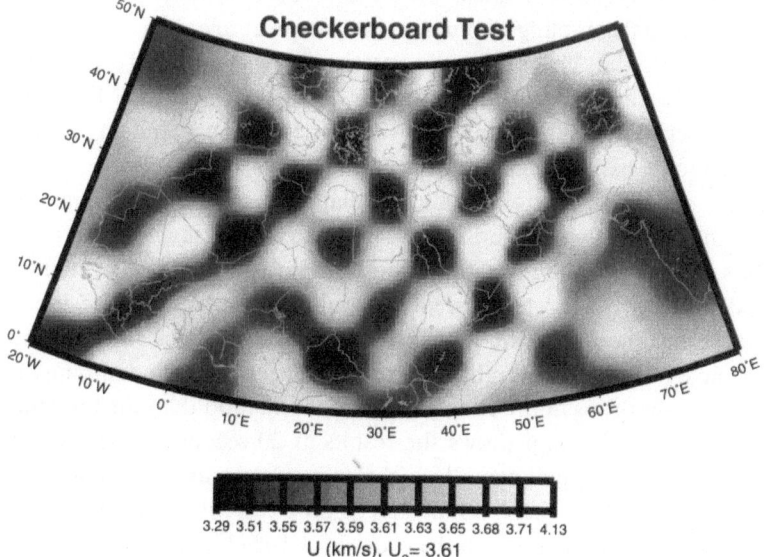

$$3.29 \ 3.51 \ 3.55 \ 3.57 \ 3.59 \ 3.61 \ 3.63 \ 3.65 \ 3.68 \ 3.71 \ 4.13$$
U (km/s), $U_0 = 3.61$

Figure 5

Distribution of paths for 50 second period Rayleigh waves (top) and results of checkerboard resolution test (bottom) of the same. The color scheme varies from slow (black) to fast (white).

solution for the damping parameter of between 1.0 and 3.0. We selected a damping parameter of $\lambda = 1.5$ for our inversions, based on the variance reduction, overall

►

Figure 6

Inversion results for 20 second Rayleigh waves performed using damping factors of 0.1 (top), 1.0 (middle), and 5.0 (bottom). The color scheme varies from slow (red) to fast (purple) with areas of poor resolution shown in white.

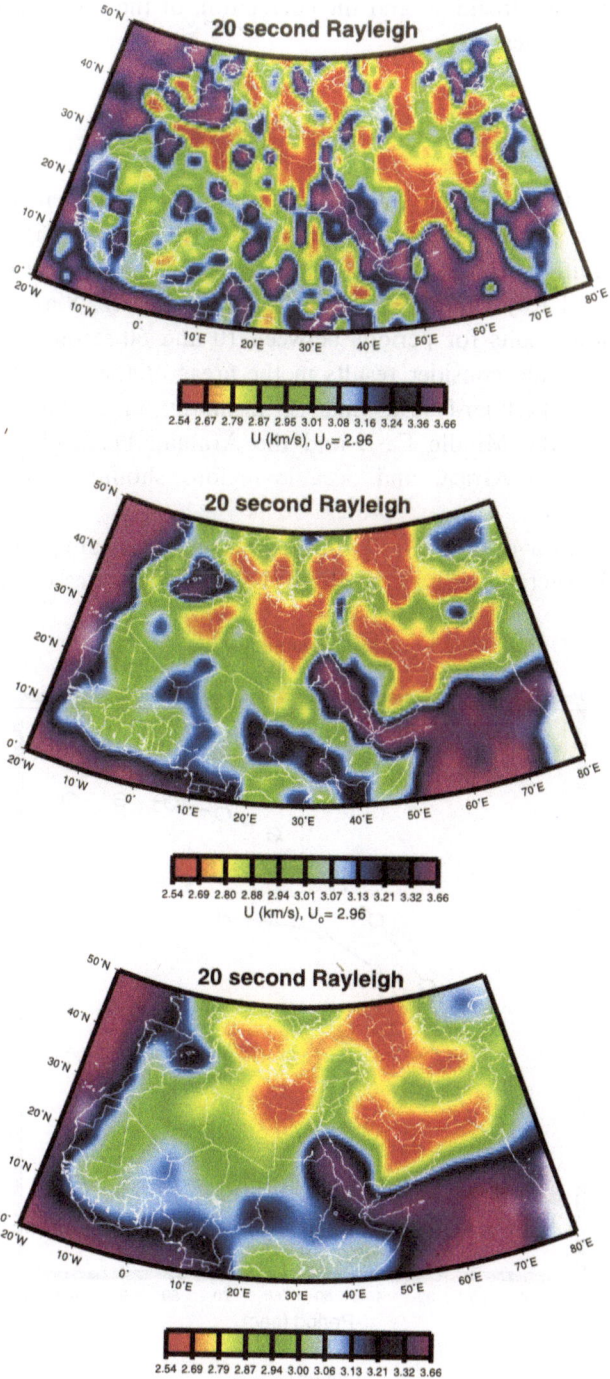

model smoothness and streaking, and the correlation of the tomography results to certain known tectonic features.

Results

Our results for the inversions of the group-velocity measurements performed using the conjugate gradient method are shown in Figures 7–9. Significant lateral group-velocity variations are apparent at all periods. Figure 7 indicates the average velocity of Love and Rayleigh waves for each period. Figures 8 and 9 present the individual inversion results for periods between 10 and 60 seconds. In looking at these figures, one should consider results in the areas of the map that were well-resolved in the checkerboard test, that is, primarily in northeast Africa, the Mediterranean Sea, the Middle East, and the Arabian Peninsula. Results from northern Eurasia, west Africa, and oceanic regions should be taken with the appropriate precautions.

For the Rayleigh waves (Fig. 8), shorter periods (< 30 sec) are primarily sensitive to shallow crustal structure, such as the relatively low velocities associated with large

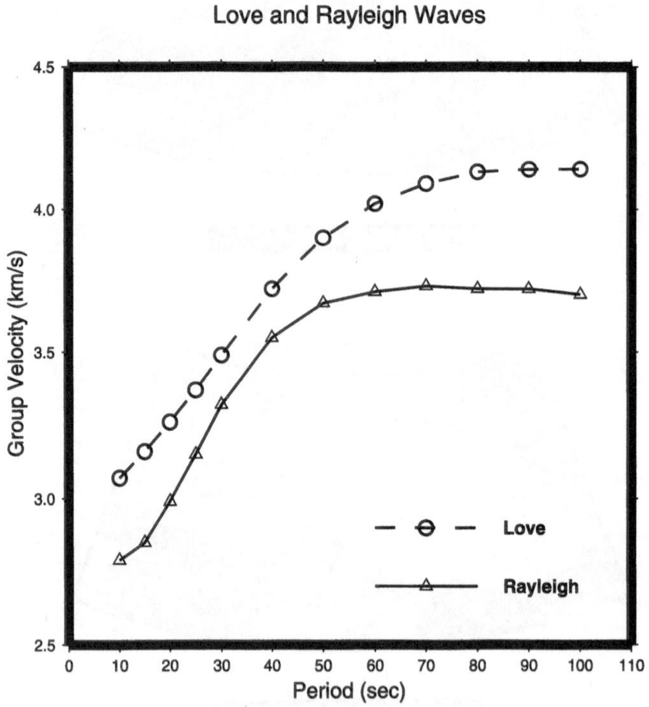

Figure 7

Average inversion velocities for Rayleigh waves (triangles and solid lines) and Love waves (circles and dashed lines) over the complete range of inversion periods.

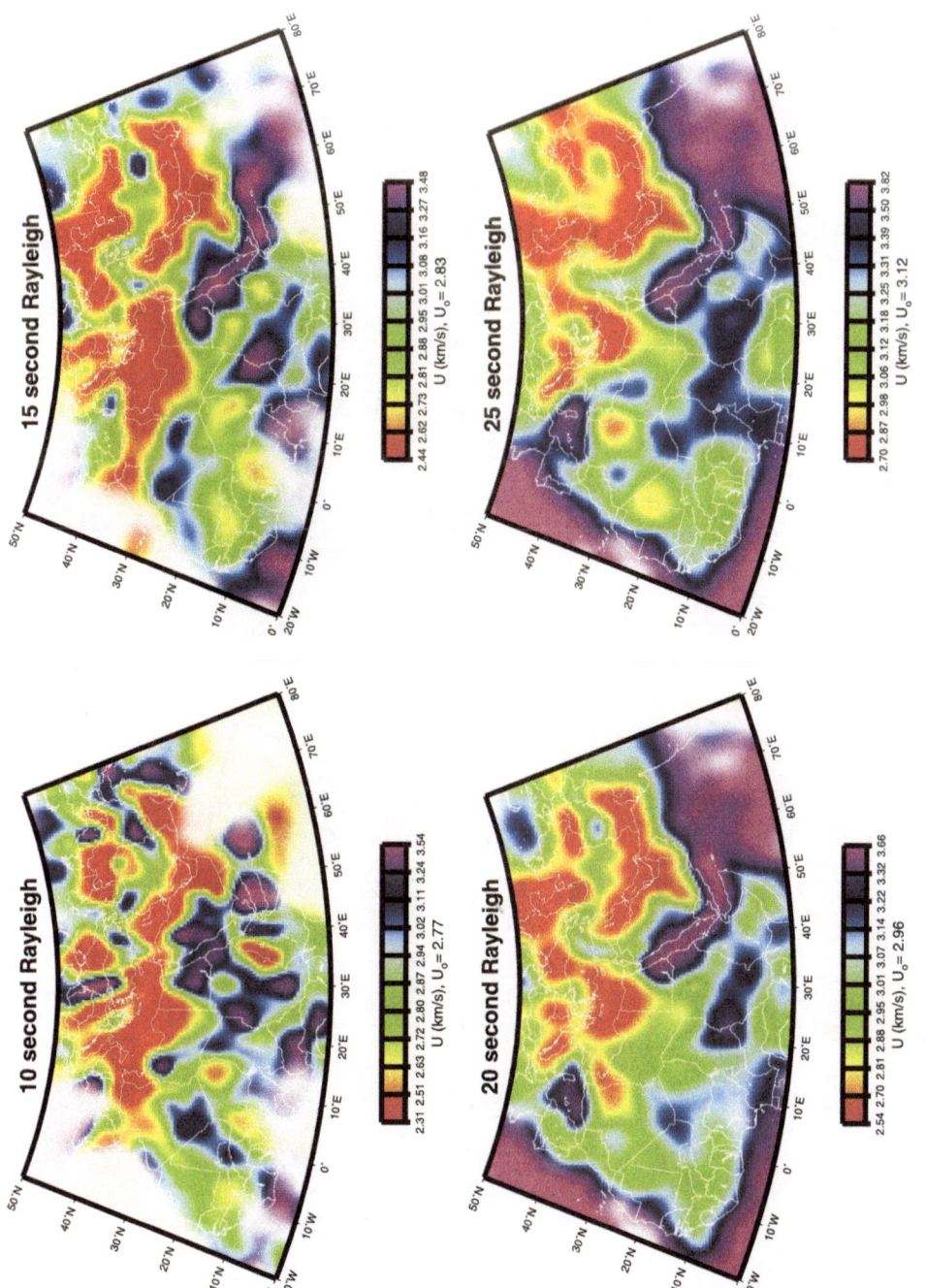

Figure 8

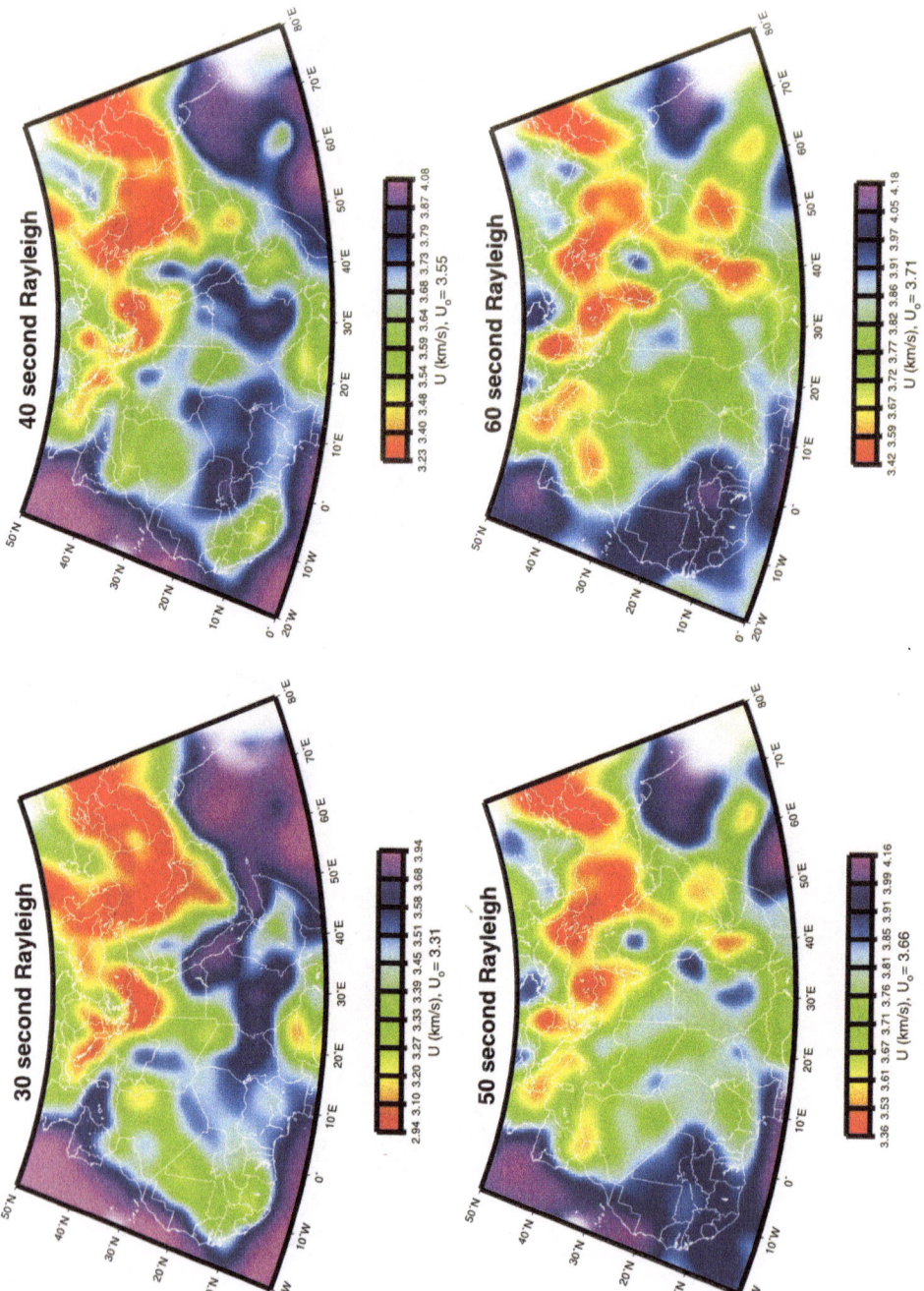

Figure 8

sedimentary features (e.g., Mesopotamian Foredeep, Persian Gulf, eastern Mediterranean, Caspian Sea, Black Sea, Tadzhyk Depression). The extent of these features across the period range gives us an indication of the depth of these sedimentary basins. For example, at the shortest periods (≤15 seconds), there are low velocity anomalies associated with the entire Mediterranean Sea. As we move to longer periods, however, we find that while the anomaly persists in the eastern Mediterranean, it disappears from the western Mediterranean. Sediments are generally considered to be considerably deeper in the eastern portion of the Mediterranean Sea (LASKE and MASTERS, 1997).

Oceanic areas like the Atlantic Ocean, Indian Ocean and the Arabian Sea have fast group velocities, in excess of 15% greater than the mean. We also find fast velocities (10–15% faster than the mean) in African shield regions, such as Tibesti, Darfour and Hoggar, as well as the Arabian and Nubian shields. The Turkish-Iranian plateau and the remainder of Africa are both moderately fast to fast (5–10% faster). We find the largest variability in the group velocities at shorter periods, where variations are generally greater than 10%, with decreasing variability at longer periods.

Rayleigh waves become more sensitive to crustal thickness and average crustal velocity at longer periods (∼50 sec). For example, at 40 seconds we can see an arc of low velocities along Asia Minor and the Zagros Mts. that continues to the east along the Hindu Kush and into Tibet. In these regions of thick crust, the group velocities are sensitive to the slow crustal velocities as opposed to the fast mantle velocities from the outlying regions. We are also starting to see slow group velocities associated with the oceanic ridges. These features become decidedly clearer with inversions performed using a smaller grid spacing, which unfortunately, is not justified for our entire data set.

By 60 seconds, we are probably starting to sample mostly the upper mantle. Group velocities from the oceanic regions continue to be fast, except for slow velocities along the oceanic ridges. There is also evidence of a velocity contrast between the western and eastern parts of North Africa. Group velocities in the west are comparable to those of the oceanic regions, while slower velocities are found in the east. There are also indications of slow mantle velocities beneath the Red Sea, Gulf of Aden, east African and mid-Indian rifts. Variations in the group velocities in this period range are generally less than 10%. The inversion results between 60 and 90 seconds are all very similar to each other. At the longest periods in our study (100 sec), we as yet have insufficient path coverage to produce reasonable maps.

The Love waves (Fig. 9), which are more sensitive to shallow structure, show somewhat similar features to the short-period Rayleigh waves. In general, however, the resolution of the Love waves is poorer than that of the Rayleigh waves, most likely attributable to the fewer number of measurements and the lower signal-to-noise ratio of the data. Unlike the Rayleigh waves, the Love waves look fairly similar over a wide range of frequencies. This is perhaps not unexpected due to the continuing sensitivity of the Love waves to shallow surface structure even at longer periods. Love waves between 10 and 20 seconds highlight shallow sedimentary

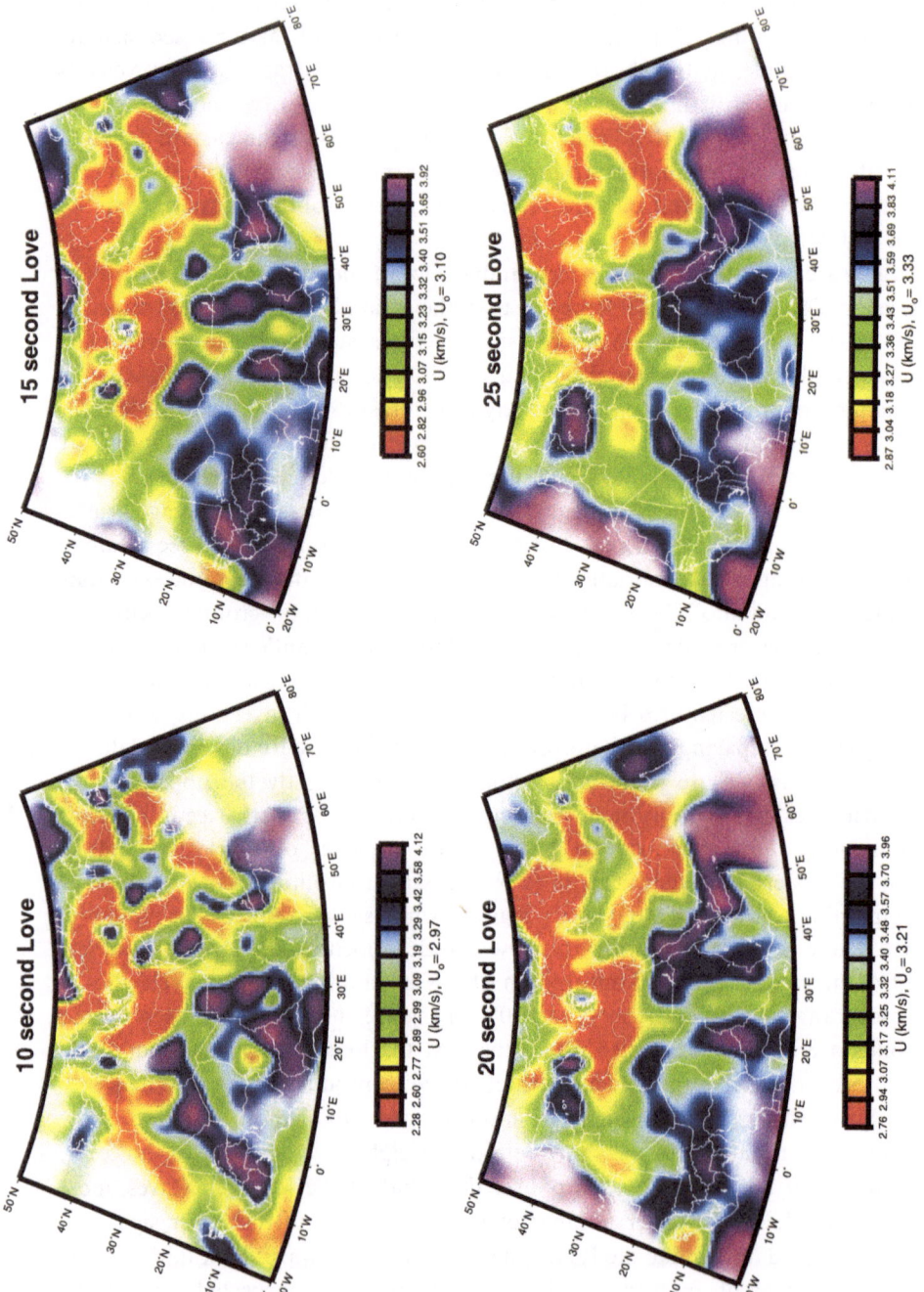

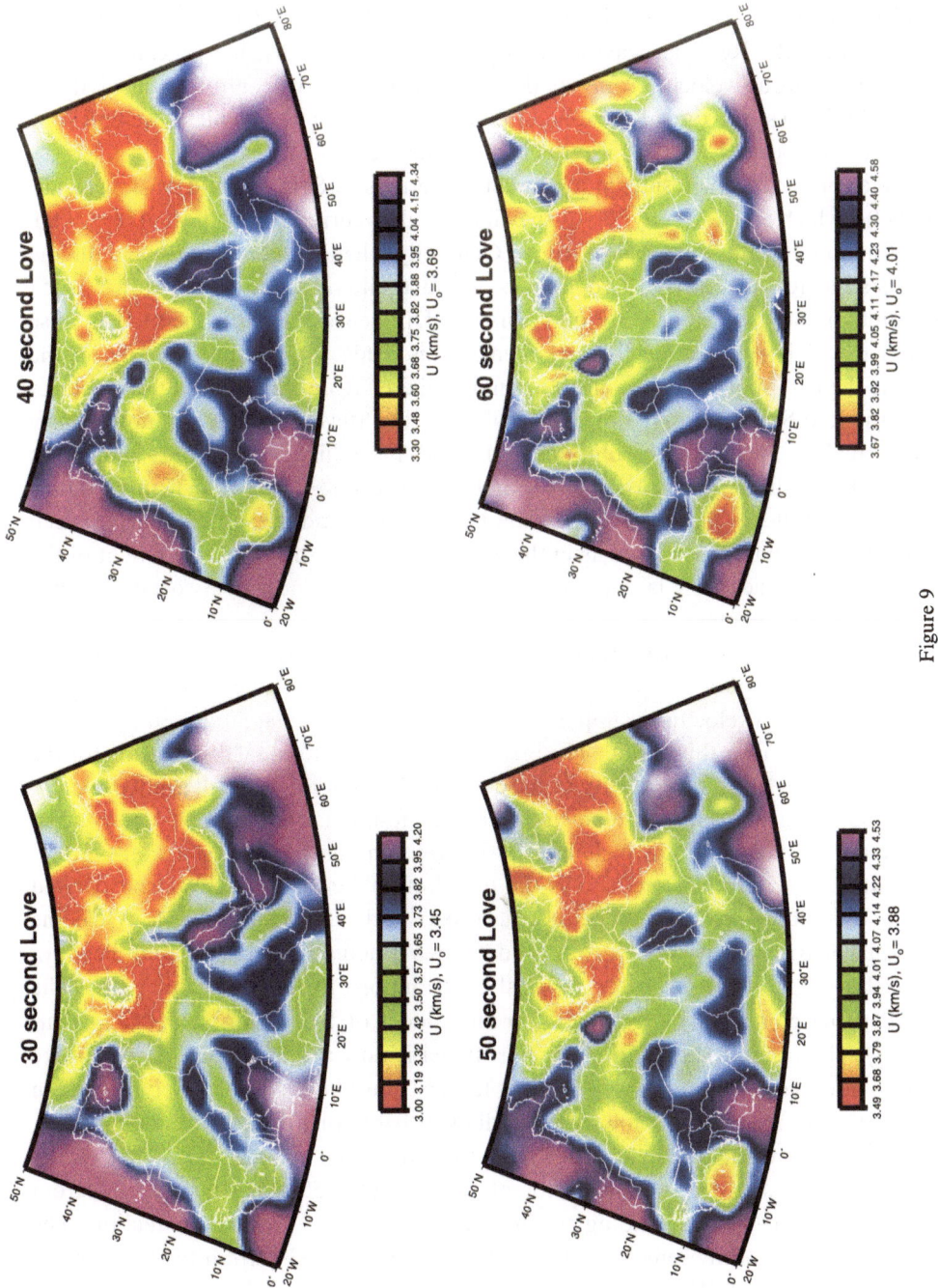

Figure 9

Inversion results for Love waves at 10, 15, 20, 25, 30, 40, 50, and 60 seconds. Color scheme same as in Figure 6.

basins. Between 25 and 40 seconds, slow group velocities are limited to the deepest basins. We find slow velocities in the eastern Mediterranean, Persian Gulf, Black Sea and Caspian Sea, and fast velocities in the Indian Ocean and throughout most of Africa. Like the Rayleigh waves, we find that the western Mediterranean has slow Love-wave group velocities only at the shortest periods. We only see sensitivity to crustal thickness starting at the longest periods shown (~60 sec), where we find slow velocities associated with the Zagros Mts., Caucasus, and Himalayas.

We compared our results at a number of periods to STEVENS and MCLAUGHLIN (1997), which is a 5° by 5° dispersion model currently used at the CTBT International Data Center in Vienna. Generally, there is good agreement in the gross-scale features of the two models. In fact, filtering our model down to 5° resolution, the two models are quite similar. Differences between the models arise from the higher resolution features seen in this study. For example, while on the whole the two maps at 20 seconds are similar, there are discrepancies. Small features like the Red Sea, which are not well-resolved in the Stevens model, are more clearly delineated in the higher-resolution study and have higher amplitude anomalies associated with them because they are not averaged into adjacent structures. There are also significant disagreements in the Persian Gulf and Mesopotamian foredeep region. While this anomaly exists in the 5° model, in our study the areal extent of the anomaly is considerably larger and the negative velocity anomaly associated with it is significantly greater.

Figure 10 displays histograms of the 20 second Rayleigh-wave group velocity residuals (in km/s) for several models and for the inversion results. The calibration of 20 second group velocities is important since surface-wave magnitudes are generally measured at or near this period. The first panel shows the residual distribution for the starting model (the mean velocity of the data), which is not unimodal, is not centered, and has a long tail. The measurements with large residuals in the figure correspond to primarily oceanic paths, which cannot be accommodated by the average group velocities. The second panel shows the distribution for the group velocity derived from the PREM velocity model (DZIEWONSKI and ANDERSON, 1981). While the residual distribution is even more skewed, the RMS actually decreases from the mean model because of the decreased residual contributions from the oceanic paths. The third panel compares the residuals for the model of STEVENS and MCLAUGHLIN (1997). The residual distribution for this model is now a single-mode Gaussian-like distribution centered at 0.0, nonetheless the distribution is still relatively broad. In contrast, the final panel shows the residual distribution for the inversion results, which manifest a markedly sharper peak than the other models. The RMS of the residuals decreases by about a factor of 2 or about a 50% RMS reduction from the mean model, corresponding to approximately a 75% variance reduction. The variance is reduced by some 73% from PREM and about 33% from the Stevens and McLaughlin model. We could choose to fit the travel time even better (up to an 80% variance reduction from the starting model) by reducing the weighting factor λ,

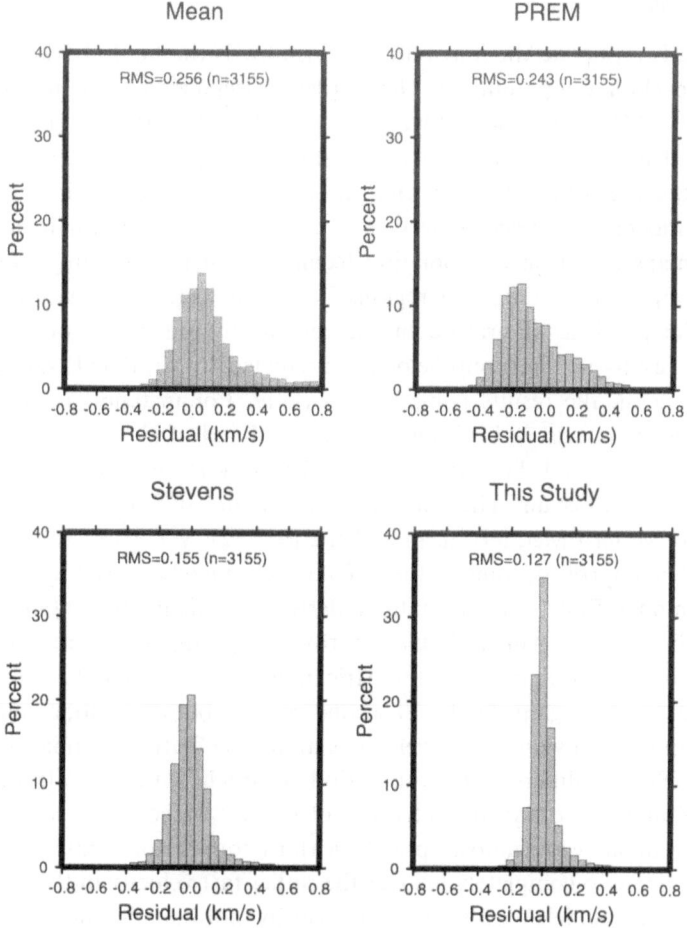

Figure 10
Histograms of residuals for 20 second period Rayleigh waves for starting (mean) model, PREM, regionalized surface waves, and group-velocity inversion (this study).

although at the expense of model smoothness. Because this study has higher resolution than global models, we expect that it will explain the data better than other existing models of the region.

Applications

This surface-wave study has many applications to CTBT monitoring and to general earth structure investigations. Below, we briefly discuss several applications we are pursuing.

Regional Models

The CTBT group at the Lawrence Livermore National Laboratory is in the process of developing a preliminary, first-order regionalization for the Middle East/ North Africa (MENA) region (SWEENEY and WALTER, 1998). The background model is used to provide estimates of calibration properties, such as attenuation, crustal thickness, crustal velocities, and phase behavior, where calibration data are sparse. The model is intended to serve as a background model for forming station calibration maps based on interpolation techniques such as kriging. This model is currently comprised of about 30 regions based on tectonics, topography, and a review of the published literature in the region. In Figure 11, we compare our tomography results to the group velocities predicted from the MENA background model. In general, the results compare favorably. For instance, in the 15 second Rayleigh-wave comparison, both figures reveal a band of slow velocities along the Mediterranean Sea and Persian Gulf, moderate velocities in Africa, and fast velocities under the oceans. There are, however, major differences between the two models in the western Mediterranean and the Red Sea. While our resolution at this period could be improved, some of the differences are likely caused by deficiencies in the MENA model. Efforts are currently underway to redefine the sedimentary layers for the MENA model, which should improve our comparisons at short periods. There are also sizable differences in the African interior that could be caused by poor path coverage in the region. In the 30 second case, in both the model and inversion results, we find the slowest group velocities along the Turkish-Iranian plateau and Zagros Mts. Notable differences at this period occur where there is a lack of crossing paths. For example, we find the intrusion of oceanic velocities onto the continent (i.e., southern India, westernmost tip of Africa). In both periods that we consider, the inversion is unable to completely recover the rather fast velocity in the oceans which is predicted by the model. We also find significantly more small-scale perturbations in the tomography than in the background model. Comparisons like these and similar comparisons of other parameters such as P_n can be used to construct, compare, modify and improve the background velocity models.

Discrimination

One immediate application of the tomography results is to construct phase-matched filters for new paths, using the predicted group velocities. Phase-matched filters can improve weak surface-wave signals by compressing the dispersed signals (HERRIN and GOFORTH, 1977). In turn, the compressed signal can be cleaned to exclude noise sources such as microseismic noise, multipathing, body waves, higher order surface waves, and coda. With this methodology it is possible to extract surface-wave signals from noisy measurements, calculate regionally determined M_s measurements, and lower the threshold of surface-wave magnitude measurements. As first demonstrated by BRUNE *et al.* (1963) and more traditionally codified by

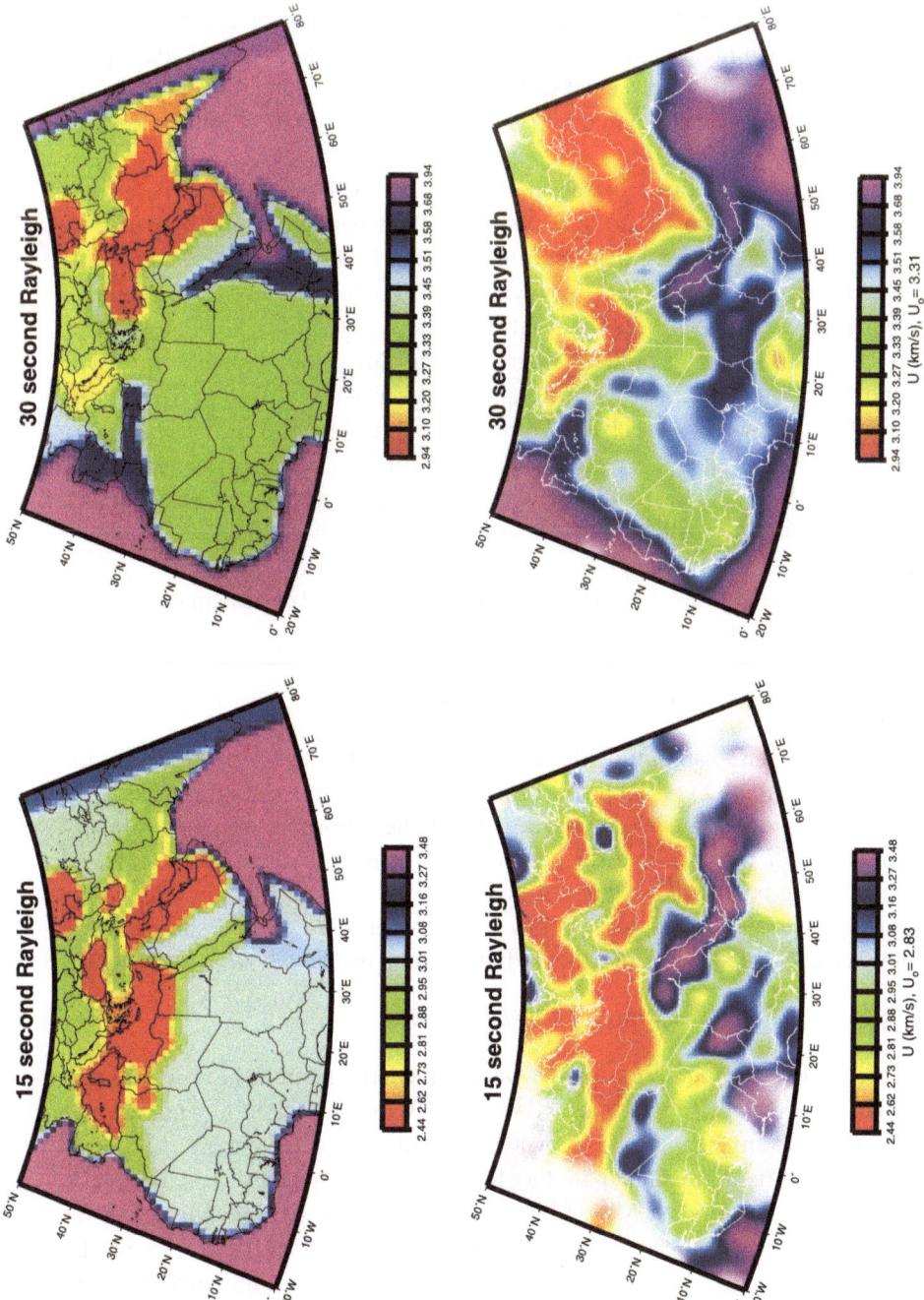

Figure 11

A comparison of inversion results (bottom) to MENA background model (top) for 15 and 30 second Rayleigh waves. The background model is not defined east of 75 degrees E.

LIEBERMANN *et al.* (1966) and LIEBERMANN and POMEROY (1967), surface-wave magnitude estimates can be combined with m_b to form one of the best known discriminants of earthquakes and explosions. While the threshold for calculating M_S depends on factors such as epicentral distance and long-period station noise, using phase-matched filters has the potential to reduce the $m_b{:}M_S$ discriminant to even lower magnitude levels.

Geophysical models can be biased in comparison to the real earth. To correct these biases, one needs to create a set of spatially-dependent model corrections, known as correction surfaces (i.e., SCHULTZ *et al.*, 1998), which are commonly provided in a station-centric format. Applying our inversion results we can construct group-velocity correction surfaces for a station, wave type (i.e., LR, LQ), and period. The correction surfaces are produced by integrating slownesses for the appropriate period and wave type from the station to all points on the grid. For a given station and source location, we can simply look up the group velocities to use in the phase-matched filter. An example of a correction surface for 20 second Rayleigh waves at station RAYN (Fig. 12a) and station KIV (Fig. 12b). Note the "wakes" of slow group velocities in back of large sedimentary features such as the Mesopotamian Foredeep and the Caspian Sea.

We use the formulation of MARSHALL and BASHAM (1972) to calculate regional surface wave magnitudes, which is given as:

$$M_{sr} = \log A + B'(\Delta) + P(T) \; , \tag{4}$$

where A is the maximum amplitude (in nm), $B'(\Delta)$ is the distance correction, Δ is the distance (in degrees), $P(T)$ is the path correction, and T is the period (in seconds). The path correction can be determined with the appropriate group-velocity dispersion curve for the path and the passband of the measurement can be selected where we have the best signal-to-noise ratio.

The discriminant is calculated as $M_{Sr} - M_{So}$ where M_{So} is the surface wave magnitude which is predicted from the body-wave magnitude m_b. M_{So} is given as:

$$M_{So} = 1.60 m_b - 4.50 \tag{5}$$

and has been derived by equilibrating the empirical energy relationships of the body and surface-wave magnitude scales (LAY and WALLACE, 1995). This should effectively remove any magnitude dependence from the discriminant. Figure 13 shows the discriminant calculated in the 10–14 second passband at station NIL (Nilore, Pakistan) for the 11 May 1998 Indian nuclear explosion (star) and 28 earthquakes (triangles) located near the explosion. The left side of the plot shows the mean (center line) and first and second standard deviations (outer lines) of the earthquake population. The explosion lies outside of the third standard deviation of the earthquake population. Because of amplitude modulations with azimuth caused by the source mechanism, a multi-station discriminant, in which single-station results

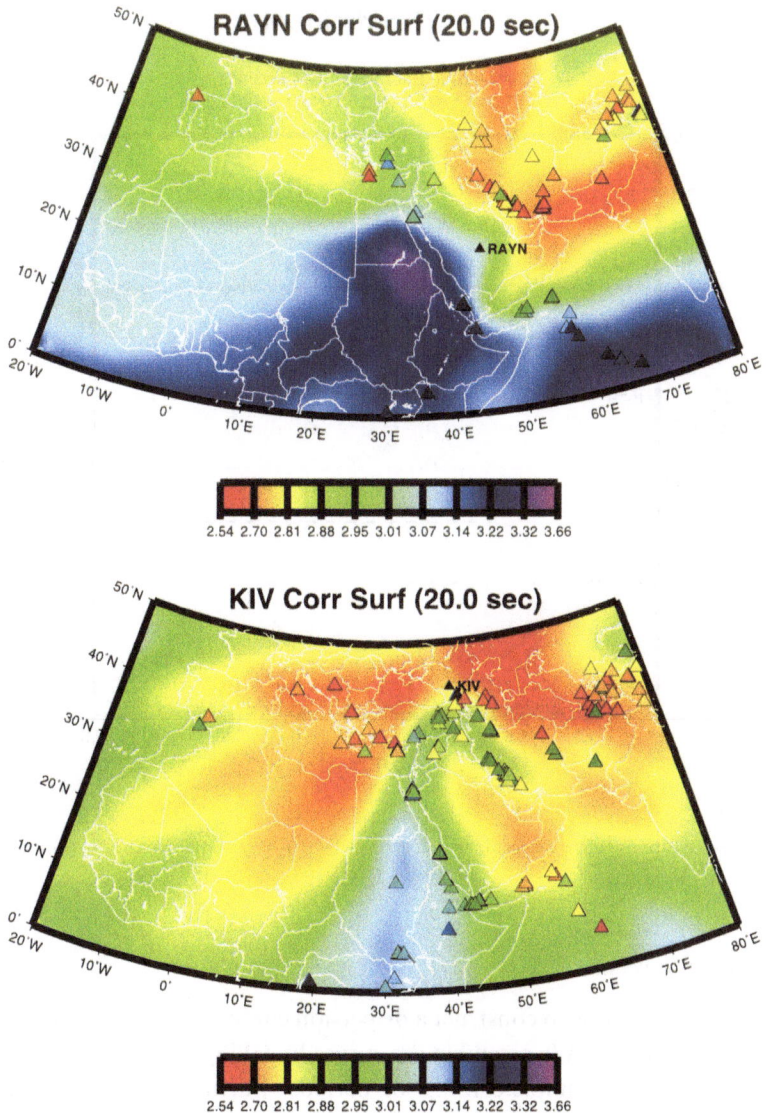

Figure 12
Correction surfaces for 20 second Rayleigh waves at stations RAYN in Saudi Arabia and KIV in Russia.
Triangles indicate group velocities measured at that station.

are averaged over several stations, would have even smaller variability in the earthquake population. In this case, these events had good signal-to-noise and the phase-matched filters had little effect on the results. The real power of the phase-matched filters will be in lowering the maximum M_S estimate threshold in order to discriminate even smaller nuclear tests from earthquakes.

M$_{sr}$ Discriminant

Station NIL - Passband 10-14 seconds
Statistics -0.30 +/- 0.53 (Nobs = 123)

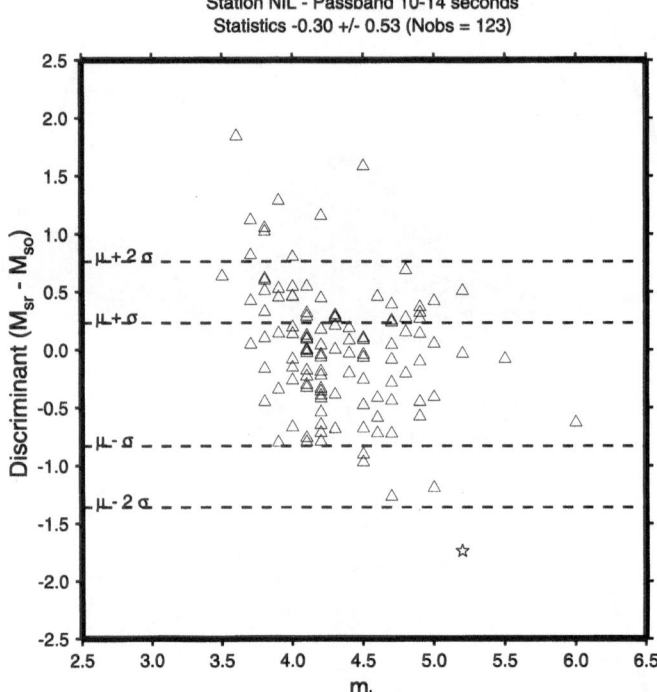

Figure 13
Regional surface-wave/body-wave discriminant at station NIL for earthquakes (triangles) and a single explosion (star).

Layered Velocity Inversion

It is now possible for us to construct a dispersion curve for any particular location within our inversion area. We assemble the curves by taking the inversion results for a single point over the complete range of periods. Furthermore, we can use the constructed curves to forward model or invert for a layered velocity model at that point. Figure 14 shows an example from Kuwait. A simple grid search was performed for three layers (sediment, crust, and upper mantle) over a mantle model (IASPEI), where sediment thickness, crustal velocity (exclusive of sediment), crustal thickness, and upper mantle velocity were allowed to vary. We use ranges of 15–50 km for crustal thickness, 0–5 km for sediment thickness, 6.0–6.8 km/s for average crustal velocity, and 7.7–8.1 km/s for upper mantle velocity. For each layer in the grid search the Poisson's ratio was held constant, fixing the P-to-S ratio. To allow for transverse anisotropy in the upper mantle, we search over two velocities in the upper mantle, which are used to construct the dispersion curves for the Love and Rayleigh

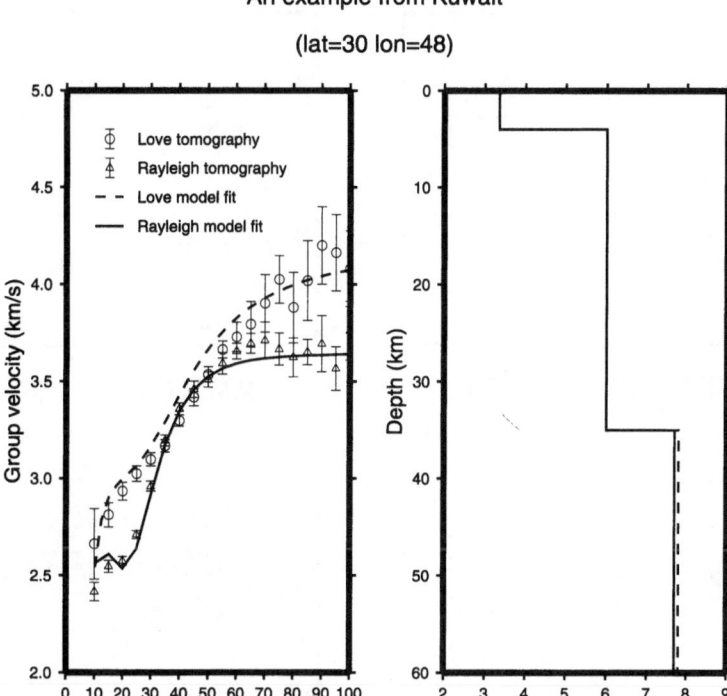

Figure 14

An example of layered velocity modelling. The figure to the left shows a fit of Rayleigh (triangle) and Love (circle) waves to curves for the best fitting model (solid and dashed). The right figure shows the best-fitting velocity model which is slightly anisotropic.

waves separately. In our example, the anisotropy is small (0.1 km/s) and the misfit would be only slightly higher for the isotropic model. We can now model Middle East/North Africa structure by performing this layered velocity modelling at a regular interval (in this case every 2 degrees), assembling the results, and observing how they correspond to known geology and tectonics, as well as other studies. This exercise is provided principally to substantiate the results of the group velocity tomography. A more comprehensive inversion for 3-D velocity structure would allow more layers and a wider range of variables and would include other data which could reduce non-uniqueness (i.e., sedimentary models, P_n velocity, phase velocity, receiver functions, etc.).

Figure 15 shows the results for crustal thickness. Owing to the large tradeoffs between crustal thickness and upper mantle velocity, surface waves are perhaps not the best way to estimate this parameter. Despite this fact, like crustal thickness estimates from other studies (NATAF and RICARD, 1996), we find significantly thicker crust in the Caucasus and along the Zagros Mts. and Turkish-Iranian Plateau and

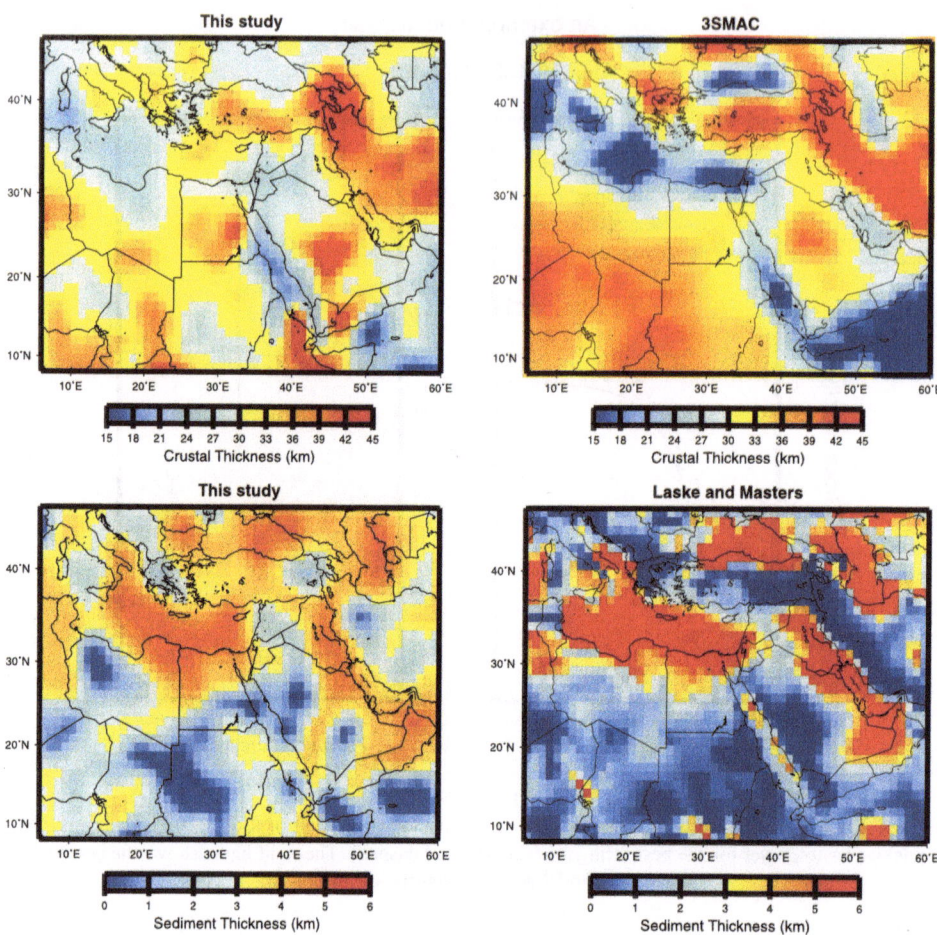

Figure 15
A comparison of crustal thickness (top) and sediment thickness (bottom) estimates determined from modelling of surface-wave data (left) and from other studies denoted in text (right).

thinner crusts under the oceanic regions of the Mediterranean Sea and Indian Ocean, with intermediate thicknesses across Africa. Figure 15 also shows similar plots for sediment thickness as compared to other estimates (LASKE and MASTERS, 1997). Both figures show the greatest sediments loads beneath the Mediterranean Sea and in the Persian Gulf – Mesopotamian foredeep area. In other respects, the two figures indicate some differences. A current limitation of our search is a single sediment layer with single compressional and shear-wave velocities. In practice, there is significant variation in sediment velocity. We are also currently unable to resolve smaller scale sedimentary features such as the Red Sea Rift. We expect that our estimates of sediment thickness will improve as we increase the currently sparse path coverage of the shortest period Love waves. Figure 16 shows a profile from Sudan to Iran along a well-resolved portion of the study area. We see a few kilometers of sediment across the Sudan, with the thickest sediments along the profile occurring under the Persian Gulf. Crustal thickness is relatively uniform (30–35 km) under Africa. Significant thinning of the crust is seen under the Red Sea rift, increases under Arabia and is the thickest under the Zagros Mts. (40–45 km).

In the future, we plan to incorporate other data (sediment thickness, crustal thickness, P_n velocity, etc.) in order to produce an integrated velocity model that satisfies all of these diverse and independently-derived data sets. This approach would reduce the non-uniqueness of the group velocity modelling, such as the tradeoffs between crustal thickness and upper mantle velocity. Finally, all of the velocity models that we develop can be tested by using the results to construct a travel-time model of the region. The corrections can then be applied to travel-time picks to determine whether locations for ground-truth events are improved using the model. As we increase our coverage of the region and improve the resolution of

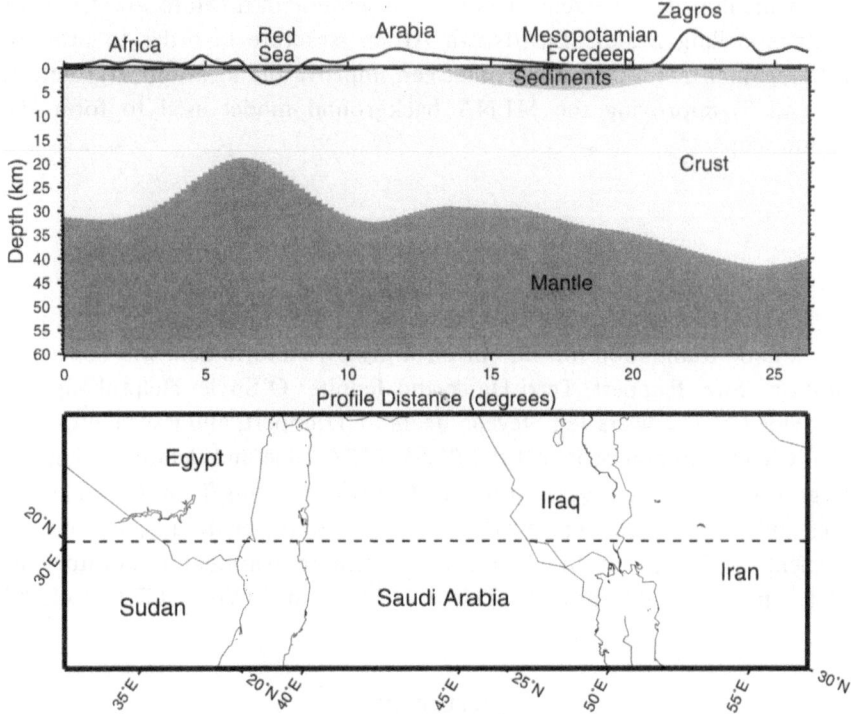

Figure 16

A profile of sediment and crustal thickness from 3-layer modelling of surface-wave data along a well-resolved segment of the study area from the Sudan to Iran (top). The topography along the segment, which is plotted at the top of the profile, has been exaggerated 3 times. A map view of the profile is shown in the bottom figure.

our group velocity inversion, we shall be performing these tests for our crustal model, in combination with several mantle models, against other crustal models of the region.

Conclusions

We find that Rayleigh-and Love-wave group-velocity models, for periods ranging from 10–60 seconds, vary laterally across the region and diverge significantly from laterally homogeneous earth models. In addition, our inversion results better predict surface-wave group velocities than other models that we tested. For this reason it is important to continue utilizing regional data to more accurately determine the lateral variation of shear-wave velocity across the region. We intend to do this by including additional raypaths in the inversion and expanding our analysis to Rayleigh and Love wave phase velocity. Current and future applications of the tomography results are: 1) using phase-matched filters to extract the surface-wave signal from noisy measurements in order to extend the $m_b:M_S$ discriminant down to lower magnitude levels; 2) modelling Middle East/North Africa structure in order to produce an accurate regional travel-time model that can improve the location of ground-truth events; and 3) improving the MENA background model used to form station calibration maps.

Acknowledgments

We thank Dan McNamara who worked on the forerunner to this project, building a good foundation for our current work. Waveform data was collected and organized by Stan Ruppert, Terri Hauk and Jennifer O'Boyle. Helpful suggestions were provided by reviewers Jeff Stevens, Jeannot Trampert, and issue editor Anatoli Levshin. Chuck Ammon provided the PGSWMFA code and details on some of its workings. Figures were generated using the Generic Mapping Tools (GMT) software (WESSEL and SMITH, 1998). This work was performed under the auspices of the U.S. Department of Energy at the Lawrence Livermore National Laboratory under contract number W-7405-ENG-48. This is LLNL contribution UCRL-JC-133973.

REFERENCES

ASUDEH, I. (1982), *Seismic Structure of Iran from Surface and Body Wave Data*, Geophys. J. R. astr. Soc. *71*, 715–730.

BOSCHI, E., GIARDINI, B., MORELLI, A., ROMEO, G., and TACCETTI, Q. (1988), *Mednet; The Italian Broadband Seismic Network for the Mediterranean*, Proceedings of a Workshop on Downhole Seismometers in the Deep Ocean, Woods Hole, MA (ed. Purdy, G. M.) pp. 116–124.

BRUNE, J., ESPINOSA, A., and OLIVER, J. (1963), *Relative Excitation of Surface Waves by Earthquakes and Underground Explosions in the California-Nevada Region*, J. Geophys. Res. *68*, 3501–3513.

DORBATH, L., and MONTAGNER, J. P. (1983), *Upper Mantle Heterogeneities in Africa Deduced from Rayleigh Wave Dispersion*, Phys. Earth Planet. Int. *32*, 218–225.

DZIEWONSKI, A. M., and ANDERSON, D. L. (1981), *Preliminary Reference Earth Model*, Phys. Earth Planet. Inter. *25*, 297–356.

DZIEWONSKI, A. M., BLOCH, J., and LANDISMAN, M. (1969), *A New Technique for the Analysis of Transient Seismic Signals*, Bull. Seismol. Soc. Am. *59*, 427–444.

EZEN, U. (1991), *Surface Wave Dispersion and Upper Crustal Structure Along N-S Direction in Western Turkey from Burdur Earthquake of 12 May 1971*, Bull. IISEE *25*, 39–59.

GUMPER, F., and POMEROY, P. W. (1970), *Seismic Wave Velocities and Earth Structure on the African Continent*, Bull. Seismol. Soc. Am. *60*, 651–668.

HADIOUCHE, O., and JOBERT, N. (1988), *Geographical Distribution of Surface Wave Velocities and Three-dimensional Upper Mantle Structure in Africa*, Geophys. J. R. Astron. Soc. *95*, 87–109.

HADIOUCHE, O., JOBERT, N., and ROMANOWICZ, B. (1986), *First Two-Station Results for Long Period Surface Waves Velocity from the Geoscope Stations in Africa*, Geophys. Res. Lett. *13*, 547–550.

HADIOUCHE, O., JOBERT, N., and MONTAGNER, J. P. (1989), *Anisotropy in the African Continent Inferred from Surface Waves*, Phys. Earth Planet. Inter. *58*, 61–81.

HADIOUCHE, O., and ZÜRN, W. (1992), *On the Structure of the Crust and Upper Mantle beneath the Afro-Arabian Region from Surface Wave Dispersion*, Tectonophysics *209*, 179–196.

HAZLER, S. E. (1998), *One-dimensional Shear Velocity Structure of Northern Africa from Rayleigh Wave Group Velocity Dispersion*, Master Thesis, University of Colorado, Boulder, Colorado, 120 pp.

HAZLER, S. E., SHEEHAN, A. F., MCNAMARA, D. E., and WALTER, W. R. (2001), *One-dimensional Velocity Structure of Northern Africa as Determined by Rayleigh Wave Group Velocity Dispersion*, Pure appl. geophy., this volume.

HERRMANN, R. B. (1973), *Some Aspects of Bandpass Filtering of Surface Waves*, Bull. Seismol. Soc. Am. *63*, 663–671.

HERRIN, E., and GOFORTH, T. (1977), *Phase-matched Filters: Applications to the Study of Rayleigh Waves*, Bull. Seismol. Soc. Am., *67*, 1259–1275.

KNOPOFF, L., and FOUDA, A. A. (1975), *Upper-mantle Structure Under the Arabian Peninsula*, Tectonophysics *26*, 121–134.

KNOPOFF, L., and Schlue, J. W. (1972), *Rayleigh Wave Phase Velocities for the Path Addis Ababa – Nairobi*, Tectonophysics *15*, 157–163.

KNOX, R. P., NYBLADE, A. A., and LANGSTON, C. A. (1998), *Upper Mantle S Velocities beneath Afar and Western Saudi Arabia from Rayleigh Wave Dispersion*, Geophys. Res. Lett. *25*, 4233–4236.

LASKE, G., and MASTERS, G. (1997), *A Global Digital Map of Sediment Thickness*, EOS Trans. AGU *78*, F483.

LAY, T., and WALLACE, T. C., *Modern Global Seismology* (Academic Press, San Diego, 1995).

LEVEQUE, J.-J., RIVERA, L., and WITTLINGER, G. (1993), *On the Use of the Checker-board Test to Assess the Resolution of Tomographic Inversions*, Geophys. J. Int. *115*, 313–318.

LEVSHIN, A., RATNIKOVA, L., and Berger, J. (1992), *Peculiarities of Surface Wave Propagation across Central Asia*, Bull. Seismol. Soc. Am. *82*, 2464–2493.

LIEBERMANN, R. C., KING, C.-Y., BRUNE, J. N., and POMEROY, P. W. (1966), *Excitation of Surface Waves by the Underground Nuclear Explosion Longshot*, J. Geophys. Res. *71*, 4333–4339.

LIEBERMANN, R. C., and POMEROY, P. W. (1967), *Excitation of Surface Waves by Events in Southern Algeria*, Science *156*, 1098–1100.

MARSHALL, P. D., and BASHAM, P. W. (1972), *Discrimination between Earthquakes and Underground Explosions Employing an Improved M_s Scale*, Geophys. J. R. Astron. Soc. *28*, 431–458.

MCNAMARA, D., WALTER, W., and HAZLER, S. (1997), *Surface Wave Group Velocity Dispersion across Northern Africa, Southern Europe, and the Middle East*, 19th Seismic Research Symposium on Monitoring a CTBT, 83–92.

MINDEVALLI, O. Y., and MITCHELL, B. J. (1989), *Crustal Structure and Possible Anisotropy in Turkey from Seismic Surface Wave Dispersion*, Geophys. J. Int. *98*, 93–106.

MOKHTAR, T. A., and AL-SAEED, M. M. (1994), *Shear Wave Velocity Structures of the Arabian Peninsula*, Tectonophysics *230*, 105–125.

NATAF, H.-C., and RICARD, Y. (1996), *3SMAC: An a priori Tomographic Model of the Upper Mantle Based on Geophysical Modeling*, Phys. Earth Planet. Inter. *95*, 101–122.

PAIGE, C. C., and SAUNDERS, M. A. (1982), *LSQR: An Algorithm for Sparse Linear Equations and Sparse Least Squares*, Assn. Comp. Mech. Trans. on Mathematical Software *8*, 43–71.

RITZWOLLER, M. H., and LEVSHIN, A. L. (1998), *Eurasian Surface Wave Tomography: Group Velocities*, J. Geophys. Res. *103*, 4839–4878.

RODGERS, A. J., WALTER, W. R., MELLORS, R. J., AL-AMRI, A. M. S., and ZHANG, Y.-S. (1999), *Lithospheric Structure of the Arabian Shield and Platform from Complete Regional Waveform Modeling and Surface Wave Group Velocities*, Geophys. J. Int. *138*, 871–878.

ROMANOWICZ, B., CARA, M., FELS, J. F., and ROULAND, D. (1984), *Geoscope: A French Initiative in Long-period, Three-component, Global Seismic Networks*, EOS, Trans. Am. Geophys. U. *65*, 753–754.

RUPPERT, S., HAUK, T., LEACH, R., and O'BOYLE, J. (1998), *LLNL Middle East and North Africa Research Database*, Proceedings of the 20th Annual Seismic Research Symposium on Monitoring a Comprehensive Test-Ban Treaty, Santa Fe, NM, Department of Defense, Nuclear Treaty Programs Report, 727–735.

SCHULTZ, C., MYERS, S., HIPP, J., and YOUNG, C. (1998), *Nonstationary Bayesian Kriging: Application of Spatial Corrections to Improve Seismic Detection, Location, and Identification*, Bull. Seismol. Soc. Am. *88*, 1275–1288.

STEVENS, J. L., and McLAUGHLIN, K. L. (1997), *Improved Methods for Regionalized Surface Wave Analysis*, Final Report PL-TR-97-2135, Phillips Laboratory, Directorate of Geophysics, AFMC, Hanscom AFB, 26 pp.

SWEENEY, J. J., and WALTER, W. (1998), *Preliminary Definition of Geophysical Regions for the Middle East and North Africa*, UCRL-ID-132899, Lawrence Livermore National Laboratory, 41 pp.

TAHA, Y. S. (1991), *Crust and Upper Mantle Structure in Egypt from Phase Velocity of Surface Waves*, Individual Studies Participants to the International Institute of Seismology and Earthquake Engineering, Tsukaba, *26*, 1–12.

VERNON, F., MELLORS, R., BERGER, J., EDELMAN, A., AL-AMRI, A., ZOLLWEG, J., and WOLFE, C. (1996), *Observations from Regional and Teleseismic Earthquakes Recorded by a Deployment of Broadband Seismometers in the Saudi Arabian Shield*, EOS, Trans. Amer. Geophys. U. *77*, 478.

WALTER, W. R., and RITZWOLLER, M. (1998), *Summary Report of the Workshop on the U.S. Use of Surface Waves for Monitoring the CTBT*, UCRL-ID-131835, Lawrence Livermore National Laboratory, 16 pp.

WESSEL, P., and SMITH, W. H. F. (1998), *New, Improved Version of Generic Mapping Tools Released*, EOS Trans. Amer. Geophys. U. *79*, 579.

(Received June 4, 1999, revised December 17, 1999 accepted January 3, 2000)

 To access this journal online:
http://www.birkhauser.ch

Pure appl. geophys. 158 (2001) 1475–1493
0033–4553/01/081475–19 $ 1.50 + 0.20/0

▌Pure and Applied Geophysics

One-dimensional Shear Velocity Structure of Northern Africa from Rayleigh Wave Group Velocity Dispersion

S. E. HAZLER,[1] A. F. SHEEHAN,[1] D. E. MCNAMARA,[2] and W. R. WALTER[3]

Abstract — Rayleigh wave group velocity dispersion measurements from 10 s to 160 s periods have been made for paths traversing Northern Africa. Data were accumulated from the IRIS DMC, GEOSCOPE, and MEDNET seismic networks covering the years 1991–1997. The group velocity measurements are made including the effects of debiasing for instantaneous period and a single-iteration, mode-isolation (phase match) filter. The curves are grouped by tectonic province and compared to tomographic model-based curves in an effort to test and validate the tomographic models. Within each tectonic category (rift, orogenic zone, or craton) group velocity curves from various provinces are similar. Between tectonic categories, however, there are marked differences. The rift related paths exhibit the lowest group velocities observed, and cratonic paths the fastest. One-dimensional shear velocity inversions are performed, and while highly nonunique, the ranges of models show significant differences in upper mantle velocities between the tectonic provinces.

This work is part of a larger project to determine group velocity maps for North Africa and the Middle East. The work presented here provides important tools for the validation of tomographic group velocity models. This is accomplished by comparing group velocity curves calculated from the tomographic models with carefully selected high-quality group velocity measurements. The final group velocity models will be used in M_s measurements, which will contribute to the $m_b{:}M_s$ discriminant important to the Comprehensive Nuclear-Test-Ban Treaty (CTBT). The improved shear wave velocity models provided by this study also contribute to the detection, location, and identification of seismic sources.

Key words: Seismology, surface waves, Africa, mantle, crust, lithosphere.

Introduction

North Africa has been only sparsely sampled both geologically and seismically; yet knowledge of regions such as this is critical for CTBT monitoring purposes. In an effort to fill in the gaps present in our knowledge of the seismic structure of Northern Africa, a comprehensive study of the shear wave velocity structure beneath this

[1] CIRES and Department of Geological Sciences, University of Colorado at Boulder, Campus Box 399, Boulder, CO 80309, USA.
[2] U.S. Geological Survey, National Earthquake Information Center, Federal Center, MS967, Denver, Co 80225, USA.
[3] Geophysics and Global Security, Lawrence Livermore National Laboratory, Livermore, CA 94550, USA.

region has been undertaken (HAZLER, 1998; PASYANOS *et al.*, 2001). North Africa includes a number of interesting tectonic provinces (Fig. 1), and the work presented here seeks to seismically characterize these provinces through a surface wave regionalization approach (Fig. 2). Orogenic domains are grouped together into larger regions where the tectonics of mountain building dominate. Similarly shield bodies have been grouped into larger cratonic bodies. Regions of recent tectonic activity have also been grouped together. Should these regions show a distinct seismic signal, an important step toward the seismic characterization of North Africa will have been made.

Most previous seismic studies concerning Africa were small in scale (LAST *et al.*, 1997; NYBLADE *et al.*, 1996; ACHAUER, 1992; CLOUSER and LANGSTON, 1990; BERKHEMER *et al.*, 1975; LONG *et al.*, 1972; KNOPOFF and SCHULE, 1972), and while providing important constraints on particular regions, do not address the large-scale structure of North Africa specifically. These regional studies contrast with global velocity models in which Africa plays only a small part (e.g. EKSTRÖM *et al.*, 1997; MASTERS *et al.*, 1996; ZHANG and LAY, 1996; SU *et al.*, 1994). In these global studies,

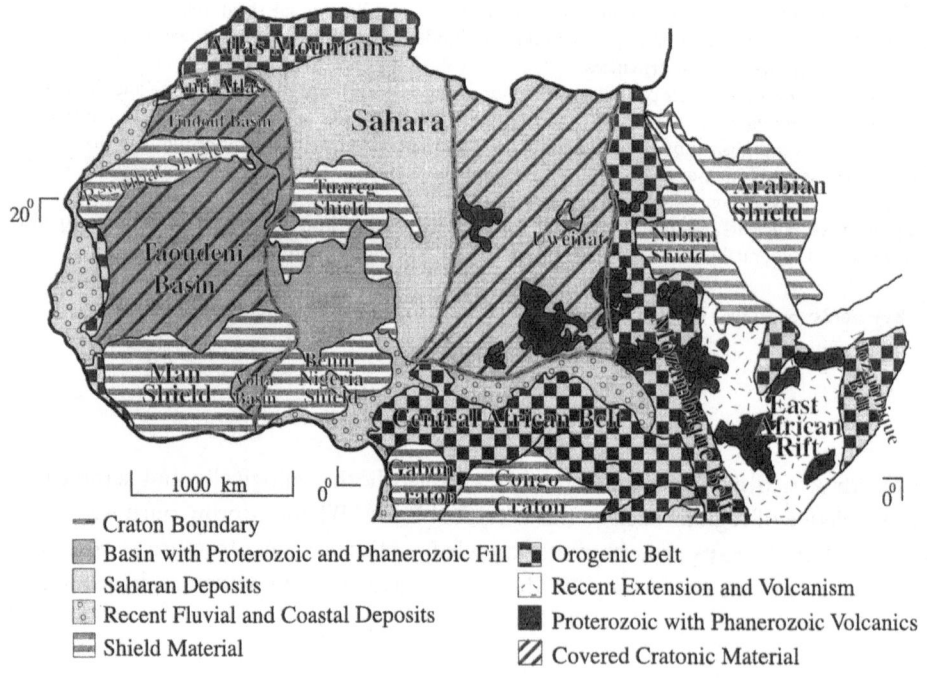

Figure 1

Generalized geologic divisions of North Africa. Major divisions include shield material, covered cratonic material, basins filled with Proterozoic and Phanerozoic sediment, Saharan deposits, recent fluvial and coastal deposits, orogenic belts, regions of recent extension, and Proterozoic regions with Phanerozoic volcanics. After GOODWIN (1996).

resolution constraints due to path coverage and wavelengths used limit the ability to distinguish the properties and interrelations between the various tectonic provinces of Africa. In an attempt to link seismic data with geologic information, CHRISTENSEN and MOONEY (1995) inferred global crustal composition and structure on the basis of seismic refraction profiles and high pressure laboratory studies of common crustal rocks. Their inferences are applied to all continents although data points for Africa were only taken in Morocco, in Southern Africa, and along the East African Rift. The recent crustal model CRUST 5.1 (MOONEY et al., 1998) also lacks sufficient coverage in Africa. Some work has also been done on a regional level in Africa (HADIOUCHE and ZÜRN, 1992; TANAKA and HAMAGUCHI, 1992; HADIOUCHE and JOBERT, 1988b; HADIOUCHE et al., 1986; GUMPER and POMEROY, 1970); but these regional studies often fail to link results to regional geology. Further, considerably more data are available now to augment this work.

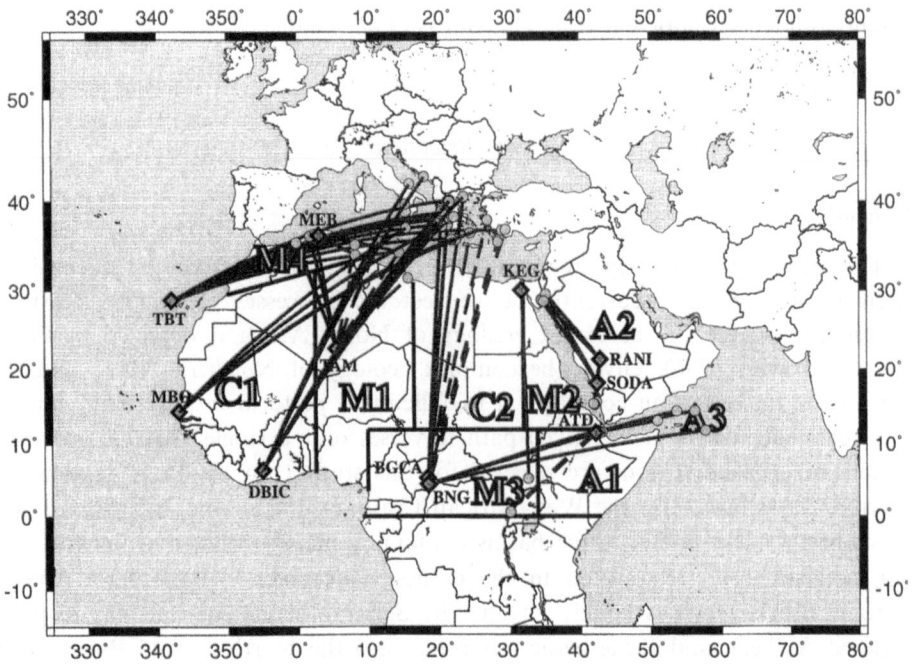

Figure 2

Tectonic provinces used in regionalization. The numbers correspond to the following regions: C1) the West African Cratonic Block, C2) the East African Cratonic Block, M1) the North-Central Pan-African Orogenic Block, M2) the Eastern Pan-African Block, M3) the Central Pan-African Block, M4) the Atlas Mountains, A1) the East African Rift Block, A2) the Arabian Shield, and A3) the Gulf of Aden Spreading Center. The tectonic regions were chosen on the basis of basement geology and current active tectonic processes (see text). Also labeled are the great circle paths, the broadband seismic stations (diamonds), and the earthquakes (circles) used in this study.

Continental scale studies are an important complement to global studies and allow us to concentrate on specific regions and processes in more detail. To do this, we make use of Rayleigh wave group velocity dispersion, and concentrate on the crust and upper mantle beneath North Africa. By examining group velocities which describe a single path, certain ambiguities found in global group velocity models such as those related to poor crossing paths can be eliminated. In this study, group velocities of regional and teleseismic Rayleigh waves in the period range of 10 to 160 s are measured from seismograms which have been recorded on broadband seismic stations. The group velocity curves are grouped according to the tectonic region they traverse. The clustered group velocity curves are averaged; and the averaged curves are inverted for one-dimensional shear velocity structure. Increased understanding of the regional tectonic setting and shear velocity structure will better define the seismic characteristics need to properly monitor the Comprehensive Test-Ban Treaty (CTBT). In particular, increased knowledge of seismic shear wave structure will lead to improved detection, location, identification capabilities in the region. Our high-quality path measurements will be used to validate 3-D models currently under development (e.g., PASYANOS *et al.*, 2001).

Data Selection

Regionalization

The North African study area is divided into tectonic provinces on the basis of basement geology and currently active tectonic processes (HAZLER, 1998). A generalized map of the basement geology of North Africa is depicted in Fig. 1 (after GOODWIN, 1996). Given the complex geology of Northern Africa and the sparse earthquake/station coverage, it is difficult to perform a 'pure-path' surface wave regionalization where a given path traverses only a single geologic province. Instead, nine broad regions are selected for this study (Fig. 2). These regions are described as follows: C1) the West African Cratonic Block which consists of two Precambrian shield bodies, three basins containing mainly lithified sediments, and four orogenic belts reactivated in the earliest Paleozoic, C2) the East African Cratonic Block which is largely covered by Saharan sediments but also displays scattered Archean and Proterozoic outcrops, M1) the North-Central Pan-African Orogenic Mobile Belt which also contains a large portion of recent sediments as well as the massively deformed Benin-Nigeria Shield and the Tuareg Shield, M2) the Eastern Pan-African Mobile Belt which splits its area between the Mozambique Orogenic Belt and the Nubian Shield (an amalgam of island arc material), M3) the Central Pan-African Mobile Belt and Pan-African Orogenic Belt, M4) the Atlas Mountains, A1) the East African Rift Block; an area affected by recent rifting and magmatism, A2) the Arabian Shield Block which consists of former island arc material once contiguous with the Nubian Shield, and A3) the Gulf of Aden Spreading Center. Figure 2 depicts these nine regions as well as the Rayleigh wave paths used to study them.

Paths Chosen

Once these boundaries were established, a search was made for earthquake/ station pairs whose connecting paths could provide insight into the chosen tectonic regions. Broadband three-component stations from the IRIS/GSN, MEDNET, and GEOSCOPE networks were included in the search. Seismograms from all earthquakes from the years 1991–1997 of magnitude 4.0 and greater located between 15°W to 65°E longitude and 5°S to 50°N latitude with paths crossing North Africa or the Middle East were examined. The bulk of the earthquakes used in this study surround the Mediterranean Sea. In addition to the Mediterranean events, several sources were located in the Gulf of Aden, several more were found in the Sinai Peninsula region, several more were centered in the southern Red Sea, and two earthquakes were located in the interior of the African continent (Fig. 2). Chosen paths are not intended to represent pure path propagation within a given tectonic block. Event/station locations in the region often preclude such pure path geometries. Thus, paths chosen in this study provide a broad comparison between chosen geologic blocks.

Analysis

Measurement Technique

Before undergoing the group velocity measurement procedure, the data were demeaned and detrended, and the instrument response was removed from all seismograms. The remainder of the measurement process follows the procedure outlined in AMMON (1998). The Rayleigh waves present in the chosen broadband records were examined using a narrow band filter centered on a suite of different periods (HERRMANN, 1973) referred to as filter periods. Measurements were made for filter periods ranging from 7 to 200 s. Typical measurements in this study range from 15 to 120 s. The peak amplitude of each energy packet is determined using an envelope function. The arrival time of the peak amplitude is recorded and converted to a velocity value assuming a known location and origin time of the source. These velocity values were then compiled into a group velocity dispersion curve for each station/event pair. Group velocity errors for each period were estimated by measuring the width of the group velocity peak.

In an effort to resolve the most accurate group velocity dispersion curve, the instantaneous period is used for the final group velocity dispersion curve rather than the filter period. The instantaneous period refers to that period which is most closely associated with the peak amplitude of the energy packet (CLAERBOUT, 1992). The filter period is often very close to the instantaneous period in cases where a good signal-to-noise ratio is present for the energy packet under examination. However, in cases where a spectral hole is present or where amplitudes are diminishing, the variation between instantaneous period and filter period can be significant.

Curve Averaging

Dispersion curves obtained from the narrow band-filtering method were averaged in an effort to minimize errors due to noise endemic to any particular path. Paths that contributed to the averaged curve satisfied the following criteria: a) the paths must be used to examine the same geologic block; and b) paths must have backazimuths which diverge by three degrees or less. The dispersion curves satisfying these criteria were collected and averaged. A total of 2529 Rayleigh wave group velocity dispersion curves were measured for the entire project, which includes the larger scale tomography work of PASYANOS *et al.* (2001). Of these 2529 curves, a subset (69 before averaging and 16 after averaging) was selected for our regionalized one-dimensional inversions. Tectonic regions are sampled by anywhere from a single dispersion curve in the case of the East African Rift to fifteen dispersion curves in the case of the Arabian Shield.

Inversion Technique

A maximum likelihood technique was used to invert the group velocity dispersion curves for earth shear velocity structure (e.g., WIGGINS, 1972; TAYLOR, 1980). Input for this process consists of an initial velocity model and the measured group velocities as well as their corresponding periods and the errors associated with each data point. A linearized system of equations is obtained by performing a Taylor series expansion to first order about the initial model, and the system of equations is solved using damped least squares. In the inversion, only shear velocity is adjusted, i.e., compressional velocity, density, and layer thicknesses remain fixed. Since errors in group velocity measurements vary with period and partial derivatives are a function of layer thickness, the solution is weighted in both model and data space. The inversion is terminated when the misfit to the observed data drops below a preset minimum; or after 40 iterations. If convergence is not achieved in 40 iterations, the final model is discarded. The partial derivatives that make up the Rayleigh wave sensitivity kernels are recalculated for each iteration. Final output consists of a final velocity model, the model resolution matrix, covariance information, and error propagation information.

The original starting model for the inversion was developed by forward modeling (MCNAMARA and WALTER, 1995; HAZLER, 1998). This model consists of four crustal layers over the mantle part of the radial earth model PREM (DZIEWONSKI and ANDERSON, 1981). In an attempt to obtain the best possible fit to the input data, velocity and thickness values in the original starting model were varied randomly by ±1% in velocity and ±2.5 km in thickness. This variation process was used to produce a suite of one hundred different starting models. Each of these different starting models was employed in the inversion process to produce a large suite of inverted models. The various models indicate the large range of possible solutions that can equally or nearly equally fit the data, illustrating the nonuniqueness of group velocity inversion. Despite the nonuniqueness, the models for each tectonic grouping do display common features, which will be discussed further in the following sections.

Error Sources

Scattering and inaccurate source information are two of the factors that contribute error or uncertainty to the group velocity curves. Scattering occurs when an incoming seismic ray encounters a geologic feature which changes the ray's path so that it deviates from the path predicted by the ray-tracing method employed in a given study (e.g., WANG and DAHLEN, 1995). The errors associated with this scattering effect are due to the observer's choice of an assumed path which deviates from the true path of the seismic wave. Errors can also appear due to the interaction of several different raypaths.

Two values that are paramount in measuring group velocity are the arrival time of the wave packet and the distance the wave packet traveled from source to receiver. These values are extremely sensitive to the reported event locations and the assigned origin time. A study of the deviation between teleseismically determined epicentral locations and locally determined epicentral locations in the Middle East and Northern Africa was developed in SWEENEY (1996). A maximum of 15 km mislocation in ISC (International Seismic Center) source locations and less than 2.5 s in origin time was found. For path lengths relevant to this study, we find that the potential errors associated with the longest path are ±0.02 km/s; while the errors associated with the shortest path are ±0.12 km/s. As a result of this work, the value of ±0.2 km/s has been assigned as a conservative error estimate for each group velocity measurement. Comparison of group velocity curves with nearly identical paths (same station, earthquakes within 10 km) has been performed, and the resulting group velocity curves are nearly identical (HAZLER, 1998).

Discussion

Since seismicity and seismic stations are sparse in North Africa, we choose to look at first order differences between regions of North Africa by grouping dispersions curves which sample regions with similar geology, and solving for one-dimensional shear velocity structure in these regions. We have identified nine different regions on the basis of their geology and surface wave path coverage. We examine the tectonic regions by comparing them in three similar groups: active, orogenic, and cratonic. The active tectonic grouping includes the East Africa Rift, Arabian Shield/Red Sea Rift, and Gulf of Aden regions (regions A1, A2, and A3 of Fig. 2, respectively). The orogenic grouping includes the North-Central Pan-African Belt, Mozambique Belt/Eastern Pan-African Block, Central Pan-African Belt, and the Atlas Mountains (regions M1, M2, M3, and M4 of Fig. 2). The cratonic grouping includes the West African Craton and East African Craton (regions C1 and C2 of Fig. 2). The chosen paths do not represent pure path propagation within a given tectonic block but rather provide a rough comparison between chosen geologic blocks. The model validation aspects of this study are not affected by the degree to which a given path is contained within a particular tectonic region, because model curves can be generated for any given path.

Tectonically Active Paths

The top-left panel in Figure 3 depicts group velocity dispersion curves belonging to paths from tectonically active rift or rift-like settings including the East African Rift, Gulf of Aden, and the Arabian Shield region. A cursory inspection of these curves reveals that the greatest degree of heterogeneity can be found in the period range from 10 to 55 s. This heterogeneity in large part reflects the differing crustal structure of the three paths.

The East African Rift path (A1) samples faulted material belonging to the Mozambique Orogenic Belt. At the uppermost crustal level, this material is interfingered with largely alkaline volcanics implaced during the rifting process (SANDVOL *et al.*, 1998; CAHEN and SNELLING, 1984; TESHA *et al.*, 1997). These extrusive volcanics are believed to be underlain by magmatic intrusions at the middle and lower crustal level (PETTERS, 1991). Like most of East Africa, this region is covered by unlithified sediments. At short periods (10–30 s) the East African Rift dispersion curve velocities are comparable to those from non-rift related regions (Fig. 3d). This is likely due to the combination of the mix of crustal components including orogenic material, rift related volcanics, and sediment cover.

The Arabian Shield/Red Sea Rift path (A2) samples a region composed of accreted island arcs and is analogous to the Nubian Shield on the western side of the Red Sea. The Nubian Shield comprises the bulk of the Eastern Pan-African Mobile Belt (Figs. 1 and 2). The metavolcanics and younger igneous bodies of the Arabian Shield should provide a fast crustal path for the Rayleigh waves examined, as was observed by RODGERS *et al.* (1999) for paths just east of those examined here. Measurements along the Arabian Shield do indeed reveal this fast crustal structure. The Arabian Shield path records nearly the highest group velocities examined in this study, slower only than the Gulf of Aden curve, between the periods of 10 and 35 s. The Gulf of Aden path is essentially an oceanic path connecting earthquakes in the Gulf of Aden to the ATD broadband station in Djibouti. This path passes along the

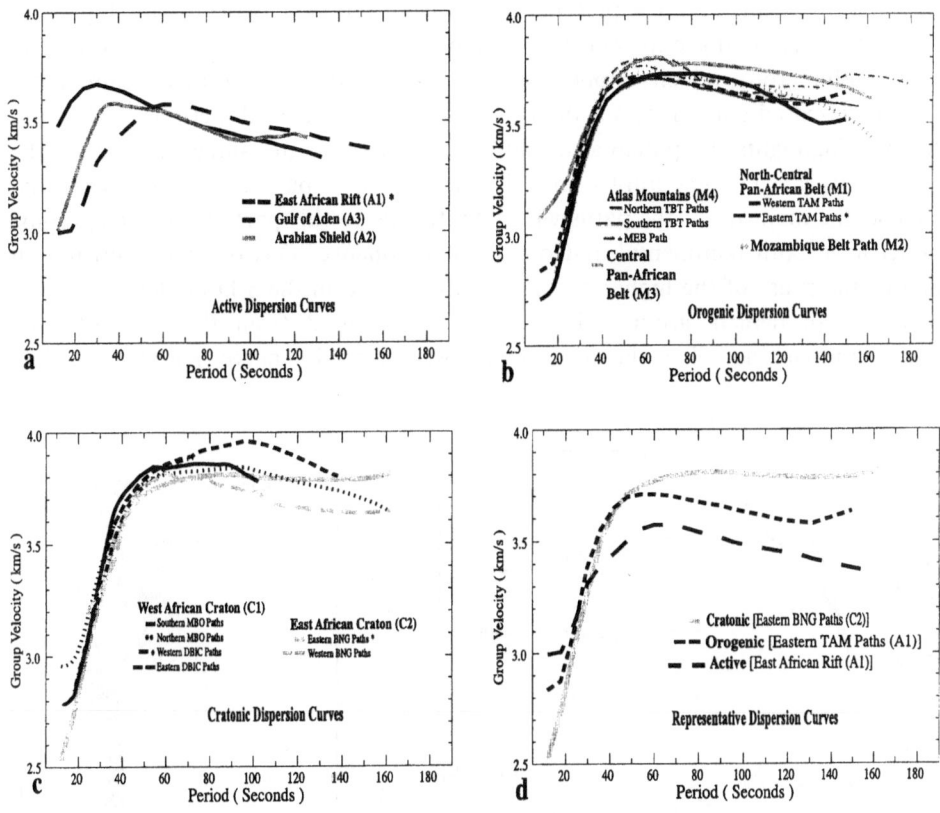

Figure 3
Rayleigh wave group velocity dispersion curves for North Africa grouped by tectonic type, including active tectonic regions (a), orogenic regions (b), and cratonic regions (c). Curves used in inversions (Figs. 4–8) denoted by *. (d) Comparison between respresentative (*) curves from (a–c).

axis of the rift which is currently expanding the Gulf of Aden. The group velocity curve corresponding to this path contains high velocities at short period (10–40 s) reflecting the volcanic material extruded from the active rift.

At periods greater than 40 s, the three group velocity curves corresponding to tectonically active are markedly slower (3–9%) than those for the orogenic or cratonic regions considered in this study (Figs. 3a, 3d), suggesting lower velocities in the upper mantle. At a period of approximately 60 s, the three curves in the tectonically active grouping converge at around 3.5 km/s. These low velocities appear to indicate the presence of high temperatures and partial melt in the upper mantle which might correspond to the presence of a plume head (EBINGER and SLEEP, 1998).

Due to the large number of paths examined in this study, inversion results for only one path from each major grouping (active tectonic, orogenic, cratonic) are

presented in detail. Figure 4 presents results from the East African Rift. Figure 4a
depicts the measured group velocity curve compared to one calculated from the 3-D
tomographic group velocity model of PAYANOS *et al.* (this issue) for the same path.
The model-based curve is faster than the measured curve at all periods >20 s, and
this plot highlights the potential deviations that a single measured path can make
from a larger 3-D tomographic model. The measured path is strictly continental
whereas the majority of the paths which contribute to this part of the tomographic
model have both continental and oceanic components. This oceanic contribution
may be the cause of the higher group velocities found in the 3-D model.

Figure 4b depicts group velocity curves calculated from the suite of final
inversion models as well as the original measured dispersion curve. The close fit of all

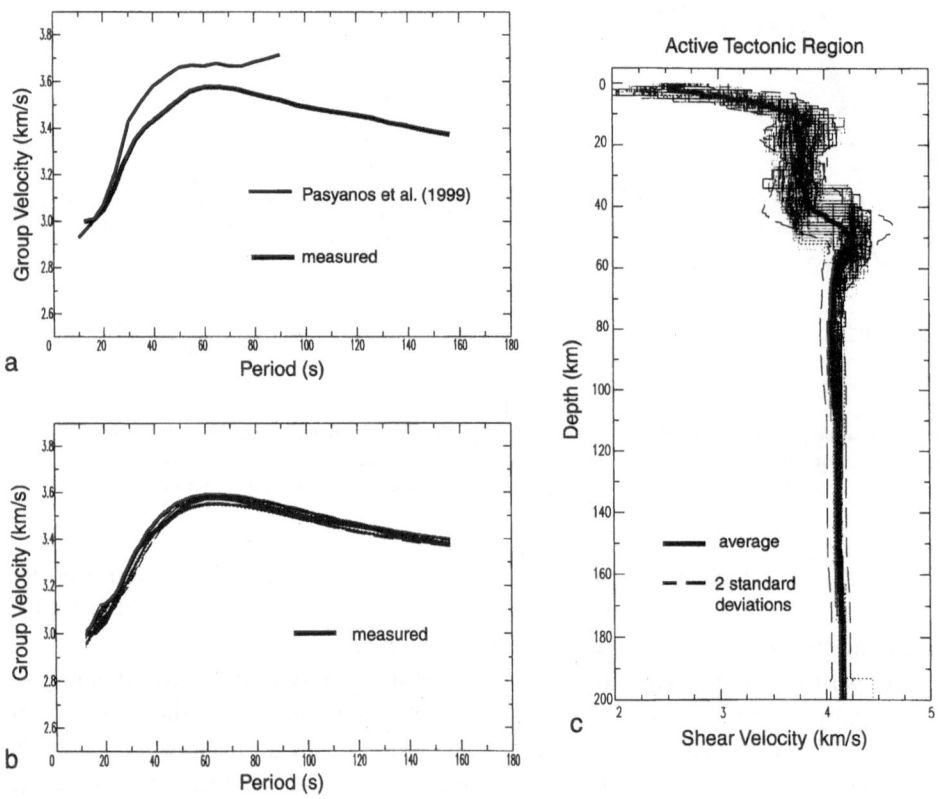

Figure 4

Inversion and forward modeling results for the East African Rift path (A1): (a) measured group velocity
curve from this study compared to one calculated from a regional 3-D tomographic velocity model
(PASYANOS *et al.*, this issue) for the same path; (b) observed group velocity curve versus group velocity
curves calculated from inversion results (shear velocity models) in panel (c); (c) inversion results for the
East African Rift path. Average of the models shown as thick solid grey line, with standard deviation
(dashed).

the calculated group velocity curves to the measured data points out the non-uniqueness of this inversion, i.e., many different models can fit the data equally well. The shear velocity structures which produced the calculated group velocity curves are displayed in Figure 4c. The inversion results are presented with their average as well as two standard deviations from that average. It is important to note that averaging shear velocity versus depth models tends to smooth out the sharp transitions likely present in the earth. Thus examining the range of inversion results is more representative of possible earth structure than merely referring to the average. Not surprisingly the shear velocity structure for this tectonically active path is uniformly lower than those reported in global radial reference earth models (e.g., PREM, DZIEWONSKI and ANDERSON, 1981). The inversion results show maximal variation in the top 60 km of the model with a Moho depth of 43±5 km. These results concur with the crustal thickness values for rift regimes found by CHRISTENSEN and MOONEY (1995). CHRISTENSEN and MOONEY report crustal thicknesses of 36±8 km in rifts. The Moho range found in the current study reflects the higher end of Christensen and Mooney's crustal thickness range perhaps because the path in question (region A1) samples the rift axis obliquely. A path which followed the rift axis more directly would be more likely to reflect a thinner crust. Other techniques, such as teleseismic receiver functions and regional waveform modeling, would be required to better constrain the crustal thickness.

Depending upon the velocity-temperature scaling relation used (e.g., BASS, 1995; SHEEHAN and SOLOMON, 1991; KARATO, 1993), the low shear wave velocities for the active tectonic paths suggest upper mantle temperature anomalies over 240 K and perhaps up to 1700 K. The high end of these temperatures is thought to be unreasonably large, as this is nearly four times the thermal variation predicted between oceans and continents, and would also elevate the temperature of the mantle rocks above their solidus and cause massive melting. The presence of high attenuation can moderate the temperature variation required (KARATO, 1993), and represents our low temperature estimate. The maximum temperature variation associated with a mantle plume is approximately 300 K (MCKENZIE and BICKLE, 1988; SCHILLING, 1991), and our models cannot be used to reject a plume hypothesis for the region (KNOX et al., 1998).

Orogenic Belts

Figure 3b depicts group velocity dispersion curves that have traveled through orogenic belts. The group velocity dispersion curves displayed in this panel show a tight grouping overall with the greatest heterogeneity at short periods, 10 to 30 s (Fig. 3). The fastest curve at these short periods corresponds to the averaged path passing through the Central Pan-African Belt. This region describes the suture zone between North Africa and the Congo Craton of Central Africa. More importantly, this region has little to no sediment cover as compared to the three other orogenic belts examined. This lack of sedimentary cover results in a faster crustal segment which in turn explains the higher velocities displayed at short periods. Longer period measurements for these four regions indicate lower group velocities than those found at similar periods in cratonic regions (Fig. 3).

Figure 5 depicts the inversion results from the eastern TAM grouping in the North-Central Pan-African Block, M1. Figure 5a displays the measured results versus those taken from the 3-D tomographic group velocity model of PASYANOS *et al.* (this issue). This plot shows a much closer fit than that in Figure 4a, indicating that the tomographic velocity model better describes orogenic belts than the actively deforming regions in this study. Figure 5b displays the group velocity

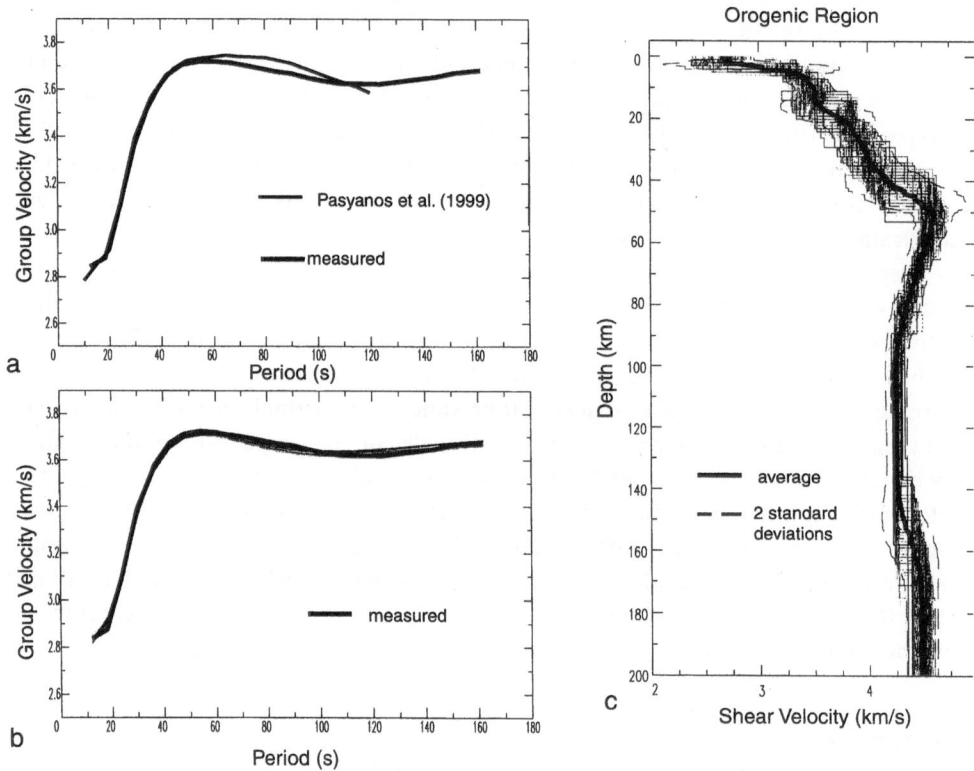

Figure 5

Inversion and forward modeling results for a sample orogenic path (the eastern TAM path of the orogenic block, M1): (a) average measured group velocity curve from this study (thick grey line) compared to one calculated from a regional 3-D tomographic velocity model (PASYANOS *et al.*, this issue) for the same path (black line); (b) average measured group velocity curve (thick grey line) compared to group velocity curves calculated from shear velocity models in panel c (thin grey lines) (c) inversion results for the eastern TAM path and the average (thick grey line) and standard deviation (dashed) of the models.

curves that have been calculated from the velocity models resulting from the suite of one hundred different starting models, in comparison to the measured dispersion curve. The close fit to the observed data highlights the non-uniqueness of the individual velocity models depicted in Figure 5c. These inversion results show maximal variation in the top 50 km of the model with Moho depths of 43 ± 5 km. Crustal thicknesses for orogenic regions as reported by CHRISTENSEN and MOONEY (1995) are found to average 46 ± 10 km. Since the orogenic belts of Northern Africa have been largely eroded and no longer show significant relief, it is not unexpected that our results reflect the thinner end of the Christensen and Mooney range. Below crustal depths, the sample orogenic shear velocity model is slightly slower than the shear velocities predicted by the PREM global model (DZIEWONSKI and ANDERSON, 1981). However, the PREM curve at these depths does fall within the region encompassed by two standard deviations from the averaged model.

Cratonic Regions

Figure 3c presents the group velocity dispersion results for the two large-scale cratonic bodies present in Northern Africa, the West African Craton and the East African Craton (Figs. 1 and 2). The West African Craton contains two major shield bodies and three major basin structures (PETTERS, 1991). The paths that represent the West African Craton are not well bounded by our generalized geologic boundaries, with just half to two-thirds of their path length within shield material. However, these paths were included in this study in the hope that they might reveal significant difference from or similarity to other paths within this study.

The extent of the East African Craton is not well defined as the bulk of the hypothesized cratonic body lies beneath a thick layer of Phanerozoic sediment, the Sahara desert. The existence of a cratonic body beneath the Sahara is proposed on the basis of scattered Precambrian outcrops in the region (CONDIE, 1982). The oldest of these outcrops is the Archean aged Uweinat Inlier. Other outcrops such as Tibesti, Kordofan, and Darfour display Proterozoic basement material mingled with Phanerozoic volcanics (GOODWIN, 1996; CAHEN and SNELLING, 1984; VINCENT, 1970) (see Fig. 1).

Despite these uncertainties, the group velocity curves corresponding to cratonic paths indicate a tight grouping between the periods of 20 s and 80 s (Fig. 3c). The divergence between the group velocity curves from 10 s to 20 s can be ascribed to differences in sedimentary thicknesses sampled by the different paths. For example, the northern MBO (a broadband station in M'bour, Senegal) paths pass through exposed basement and lithified, Paleozoic, basin material; however, the southern MBO paths cross twice the distance through thick unlithified sediments that the northern MBO paths cross. This difference is demonstrated in the short-period segments of the two averaged group velocity curves. Between 10 s and 20 s we can easily see the faster group velocities of the northern MBO path, less sampling of unlithified sediments, and the slower group velocities of the southern MBO path, greater sampling of unlithified sediments.

Overall, the group velocity curves belonging to the East African Craton record slower group velocities at the same periods than the group velocity curves belonging to the West African Cratonic Block. This west to east velocity gradient (faster in west, slower in east) is also present in the African portion of global velocity models (e.g., EKSTRÖM *et al.*, 1997; MASTERS *et al.*, 1996). This velocity gradient might be the result of the proposed plume head beneath Eastern Africa (e.g., KNOX *et al.*, 1998; LITHGOW-BERTELLONI and SILVER, 1998; NYBLADE and ROBINSON, 1994; HADIOUCHE, 1990; HADIOUCHE and JOBERT, 1988a).

As with the tectonically active regions and the orogenic regions, a sample path has been chosen to represent the cratonic group. Figure 6 depicts the inversion results from the eastern BNG grouping in the East African Craton. Figure 6a demonstrates the very close fit of the measured data to the dispersion curve taken from the 3-D tomographic group velocity model (Pasyanos *et al.*, this issue), indicating that this region is well resolved by the 3-D velocity model. Figure 6b shows the measured group velocity dispersion curve and the group velocity curves calculated using the suite of velocity models resulting from the inversion process. These velocity models are displayed in Figure 6c. These inversion results show maximal variation in the top 50 km of the model with Moho 43 ± 5 km which compares to the global average of crustal thickness for shields and platforms of 41.5 ± 6 km (CHRISTENSEN and MOONEY, 1995). Below the crust, shear velocities reflected in this averaged model are slightly faster than those found in PREM (DZIEWONSKI and ANDERSON, 1981). However, the PREM curve at these depths does fall within the region encompassed by two standard deviations from the averaged model.

Comparison Between Regions

Figure 3d shows observed group velocity curves for the different tectonic regions, and Figure 7 depicts the corresponding velocity inversion results. The hashed regions in Figure 7 represent the area encompassed by two standard deviations from the mean value at each depth. All curves overlap at the 2 sigma level from 0 to 50 km depth. Below 50 km, the active curve is markedly slower and completely separated from the cratonic path. The active curve is also slower than the orogenic curve below 50 km and indicates minimal overlap within two standard deviations of the average. The cratonic curve displays the fastest shear velocities below 60 km. These results are not particularly surprising, but are useful tectonic constraints in this region of sparse seismic coverage. Given the range of acceptable values for cratonic mantle velocities, we cannot establish the degree to which North African cratonic bodies do or do not differ from other cratonic bodies.

Summary

In this study we have measured Rayleigh wave group velocity dispersion across nine regions based upon tectonics and geology. We have performed inversion tests for three representative tectonic regions. Groupings of active tectonic regions, orogenic regions, and cratons reflect mantle velocity structures comparable to those from similar tectonic regions elsewhere in the world. Shallow crustal velocities are strongly controlled by the presence or absence of sedimentary basins. Our results from active tectonic paths suggest the presence of a large low velocity body apparently centered on the Afar triple junction. This region has been subject to significant volcanic activity in the recent geologic past, and it is likely that the low

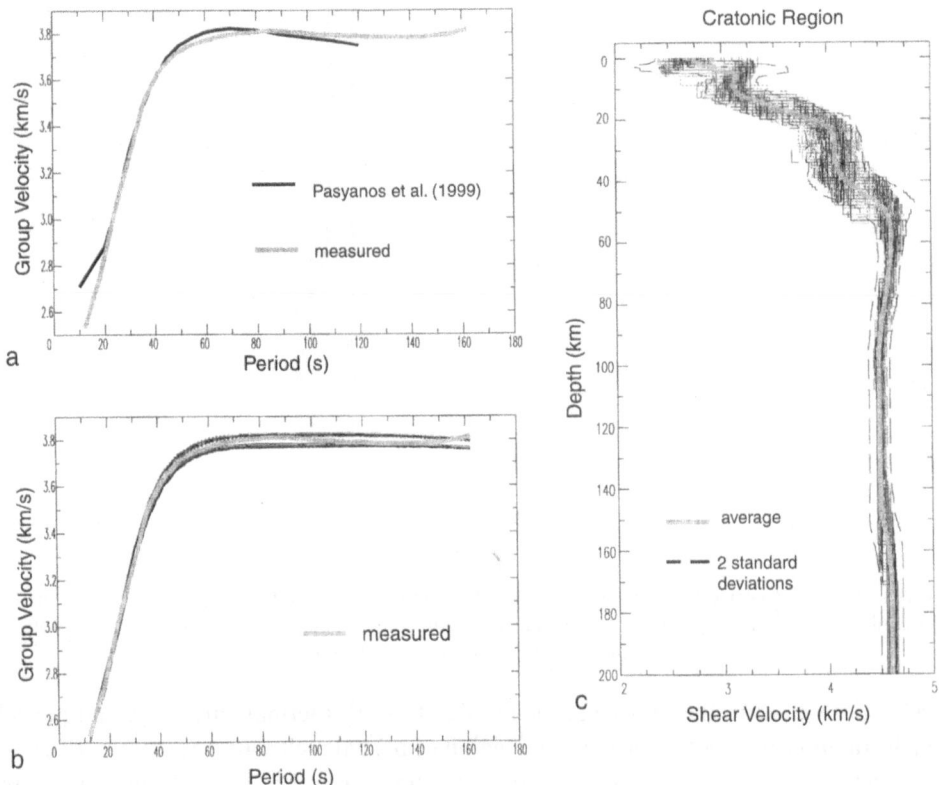

Figure 6

Inversion and forward modeling results for a sample cratonic path (the eastern BNG path of the cratonic block, C2): (a) average measured group velocity curve from this study (thick grey line) compared to one calculated from a regional 3-D tomographic velocity model (PASYANOS et al., this issue) for the same path (black line); (b) average measured group velocity curve (thick grey line) versus group velocity curves (thin grey lines) calculated from shear velocity models (inversion results) in panel c; (c) inversion results for the eastern BNG path and the average (thick grey line) and standard deviation (dashed) of the shear velocity models.

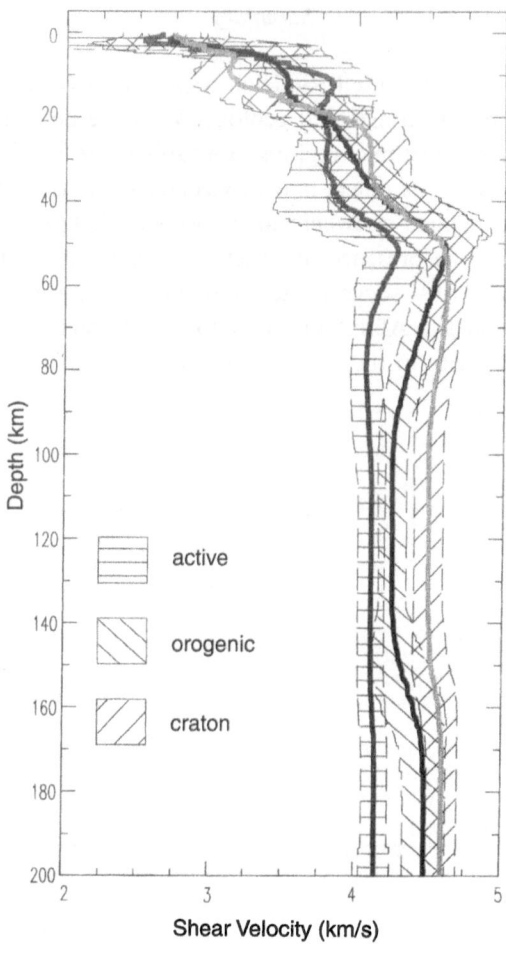

Figure 7
Inversion results (shear velocity models) from the East African Rift, the M1 orogenic block, and the C2 cratonic block. The average (thick solid line) as well as the area encompassed by two standard deviations (dashed) are shown for each region.

velocities we see in the upper mantle are due to both thermal variations and partial melt. In order to produce such a large region of partial melt, one might well appeal to the action of a plume. However, the one-dimensional nature of the inversions produced in this study as well as the depth penetration limitations imposed by the Rayleigh wave sensitivity kernels preclude any definitive imaging of a plume structure.

Group velocity curves are influenced by crust and upper mantle velocities and the thickness of unlithified sediments through which the surface wave passes. In terms of CTBT monitoring objectives, variations in crustal and upper mantle structure affect event locations; while the amount of unlithified sediment through which a surface wave passes affects M_s values. On the basis of these observations, a good regionalization scheme of Northern Africa for CTBT monitoring purposes should rely primarily on sediment thickness for shallow structure and on the basement geology (orogenic, cratonic, or rift regions) for the upper mantle velocities. The very low velocities associated with the broad Afar/Djibouti plume need to be better mapped and paths which traverse the region identified as potentially anomalously slow. By comparing direct group velocity measurements and/or averaged group velocity measurements to those generated from tomographic models, important insight can be gained into the accuracy of 3-D tomographic studies which will be vital in the CTBT monitoring effort. These observed curves serve as a 'ground truth' by which to validate the tomographic models. Such tomographic models used in conjunction with look-up tables of observed group velocity curves could allow for rapid, and potentially automate, comparison of a suspect waveform with the waveform one would expect from an earthquake of similar magnitude occurring within the region.

Acknowledgements

We thank C. Ammon for providing codes for measuring group velocities and S. Taylor for providing his surface wave inversion code. H. Patton, C. Jones, M. Pasyanos, A. Levshin, M. Ritzwoller, and an anonymous reviewer provided useful advice and comments. This work was performed in part under the auspice of the U.S. Department of Energy by the Lawrence Livermore National Laboratory under contract W-7405-ENG-48 and LLNL subcontract B334420 to the University of Colorado.

REFERENCES

ACHAUER, U. (1992), *A Study of the Kenya Rift Using Delay-time Tomography Analysis and Gravity Modeling*, Tectonophysics *209*, 197–207.

AMMON, C. J., Notes on Surface-wave Tomography. Part I - Surface-wave Processing, unpublished notes, 44 pages, 1998.

BASS, J. D., *Elasticity of minerals, glasses, and melts.* In *Mineral Physics and Crystallography* (ed. AHRENS, T. J.), Am. Geophys. Union (Washington, D.C., 1995), pp. 45–63.

BERKHEMER, H., BAIER, B., BARTELSEN, H., BEHLE, A., BURKHARDT, H., GERBRANDE, H., MAKRIS, J., MENZEL, H., MILLER, H., and VEES, R., *Deep seismic soundings in the Afar region and on the highland of Ethiopia.* In *Afar Depression of Ethiopia* (eds. PILGER, A., and RÖSLER, A.) (Stuttgart, Germany: E. Schweizerbart'sche Verlagsbuchhandlung, 1976), pp. 89–106.

CAHEN, L., and SNELLING, N. J., *The Geochronology and Evolution of Africa* (Clarendon Press, Oxford 1984).

CHRISTENSEN, N. I., and MOONEY, W. D. (1995), *Seismic Velocity Structure and Composition of the Continental Crust: A Global View*, J. Geophys. Res. *100*, 9761–9788.

CLAERBOUT, J. F., *Earth Soundings Analysis: Processing versus Inversion* (Blackwell Scientific Publications, Boston, MA, 1992) 304 pp.

CLOUSER, R. H., and LANGSTON, C. A. (1990), *Upper Mantle Structure of Southern Africa from P_{nl} Waves*, J. Geophys. Res. *95*, 17,403–17,415.

CONDIE, K. C., *Plate Tectonics and Crustal Evolution* (Pergamon Press, New York 1982).

DZIEWONSKI, A. M., and ANDERSON, D. L. (1981), *Preliminary Reference Earth Model*, Phys. Earth and Planet. Inter. *25*, 297–356.

EBINGER, C. J., and SLEEP, N. H. (1998), *Cenozoic Magmatism Throughout East Africa Resulting from Impact of a Single Plume*, Nature *395*, 788–791.

EKSTRÖM, G., TROMP, J., and LARSON, E. (1997), *Measurements and Global Models of Surface Wave Propagation*, J. Geophys. Res. *102*, 8137–8157.

GOODWIN, A. M., *Principles of Precambrian Geology* (Academic Press Limited, Great Britain 1996).

GUMPER, F., and POMEROY, P. W. (1970), *Seismic Wave Velocities and Earth Structure on the African Continent*, Bull. Seismol. Soc. Am. *60*, 651–668.

HADIOUCHE, O. (1990), *First Evidence for High Anelastic Attenuation Beneath the Red Sea from Love Wave Analysis*, Geophys. Res. Lett. *17*, 1973–1976.

HADIOUCHE, O., and JOBERT, N. (1988a), *Geographical Distribution of Surface-wave Velocities and 3-D Upper-mantle Structure in Africa*, Geophys. J. *95*, 87–109.

HADIOUCHE, O., and JOBERT, N. (1988b), *Evidence for Anisotropy in Northeast Africa, from Geographical and Azimuthal Distribution of Rayleigh Wave Velocities, and Average Upper-mantle Structure*, Geophys. Res. Lett. *15*, 365–368.

HADIOUCHE, O., and ZÜRN, W. (1992), *On the Structure of the Crust and Upper Mantle Beneath the Afro-Arabian Region from Surface Wave Dispersion*, Tectonophysics *209*, 179–196.

HADIOUCHE, O., JOBERT, N., and RAMANOWICZ, B. (1986), *First Two-station Results from Long-period Surface Waves Velocity from the GEOSCOPE Stations in Africa*, Geophys. Res. Lett. *13*, 547–550.

HAZLER, S. E. (1998), *One-dimensional Velocity Structure of Northern Africa as Determined by Rayleigh Wave Group Velocity Dispersion*, Masters Thesis, University of Colorado, Boulder, Colorado.

HERRMANN, R. B. (1973), *Some Aspects of Surface Waves*, Bull. Seismol. Soc. Am. *63*, 663–671.

KARATO, S. (1993), *The Importance of Anelasticity in the Interpretation of Seismic Tomography*, Geophys. Res. Lett. *20*, 1623–1626.

KNOPOFF, L., and SCHLUE, J. W. (1972), *Rayleigh Wave Phase Velocities for the Path Addis Ababa-Nairobi*, Tectonophysics *15*, 157–163.

KNOX, R. P., NYBLADE, A. A., and LANGSTON, C. A. (1998), *Upper Mantle S Velocities Beneath Afar and Western Saudi Arabia*, Geophys. Res. Lett. *25*, 4233–4236.

LAST, R. J., NYBLADE, A. A., LANGSTON, C. A., and OWENS, T. J. (1997), *Crustal Structure of the East African Plateau from Receiver Functions and Rayleigh Wave Phase Velocities*, J. Geophys. Res. *102*, 24,469–24,483.

LITHGOW-BERTELLONI, C., and SILVER, P. G. (1998), *Dynamic Topography, Plate Driving Forces and the African Superswell*, Nature *395*, 269–272.

LONG, R. E., BACKHOUSE, R. W., MAGUIRE, P. K. H., and SUNDARLINGHAM, K. (1972), *The Structure of East Africa Using Surface Wave Dispersion and Durham Seismic Array Data*, Tectonophysics *15*, 165–178.

MASTERS, G., JOHNSON, S., LASKE, G., and BOLTON, H. (1996), *A Shear-velocity Model of the Mantle*, Philosoph. Transact. Royal Soc. London *354*, 1385–1411.

MCKENZIE, D., and BICKLE, M. J. (1988), *The Volume and Composition of Melt Generated by Extension of the Lithosphere*, J. Petrol. *29*, 625–679.

MCNAMARA, D. E., and WALTER, W. R. (1995), *Regional Waveform Modeling and Phase Propagation in Northern Africa*, EOS, Transact. Am. Geophys. Union *76*, 428.

MOONEY, W. D., LASKE, G., and MASTERS, T. G. (1998), *CRUST 5.1:A Global Crustal Model at $5° \times 5°$*, J. Geophys. Res. *103*, 727–747.

NYBLADE, A. A., VOGFJORD, K. S., and LANGSTON, C. A. (1996), *P-wave Velocity of the Proterozoic Upper Mantle Beneath Central and Southern Africa*, J. Geophys. Res. *101*, 11,159–11,171.

NYBLADE, A. A., and ROBINSON, S. W. (1994), *The African Superswell*, Geophys. Res. Lett. *21*, 765–768.

PASYANOS, M. E., WALTER, W. R., and HAZLER, S. E. (2001), *A Surface Wave Dispersion Study of the Middle East and North Africa for Monitoring the Comprehensive Nuclear-Test-Ban Treaty*, Pure appl. geophys., this issue.

PETTERS, S. W., *Regional Geology of Africa* (Springer-Verlag, Germany 1991).

RODGERS, A. J., WALTER, W. R., MELLORS, R. J., AL-AMRI, A. M. S., and ZHANG, Y. S. (1999), *Lithospheric Structure of the Arabian Shield and Platform from Complete Regional Waveform Modelling and Surface Wave Group Velocities*, Geophys. J. Int. *138*, 871–878.

SANDVOL, E., SERBER, D., CALVERT, A., and BARAZANGI, M. (1998), *Grid Search Modeling of Receiver Functions: Implications for Crustal Structure in the Middle East and North Africa*, J. Geophys. Res. *103*, 26,899–26,917.

SCHILLING, J-G. (1991), *Fluxes and Excess Temperatures of Mantle Plumes Inferred from their Interaction with Migrating Ocean Ridges*, Nature *352*, 397–403.

SHEEHAN, A. F., and SOLOMON, S. C. (1991), *Joint Inversion of Shear Wave Travel Time Residuals and Geoid and Depth Anomalies for Long-wavelength Variations in Upper Mantle Temperature and Composition along the Mid-Atlantic Ridge*, J. Geophys. Res. *96*, 19,981–20,009.

SU, W., WOODWARD, R., and DZIEWONSKI, A. (1994), *Degree 12 Model of Shear Velocity Heterogeneity in the Mantle*, J. Geophys. Res. *99*, 6945–6980.

SWEENEY, J. J. (1996), *Accuracy of Teleseismic Event Location in the Middle East and North Africa*, Lawrence Livermore National Laboratory, UCRL-ID-125868.

TANAKA, S., and HAMAGUCHI, H. (1992), *Heterogeneity in the Lower Mantle Beneath Africa, as Revealed from S and ScS Phases*, Tectonophysics *209*, 213–222.

TAYLOR, S. R. (1980), *Crust and Upper Mantle Structure of the Northeastern United States*, Ph.D. Thesis, Massachusetts Institute of Technology, Cambridge, Massachusetts.

TESHA, A. L., NYBLADE, A. A., KELLER, G. R., and DOSER, D. I. (1997), *Rift Localization in Suture-thickened Crust: Evidence from Bouguer Gravity Anomalies in Northeastern Tanzania, East Africa*, Tectonophysics *278*, 315–328.

VINCENT, P. M. (1970), *The evolution of the Tibesti Volcanic Province, eastern Sahara*. In *African Magmatism and Tectonics* (eds. CLIFFORD, T. N., and GASS, I. G.) (Hafner Publishing Co., Darien, Conn., 1970), pp. 301–319.

WANG, Z., and DAHLEN, F. A. (1995), *Validity of a Surface-wave Ray Theory on a Laterally Heterogeneous Earth*, Geophys. J. Int. *123*, 757–773.

WIGGINS, R. A. (1972), *The General Linear Inverse Problem: Implication of Surface Waves and Free Oscillations for Earth Structure*, Rev. Geophys. Space Phys. *10*, 251–285.

ZHANG, Y., and LAY, T. (1996), *Global Surface Wave Phase Velocity Variations*, J. Geophys. Res. *101*, 8415–8436.

(Received April 10, 1999, revised December 20, 1999, accepted January 15, 2000)

To access this journal online:
http://www.birkhauser.ch

Surface Wave Identification, Measurement, and Source Characterizations

Pure appl. geophys. 158 (2001) 1497–1515
0033–4553/01/1497–19 $ 1.50 + 0.20/0

Pure and Applied Geophysics

Isotropic and Nonisotropic Components of Earthquakes and Nuclear Explosions on the Lop Nor Test Site, China

B. G. Bukchin,[1] A. Z. Mostinsky,[1] A. A. Egorkin,[1] A. L. Levshin,[2]
and M. H. Ritzwoller[2]

Abstract—We test the hypothesis that the existence of an observable non-zero isotropic component of seismic moment can be used as a discriminant to distinguish nuclear explosions from shallow earthquakes. We do this by applying the method described herein to a small set of data recorded between 1990 and 1996 following events (seven nuclear explosions, three earthquakes) that occurred on the Lop Nor test site in Western China. We represent each source as a sum of an isotropic component at the surface and a nonisotropic, double-couple component at an estimated depth. The explosions all possess a significant non-zero isotropic component and the estimated depth of the double-couple component of the moment tensor, presumably the result of tectonic release, lies between about 0 and 3 km. For the earthquakes studied, the isotropic component is indistinguishable from zero and the depths of the sources are estimated at 3, 17 and 31 km. The data set we have studied, although still very small, suggests that certain source characteristics (namely, double-couple depth and the ratio of the isotropic to nonisotropic components of seismic moment) may prove useful in discriminating explosions from shallow earthquakes. Further work is needed to determine whether these observations hold for explosions at other test sites, to investigate a much larger set of shallow earthquakes located in regions of interest, and to study the robustness of the estimated source parameters as source magnitude and the number of observing stations decrease.

Key words: Tectonic release, seismic moment tensor, nuclear explosions, earthquakes, surface waves.

1. Introduction

It has long been known that nearly all underground nuclear explosions have a significant component of nonisotropic seismic radiation (e.g., PRESS and ARCHAMBEAU, 1962; BRUNE and POMEROY, 1963; AKI and TSAI, 1972; TOKSÖZ and KEHRER, 1972; HELLE and RYGG, 1984; WALLACE *et al.*, 1985; WALTER and PATTON, 1990). There appears to be general agreement that the long-period part of the nonisotropic radiation is caused by tectonic release due to existing tectonic stresses in the source region (MUELLER and MURPHY, 1971; ARCHAMBEAU, 1972; MASSE, 1981; WALLACE

[1] International Institute of Earthquake Prediction Theory & Mathematical Geophysics, Moscow, Russia. E-mail: bukchin@mitp.ru
[2] Department of Physics, University of Colorado, Boulder, Boulder, CO 80309-0390, USA. E-mail: levshin@lemond.colorado.edu, web: ciei.colorado.edu

et al., 1985; DAY and STEVENS, 1986; STEVENS, 1986; DAY *et al.*, 1987; PATTON, 1991; STEVENS *et al.*, 1991; HARKRIDER *et al.*, 1994; EKSTRÖM and RICHARDS, 1994; LI *et al.*, 1995).

Most of the experimental studies of tectonic release have concentrated on explosions at the Nevada test site and the test sites in Kazakhstan. BURGER *et al.* (1986) described the long-period *S*-wave radiation by tectonic release of explosions at Novaya Zemlya. ZHANG (1994), LEVSHIN and RITZWOLLER (1995), and WALLACE and TINKER (1996) studied Chinese nuclear explosions and found that most are accompanied by tectonic release. However, we are unaware of any detailed analysis of source mechanisms of these events.

Several authors have estimated the relative contributions of the nonisotropic (tectonic release) and isotropic (explosion) components to the resulting seismic moment tensor of explosions (see, for example, EKSTRÖM and RICHARDS, 1994). As a rule these estimates have been performed using the assumption that both phenomena occur in the same small volume inside the earth.

Although the determination of the source mechanism of earthquakes is a routine operation, only a few studies have evaluated the isotropic component of earthquake radiation. This is especially true for weak events that are of the primary interest for discrimination purposes. As is well known, a pure shear dislocation does not radiate an isotropic component. Subsequently, the natural tendency is to constrain the moment tensor solution with the assumption of a zero trace. However, from the point of view of discrimination it is important to determine if a significant isotropic component could appear in unconstrained moment tensor solutions due to physical phenomena in the source region or due to inaccuracy of measurements and underlying assumptions.

In this paper we will describe an approach to estimate the isotropic and nonisotropic components of the source mechanism and will apply this method to a small set of data for several earthquakes and explosions on the Lop Nor test site. The purpose is to determine whether the existence of a strong isotropic moment is consistent with the hypothesis that the event is an explosion. For each of the ten events we consider (7 explosions and 3 earthquakes, Fig. 1), we obtain estimates of the isotropic and nonisotropic moments, characterize the tectonic release by estimating the moment tensor of the nonisotropic source, and estimate the depth of the equivalent double-couple of the nonisotropic source. Our approach differs from those of previous studies in three main ways which are designed to focus the method on the discrimination problem: (1) we allow the depth of tectonic release for all events (earthquakes and explosions) to be variable and fix only the depth of the isotropic source near the surface, (2) we fit only the amplitude of the surface wave spectrum so the method, in principal, can be applied to considerably smaller events for which the interpretation of phase may be ambiguous, and (3) we minimize the misfit between synthetic and observed surface wave amplitude spectra jointly with fitting polarities of *P*-wave first arrivals.

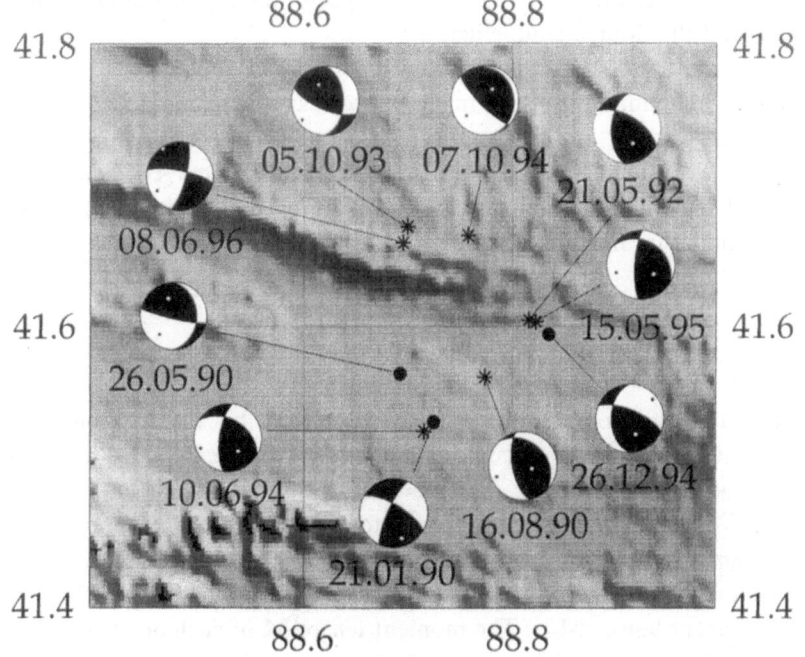

Figure 1
Epicenters and focal mechanisms of earthquakes (circles) and tectonic releases accompanying explosions
(stars).

Our approach is, therefore, based on the combination of an isotropic moment
tensor at the surface, which models an explosive source, and a pure double-couple at
an estimated depth, which models tectonic release. We find it useful to express the
seismic moment of the double-couple component, M_{0qu}, and the seismic moment of
the isotropic component, M_{0ex}, as the total seismic moment M_0 and an angle φ which
determines the ratio of the seismic moments of the isotropic and double-couple
components:

$$M_{0qu} = M_0 \cos \varphi \qquad (1)$$

$$M_{0ex} = M_0 \sin \varphi \qquad (2)$$

$$\tan \varphi = \frac{M_{0ex}}{M_{0qu}}. \qquad (3)$$

We call φ the isotropic angle, for want of a better term, because $\varphi = 0$ corresponds to
a pure earthquake and $\varphi = 90°$ corresponds to a pure explosion. With this notation
and the assumption that the isotropic component of the source occurs at a known
depth (usually at or near the surface), the source can then be characterized by six
parameters: the seismic moment M_0, the isotropic angle φ, three angles (or two

principal axes, **T** (tension) and **P** (compression)) determining the double-couple focal mechanism, and the double-couple depth h.

We determine these source parameters by fitting the polarities of P-wave first arrivals jointly while minimizing the misfit between synthetic and observed Love and Rayleigh surface wave amplitude spectra. We search for the optimal set of parameters by a systematic exploration of parameter space. To characterize the uncertainty of the source parameters, we calculate partial residual functions which describe the minimum misfit as a function of each varying parameter.

2. The Algorithm

An instantaneous point source can be described by the moment tensor, a symmetric 3×3 matrix **M**. Seismic moment, M_0, is defined by the equation $M_0 = \left(1/2 \operatorname{tr}(\mathbf{M}^T \mathbf{M})\right)^{1/2}$, where $\mathbf{M}^T$ is the transposed moment tensor and $\operatorname{tr}(\mathbf{M}^T \mathbf{M}) = \sum_{i,j=1}^{3} M_{ij}^2$. The moment tensor for any event can be presented in the form $\mathbf{M} = M_0 \mathbf{m}$, where the matrix **m** is normalized by the condition $\operatorname{tr}(\mathbf{m}^T \mathbf{m}) = 2$. We consider each event as a sum of an earthquake (0-trace moment tensor $\mathbf{M}_{qu}$) and an explosion (moment tensor $\mathbf{M}_{ex}$). The moment tensor **M** of such an event is given by the sum $\mathbf{M} = \mathbf{M}_{qu} + \mathbf{M}_{ex}$. If **I** be the 3×3 identity matrix, then $\mathbf{M}_{ex} = \sqrt{2/3}\, M_{0ex} \mathbf{I}$, where M_{0ex} is the seismic moment of the explosion. For the earthquake, $\mathbf{M}_{qu} = M_{0qu} \mathbf{m}$, where M_{0qu} is the seismic moment of the earthquake, and **m** is a normalized moment tensor that we will assume has the following characteristics: $\operatorname{tr} \mathbf{m} = 0$ and $\operatorname{tr}(\mathbf{m}^T \mathbf{m}) = 2$. Thus, **m** represents a normalized double-couple mechanism and we assume that $\mathbf{M}_{qu}$ represents a double-couple source.

The tensors $\mathbf{M}_{ex}$ and $\mathbf{M}_{qu}$ are orthogonal in the sense we describe in the next paragraph. Thus, the total moment tensor is the sum of two orthogonal components, and the seismic moments of the explosion and earthquake components of the total moment can be expressed uniquely as total seismic moment M_0 and the isotropic angle φ between **M** and $\mathbf{M}_{qu}$, as given by equations (1)–(3) above.

To show that $\mathbf{M}_{ex}$ and $\mathbf{M}_{qu}$ are orthogonal, consider the six dimensional linear Euclidean space of symmetric 3×3 matrices **M**, and let the scalar product of two vectors $(\mathbf{M}, \mathbf{N})$ be defined as $(\mathbf{M}, \mathbf{N}) = \sum_{i,j} M_{ij} N_{ij} = \operatorname{tr}(\mathbf{M}^T \mathbf{N})$. The isotropic tensors $\mathbf{M}_{ex}$ form a 1-D subspace which is orthogonal to the 5-D linear subspace of zero trace tensors $\mathbf{M}_{qu}$:

$$(\mathbf{M}_{qu}, \mathbf{M}_{ex}) = \sqrt{2/3}\, M_{0qu} M_{0ex} \operatorname{tr}(\mathbf{m}^T \mathbf{I}) = \sqrt{2/3}\, M_{0qu} M_{0ex} \operatorname{tr} \mathbf{m} = 0 \ . \qquad (4)$$

Finally, the seismic moment of the combined event can be expressed by the formula:

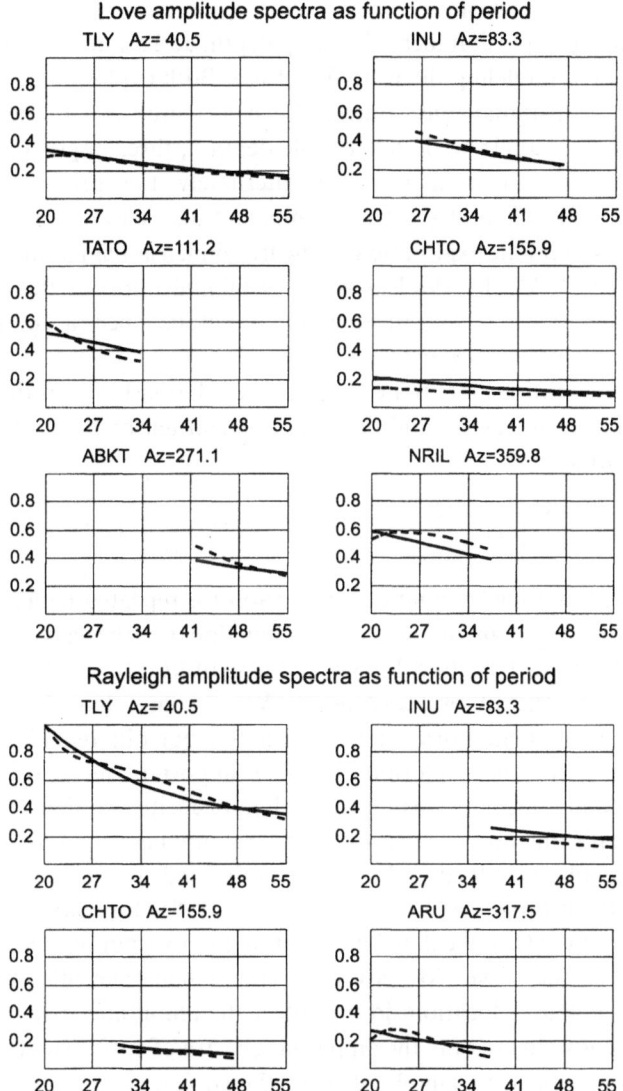

Figure 2

Comparison of observed (dashed lines) and computed (solid lines) Rayleigh and Love wave amplitude spectra for the explosion on 5 October, 1993. Stations' names and azimuths from the epicenter to the stations are provided. All spectra are normalized by the same value.

$$M_0 = \sqrt{1/2(\mathbf{M}, \mathbf{M})} = \sqrt{1/2((\mathbf{M}_{qu} + \mathbf{M}_{ex}), (\mathbf{M}_{qu} + \mathbf{M}_{ex}))}$$

$$= \sqrt{1/2(\mathbf{M}_{qu}, \mathbf{M}_{qu}) + 1/2(\mathbf{M}_{ex}, \mathbf{M}_{ex})}$$

$$= \sqrt{M_{0qu}^2 + M_{0ex}^2} \; . \tag{5}$$

Our appoach then is to consider a seismic source as the combination of an isotropic tensor, modeling an explosion located at zero depth, and a pure double-couple point source at a depth h, modeling the tectonic release. Both explosions and earthquakes are considered as instantaneous sources. Such a source can be described by six parameters: isotropic angle φ, double-couple depth h, three angles (strike, dip, slip) characterizing the focal mechanism or alternately the principal tension and compression axes (**T**, **P**), and the seismic moment M_0. We determine five of these parameters by a systematic exploration of the five-dimensional parametric space, and the sixth parameter, M_0, by the least-squares minimization of the misfit between the observed and calculated surface wave amplitude spectra for every current combination of the other parameters.

Under the assumptions mentioned above, the relation between the surface wave displacement spectrum $u_i(\mathbf{x}, \omega)$ and the total moment tensor **M** can be expressed by the following formula

$$u_i(\mathbf{x}, \omega) = \frac{1}{i\omega}\left[M_{qu_{il}} \frac{\partial}{\partial y_{qu_l}} G_{ij}(\mathbf{x}, \mathbf{y}_{qu}, \omega) + M_{ex_{jl}} \frac{\partial}{\partial y_{ex_l}} G_{ij}(\mathbf{x}, \mathbf{y}_{ex}, \omega) \right] , \qquad (6)$$

where $i, j = 1, 2, 3$ and the summation convention for repeated indices is used. Here $G_{ij}(\mathbf{x}, \mathbf{y}, \omega)$ is the spectrum of the Green's function for the chosen model of the medium and the wave type (see LEVSHIN, 1985; BUKCHIN, 1990), **y** is the source location, and **x** is the observation point. We assume that the explosion and earthquake have the same horizontal coordinates, but different depths: h for the earthquake and 0 for the explosion. Another important assumption is that the propagation medium has only weak lateral inhomogeneities so that the surface wave part of the Green's function is determined by structure near the source and receiver, by the average phase velocity along the path, and by geometrical spreading (WOODHOUSE, 1974; BABICH *et al.*, 1976; LEVSHIN, 1985; LEVSHIN *et al.*, 1989). The amplitude spectrum $|u_i(\mathbf{x}, \omega)|$ defined by equation (6) does not depend on the average phase velocity of the wave. Moreover, if lateral heterogeneities are sufficiently smooth, errors in source location do not affect the amplitude spectrum (BUKCHIN, 1990). Consequently we use only the amplitude spectra of surface waves to determine the source parameters. An example of fit to the surface wave amplitude spectra is shown in Figure 2.

Assuming that we know the relevant propagation characteristics of the medium, equation (6) provides a system of equations for the five parameters defined above (φ, h, **T**, **P**). We define a grid in the space of these five parameters and use equation (6) to calculate the amplitude spectra of the surface waves at the points of observation for every possible combination of values of the variable parameters. Comparison of the calculated and observed amplitude spectra gives us a surface wave residual $\varepsilon_{sw}^{(i)}$ for both Rayleigh and Love waves at each frequency ω. The index i runs through the entire set of observations. Specifically, it corresponds to a certain station, wave type, and frequency. Let $u^{(i)}(\mathbf{x}, \omega)$ be any

Table 1

Observed seismograms and models used for surface wave calculations for each station

Station	Azimuth (o)	Distance (o)	Model
ABKT	271	24	Tectonic
ANTO	287	42	Tectonic
ARU	317	25	Stable
BNG	260	73	Stable
BRVK	319	17	Stable
CHTO	156	24	Stable
DPC	306	49	Stable
ENH	118	20	Stable
ERM	71	40	Stable
HIA	60	23	Stable
HYB	204	26	Stable
GNI	282	33	Tectonic
INU	83	38	Stable
KEV	333	42	Stable
KIV	290	34	Tectonic
KMI	140	20	Tectonic
KONO	320	50	Stable
LZH	111	13	Tectonic
MA2	43	42	Stable
MAJO	81	38	Stable
MDJ	70	29	Stable
NIL	242	15	Tectonic
NRIL	0	28	Stable
OBN	310	37	Stable
SEY	38	42	Stable
TATO	111	31	Stable
TLY	40	14	Tectonic
YAK	36	32	Stable

observed value of the spectrum, $i = 1, \ldots, N$, and $\varepsilon_{sw}^{(i)}$ be the corresponding residual of $|u^{(i)}(\mathbf{x}, \omega)|$. We define the normalized amplitude residual by the formula

$$\varepsilon_{sw}(h, \varphi, \mathbf{T}, \mathbf{P}) = \left[\left(\sum_{i=1}^{N} \varepsilon_{sw}^{(i)2} \right) \bigg/ \left(\sum_{i=1}^{N} |u^{(i)}(\mathbf{x}, \omega)|^2 \right) \right]^{1/2}. \tag{7}$$

M_0 is found by the least-squares minimization of ε_{sw} for each combination of values of the five other variable parameters.

A well-known problem in estimating moment tensors from surface waves is that the radiation pattern of surface wave amplitude spectra is symmetric with respect to the epicenter. Another uncertainty in moment tensor estimation appears when source depth is much less than the wavelength of the observed surface waves, in which case the surface waves provide only three linear

Table 2

Crustal and upper mantle models used for surface wave calculations

h(Km)	V_p(km/s)	V_s(km/s)	ρ(g/cm³)
Lop Nor			
1.1	5.08	2.9	2.26
5.8	5.90	3.5	2.59
14.1	6.06	3.57	2.65
30.9	6.51	3.77	2.83
8.1	8.09	4.47	3.38
20.0	8.07	4.46	3.38
35.0	7.99	4.37	3.37
35.0	7.97	4.36	3.37
35.0	7.95	4.34	3.36
35.0	7.93	4.33	3.36
45.0	8.56	4.60	3.45
45.0	8.65	4.64	3.48
45.0	8.74	4.67	3.50
45.0	8.82	4.70	3.53
50.0	9.22	4.95	3.76
50.0	9.47	5.09	3.82
50.0	9.73	5.23	3.88
50.0	9.98	5.38	3.94
50.0	10.14	5.46	3.98
Stable			
19.0	6.14	3.55	2.74
19.0	6.58	3.80	3.00
12.0	8.20	4.65	3.32
10.0	8.17	4.62	3.34
10.0	8.14	4.57	3.35
10.0	8.10	4.51	3.36
10.0	8.07	4.46	3.37
10.0	8.02	4.41	3.38
25.0	7.93	4.37	3.39
25.0	7.85	4.35	3.41
25.0	7.89	4.36	3.43
25.0	7.98	4.38	3.46
25.0	8.10	4.42	3.48
25.0	8.21	4.46	3.50
50.0	8.38	4.54	3.53
50.0	8.62	4.68	3.58
50.0	8.87	4.85	3.62
50.0	9.15	5.04	3.69
50.0	9.45	5.21	3.82
Tectonic			
20.0	6.14	3.55	2.74
20.0	6.80	3.93	3.00
10.0	8.10	4.51	3.32
10.0	8.10	4.51	3.34
10.0	8.10	4.51	3.35
10.0	8.10	4.51	3.36

h is the layer thickness

Table 3

Information on the 10 events used in the present study

Date	Event type	m_b	m_s	Number of records used	Double-couple depth (km)	M_0 (N · m)	ψ (deg.)	φ (deg.)
21.01.90	earthquake	4.6	–	10	31	$0.15 \cdot 10^{17}$	72	0
26.05.90	earthquake	5.4	–	5	3	$0.42 \cdot 10^{16}$	52	0
26.12.94	earthquake	4.6	–	10	17	$0.15 \cdot 10^{16}$	56	0
16.08.90	explosion	6.2	–	7	1	$0.19 \cdot 10^{17}$	62	10–30
21.05.92	explosion	6.5	5.0	16	3	$0.62 \cdot 10^{17}$	64	10–15
05.10.93	explosion	5.9	4.7	10	0	$0.75 \cdot 10^{16}$	56	0–15
10.06.94	explosion	5.8	–	13	1	$0.52 \cdot 10^{16}$	64	20–40
07.10.94	explosion	6.0	–	11	1	$0.18 \cdot 10^{17}$	60	25–35
15.05.95	explosion	6.1	5.0	9	3	$0.15 \cdot 10^{17}$	62	40–60
08.06.96	explosion	5.9	4.3	12	3	$0.59 \cdot 10^{16}$	58	30–60

ψ is the azimuth of the horizontal compression axis, φ is the isotropic angle which determines the ratio of the isotropic and nonisotropic (double-couple) moments.

constraints on the source parameters. To improve the resolution of the source parameters, we also use the polarities of P-wave first arrivals. In calculating the radiation pattern of P waves for a set of source parameters, we assume that the waves radiated by the isotropic (i.e., explosion) and nonisotropic (i.e., tectonic release or earthquake) source arrive simultaneously. This assumption can be abolished if the observed signs of the P-first arrivals would be substituted by polarities measured from long-period P-wave spectra (BUKCHIN et al., 1997). Before inversion we apply a smoothing procedure to the observed polarities as follows. A set of observed polarities (+1 for compression and −1 for dilatation) radiated in nearly the same direction is combined to form a single cluster and is assigned the mean of the cluster direction. If the number of one polarity in the cluster is significantly larger than the other, we prescribe this polarity to the mean direction. If neither of the two polarities dominate, then the entire cluster is discarded.

Using the misfit between the calculated and observed polarities, we calculate a joint residual of surface wave amplitude spectra and polarities of P-wave first arrivals. Let ε_p be the residual of P-wave first arrival polarities (the number of wrong polarities divided by the full number of observed polarities), we then define the joint residual ε as the sum of the individuals

$$\varepsilon = 1 - (1 - \varepsilon_p)(1 - \varepsilon_{sw}). \tag{8}$$

If both residuals ε_p and ε_{sw} are small, then ε is equal approximately to their sum. We consider the values of parameters that minimize ε as estimates of these parameters.

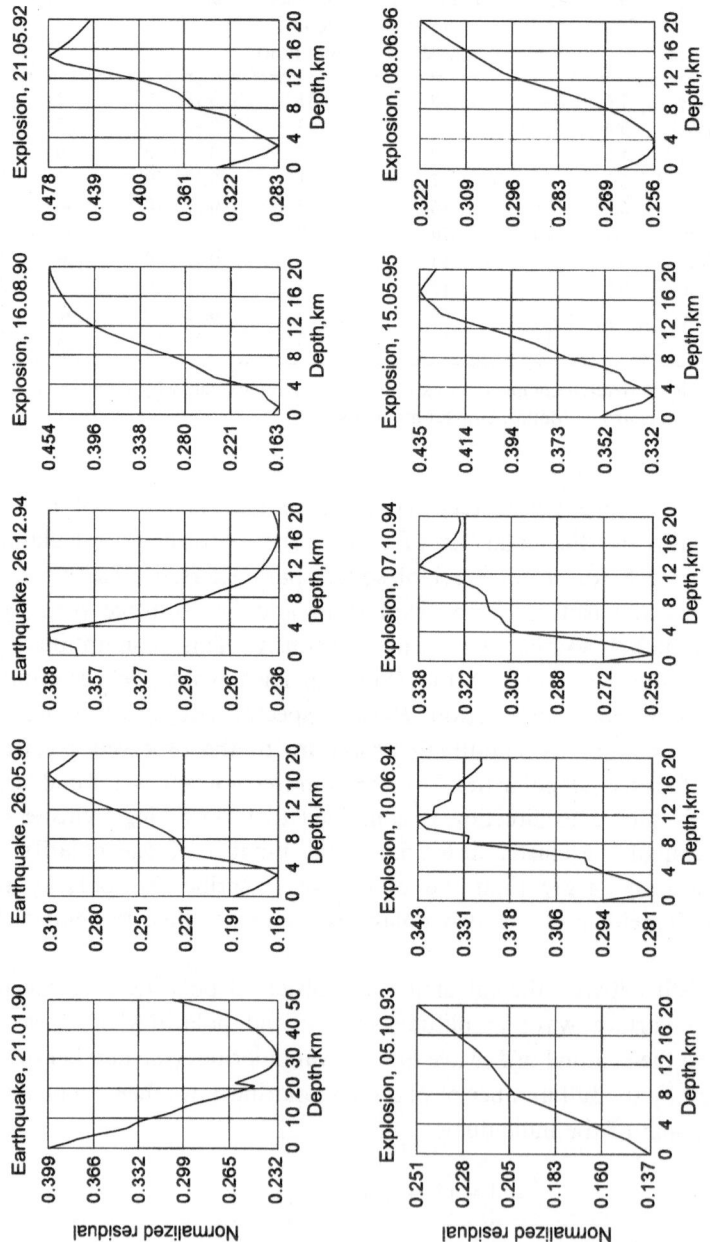

Figure 3

Partial residual functions of the double-couple depth *h*.

Main tension axis

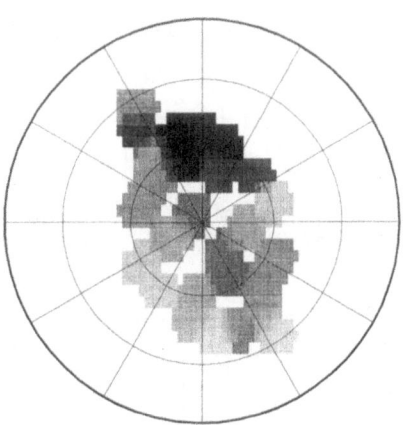

Main compression axis

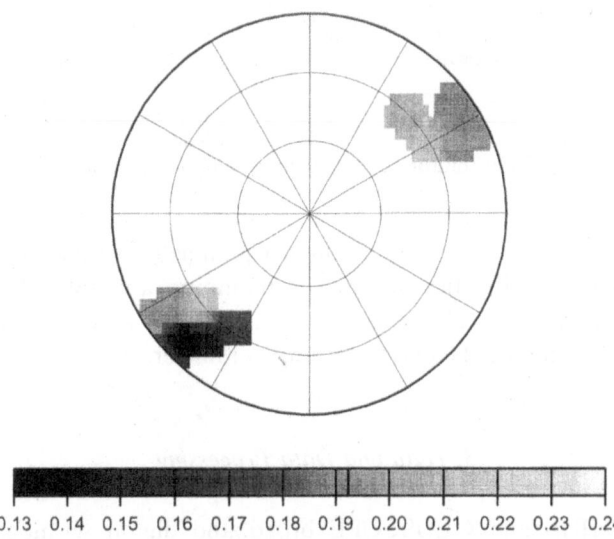

0.13 0.14 0.15 0.16 0.17 0.18 0.19 0.20 0.21 0.22 0.23 0.24

Figure 4
Partial residual functions of the orientation of the principal tension and compression axes for the double-couple focal mechanism of the explosion on 5 October, 1993. The orientation of every axis is shown using a stereographic projection on the lower hemisphere.

To evaluate the uncertainty of each of these source characteristics we calculate partial residual functions $\varepsilon_h(h), \varepsilon_\varphi(\varphi), \varepsilon_T(\mathbf{T}), \varepsilon_P(\mathbf{P})$ in the following way. For each value of a given parameter we find among all possible combinations of the other parameters such a value which provides the minimum of this function. Repeating the

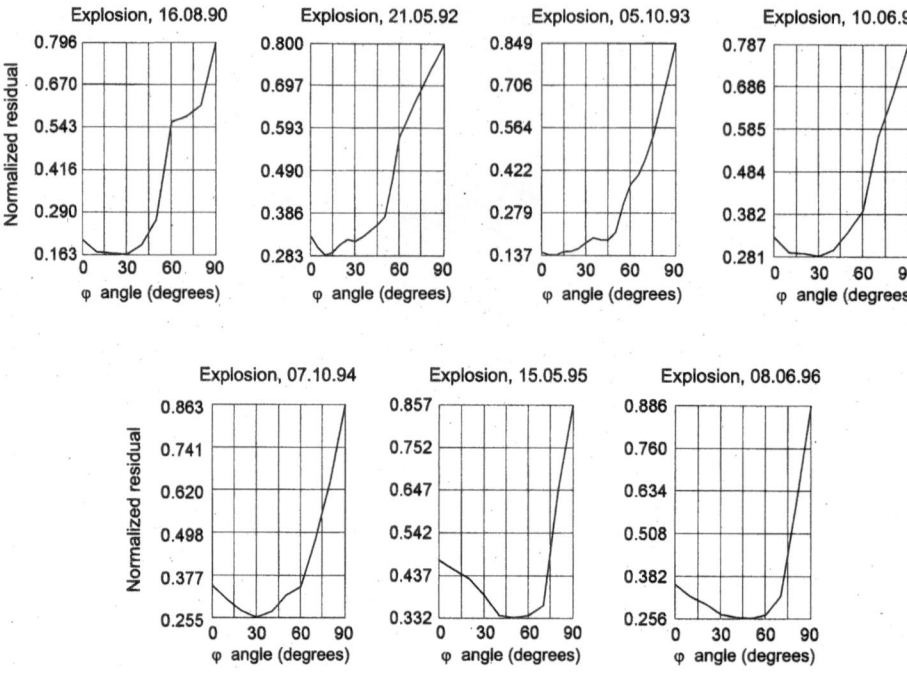

Figure 5

Partial residual functions of isotropic angle φ for the explosions. This angle determines the ratio of seismic moments of isotropic and double-couple components.

search for all possible values of the evaluated parameter we find one at which the corresponding residual function attains its minimum and take this value as an estimate of this parameter. The behavior of the residual function characterizes the uncertainty in determining the corresponding parameter.

3. Data and Data Processing

We utilized IRIS and GEOSCOPE broadband digital seismograms and the ISC bulletins for 14 events that occurred on the Lop Nor test site in China from 1990 through 1996. Eight of these events are nuclear explosions, the other six are earthquakes. However, only 10 events with high signal-to-noise surface waves recorded at several stations were selected for study. The location of the selected events (7 explosions and 3 earthquakes) is given in Figure 1. The list of stations, with their distances and azimuths from Lop Nor test site, are given in Table 1.

In our calculations of the theoretical surface wave spectra we used three different models for (1) the structure for the source region, (2) the stations deployed in the

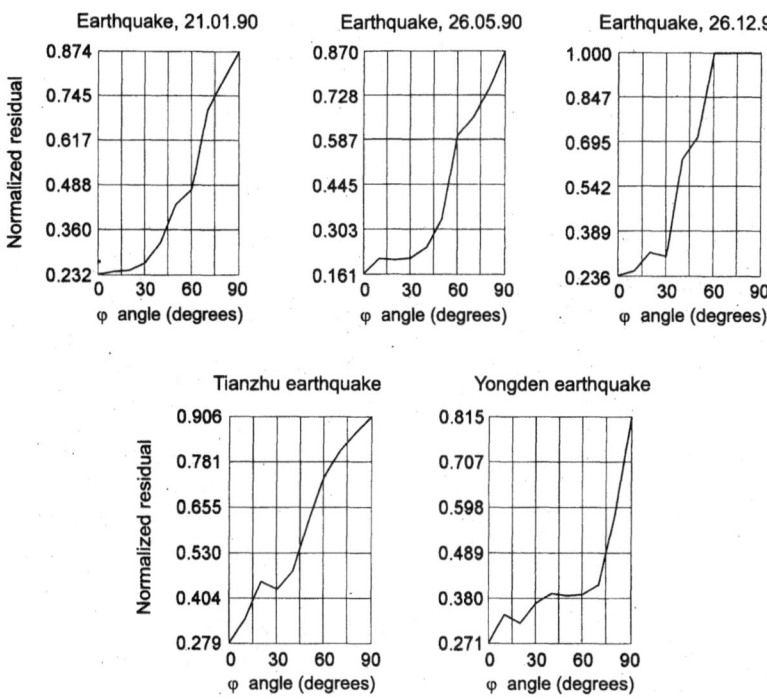

Figure 6
Same as Figure 5, but for the earthquakes.

stable continental regimes, and (3) the stations in the tectonic regions. For brevity we will call the corresponding models "Lop Nor," "Stable," and "Tectonic." The parameters for each model are presented in Table 2. For the Lop Nor model we used a regional velocity model for the four upper layers and the PREM model for deeper structure (see LASSERRE et al., 2000). The Tectonic and Stable models are identical below the first five layers. Column 4 in Table 1 indicates the model used for each station.

We estimated the source parameters using the spectra of Love and Rayleigh fundamental waves for periods ranging from 20 s to 70 s. Love and Rayleigh fundamental modes were extracted by using frequency-time analysis (FTAN) and floating filtering (e.g., LANDER, 1989, and LEVSHIN et al., 1994). We also analyzed the polarization of filtered surface waves using the technique described in LANDER (1989), PAULSSEN et al., (1990), LEVSHIN et al. (1992, 1994), and LEVSHIN and RITZWOLLER (1995). Only records in which the surface wave polarization pattern did not exhibit significant azimuthal anomalies ($\leq 15°$) were used for further analysis. Examples of normalized amplitude spectra are shown in Figure 2. Note that the amplitude of the Love waves, which cannot be excited by a pure isotropic source, is comparable with the amplitude of the Rayleigh waves.

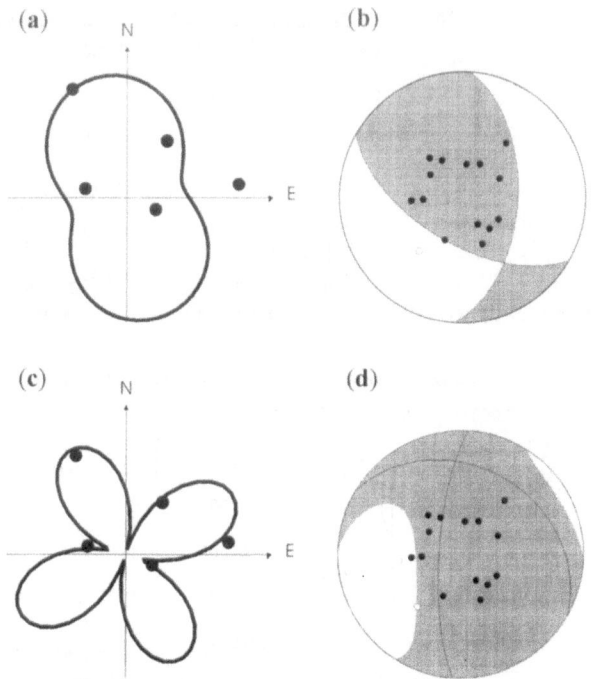

Figure 7

Comparison of observed Rayleigh wave spectral amplitudes at 26 s period and polarities of first arrivals for the explosion on June 10, 1994 with those predicted for two different source models: (a, b) – pure double-couple; (c, d) – non-zero isotropic component is added. (a, c) – observed (filled circles) and predicted (solid line) surface wave spectral amplitudes; (b, d) – observed positive (dark dots) and negative (white dots) *P*-wave polarities; predicted positive (shaded areas) and negative (white areas) *P*-wave polarities.

4. Results of Inversions

A summary of the inversion results is given in Table 3.

4.1 Double-couple Depths and Focal Mechanisms

The focal mechanisms for the earthquakes and for the double-couple components of the explosions are shown in Figure 1.

The double-couple depth is well resolved for all events studied and varies from 0 to 3 km for the explosions and for the earthquakes we obtained depths of 3 km, 31 km, and 17 km. The partial residual function of double-couple depth is presented in Figure 3 for all events.

Figure 4 shows an example of the resolution of the double-couple focal mechanism for the large explosion that occurred on October 5, 1993 with $m_b = 5.9$ and $m_s = 4.7$. This figure displays the partial residual function of principal axes

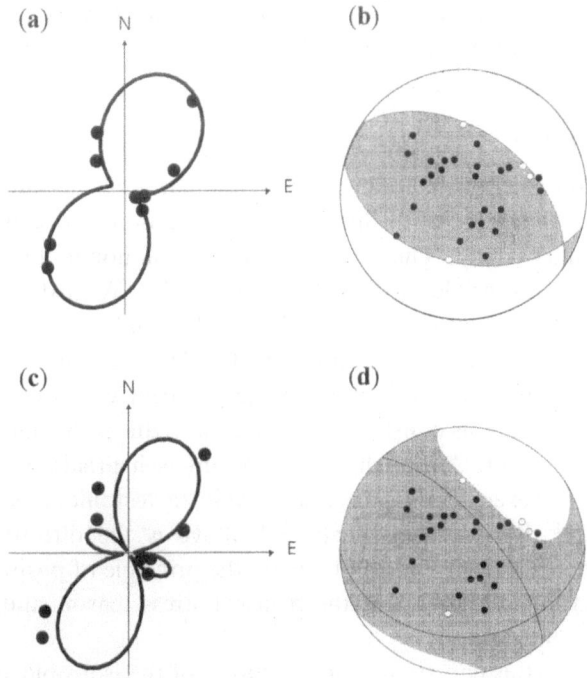

Figure 8

Same as Figure 7, but for the Yongden earthquake in Gansu province, China, on July 21, 1995 and 50 s period.

orientation. The principal compression axis is resolved better than the principal tension axis.

The number of surface wave records that we selected for the studied events was admittedly very small (see column 5 in Table 3), particularly for the earthquakes on the test site. For some events, only a few polarities of P-wave first arrivals were available. As a result, the double-couple focal mechanism was very well resolved not only for the earthquake on 21.01.90, but also for half of the events the principal compression axis, which trends in the SE-NW direction, was resolved well. Fortunately, GAO and RICHARDS (1994) estimated focal mechanisms of several earthquakes in the vicinity of Lop Nor. Considering the focal mechanisms of the studied events, and the solutions obtained by Gao and Richards, we found that while focal mechanisms vary widely, there is a moment tensor characteristic which is quite stable for all events. This characteristic is the orientation of the principal axes of the 2×2 minor of the moment tensor corresponding to the horizontal coordinates $\{M_{xx}, M_{xy}, M_{yx}, M_{yy}\}$. This minor describes the horizontal deformations on the horizontal plane. We found that for all events, horizontal compression is dominant, and the principal horizontal compression axis deviates from the average direction (with an azimuth of 60°) by no more than 12°. The direction of this axis for every event is given in the eighth column in Table 3.

4.2 Isotropic Component

These observations allow us to improve the estimates of the isotropic angle φ by assuming that the orientation of the horizontal deformations is a stable characteristic for the region under study. The average direction of horizontal compression is deviated from the fault visible in Figure 1 by about 45°. With this assumption, we applied an *a priori* constraint on the possible double-couple focal mechanisms. This constraint was formulated as a condition that the azimuth of the principal compression axis of the horizontal minor of the moment tensor cannot differ from 60° by more than ±45°. This constraint is weak, but reduces the number of possible focal mechanisms in half. Although the constraint is heuristic, we can present a special case when it is exact. This is when the deviatoric tectonic stresses in the region are horizontal and the direction mentioned above is the direction of principal compression. Then the constraint follows from the principle of positive deformation energy: $\sum_{i,j=1}^{3} S_{ij}M_{ij} \geq 0$, where **S** is the regional stress tensor, and **M** is moment tensor of event in the region.

When we applied this constraint, the resolution of the isotropic angle φ defining the seismic moments ratio was improved for three of the seven studied explosions (16.08.90, 05.10.93, and 08.06.96) but the mean of the curves did not vary appreciably. The curves of partial residual functions of the isotropic angle φ for all explosions are shown in Figure 5. Those for the three earthquakes are displayed in Figure 6. We also applied this method to two earthquakes that occurred in the Gansu province, China (Tianzhu, 1996, $M_s = 4.9$, depth 12 km, and Yongden, 1995, $M_s = 5.4$, depth 6 km) studied by LASSERRE *et al.* (2000) under the assumption of a pure double-couple source. The partial residual functions of the isotropic angle φ are also shown in Figure 6. The explosions, with perhaps one exception (05.10.93), all display substantial non-zero isotropic angles φ ranging from about 10° to 50°, corresponding to M_{0ex}/M_{0qu} ranging from about 20% to 120%. In contrast, the isotropic angles for the earthquakes are all estimated to be approximately 0.

Figure 7 illustrates the effect of adding an isotropic component to the double-couple source model for an explosion (on June 10, 1994). Theoretical Rayleigh wave amplitude spectra and *P*-wave first arrival polarities are compared here with observations for two different moment tensors. One (a, b) was obtained by minimizing the joint residual for a fixed zero isotropic angle (double-couple source model). The other (c, d) is a result of a similar minimization, but for a varying isotropic angle (30° is the estimated optimal value). The fit to *P*-wave first arrival polarities is good for both source models, however the fit to Rayleigh wave amplitude spectra is decidedly better in the case of a non-zero isotropic component.

Figure 8 illustrates the effect of adding an isotropic component for the Youngden earthquake. Similar to Figure 7, the results for two different moment tensors are presented here. One (a,b) was obtained by minimizing the joint residual for a varying isotropic angle. The other (c, d) is a result of a similar minimization but for a fixed (30°) isotropic angle. In contrast with the previous case the fit is better for a pure double-couple source.

5. Conclusions

The results above are consistent with the hypothesis that motivates this study. Namely, that for the events we analyzed on the Lop Nor test site, surface wave amplitude spectra combined with polarities of P-wave first arrivals can be used to discriminate explosions from earthquakes, based on source characteristics alone (a combination of the double-couple depth and the ratio of the isotropic to nonisotropic moments). Considerably more work remains to be done, however, to determine if the method can be applied to events with much smaller moments and if the method is transportable to other regions.

Acknowledgments

This research was supported by a NATO Linkage Grant N 9500775 between the University of Colorado at Boulder and the International Institute of Earthquake Prediction Theory and Mathematical Geophysics, Moscow, by a subcontract with the Department of Geological Sciences at Cornell University under EAR-9804859 from the US National Science Foundation, by the Russian Foundation of Fundamental Research Project 99-05-64964, the International Science and Technology Center Project 1293-99, by DSWA contract DSWA01-97-C-0157 to the University of Colorado at Boulder, by the US Arms Control and Disarmament Agency. We thank Alexander Lander and Howard Patton for helpful discussions and Istvan Bondar and Nikolai Shapiro for valuable reviews. Thanks to the IRIS-DMC and GEOSCOPE staffs for providing digital data.

REFERENCES

AKI, K., and TSAI, Y. (1972), *The Mechanism of Love Wave Excitation by Explosive Sources*, J. Geophys. Res. *77*, 1452–1475.

ARCHAMBEAU, C. B. (1972), *The Theory of Stress Wave Radiation from Explosions in Prestressed Media*, Geophys. J. *29*, 329–366.

BABICH, V. M., CHIKACHEV, B. A., and YANOVSKAYA, T. B. (1976), *Surface Waves in a Vertically Inhomogeneous Elastic Half-space with Weak Horizontal Inhomogeneity*, Izv. Akad. Nauk SSSR, Fizika Zemli *4*, 24–31.

BRUNE, J. N., and POMEROY, P. W. (1963), *Surface Wave Radiation Patterns for Underground Nuclear Explosions and Small-magnitude Earthquakes*, J. Geophys. Res. *68*, 5005–5028.

BUKCHIN, B. G. (1990), *Determination of Source Parameters from Surface Waves Recordings Allowing for Uncertainties in the Properties of the Medium*, Izv. Akad. Nauk SSSR, Fizika Zemli *25*, 723–728.

BUKCHIN, B. G., LANDER, A. V., MOSTINSKY, A. Z., and MAKSIMOV, V. I. (1997), *Determination of Seismic Source Parameters by Analysis of Coherence of Body Wave Phases*. Theoretical Problems in Geophysics. 29, Moscow (Comput. Seismol.) In (eds. V. I. KEILIS-BOROK, and G. M. MOLCHAN,) 3–17, 1997 (in Russian). English translation: Comput. Seismology and Geodynamics *4*, AGU, 2001, in press.

BURGER, R., LAY, T., WALLACE, T., and BURDICK, L. (1986), *Evidence of Tectonic Release in Long-period S Waves from Underground Nuclear Explosions at the Novaya Zemlya Test Sites*, Bull. Seismol. Soc. Am. *76*, 733–755.

DAY, S. M., and STEVENS, J. L. (1986), *An Explanation for Apparent Time Delays in Phase-reversed Rayleigh Waves from Underground Nuclear Explosions*, Geophys. Res. Lett. *13*, 1423–1425.

DAY, S. M., CHERRY, J. T., RIMER, N., and STEVENS, J. L. (1987) *Nonlinear Model of Tectonic Release from Underground Explosions*, Bull. Seism. Soc. Am. *77*, 996–1016.

EKSTRÖM, G., and RICHARDS, P. G. (1994), *Empirical Measurements of Tectonic Moment Release in Nuclear Explosions from Teleseismic Surface Waves and Body Waves*, Geophys. J. Int. *117*, 120–140.

GAO, L.-P., and RICHARDS, P. G. *Studies of Earthquakes on and near the Lop Nor, China, nuclear test site*, In *Proc. of the 16th Annual Seismic Research Symposium*, 7–9 Sept. 1994, 106–112, Phillips Lab., Directorate of Geophysics, 1994.

HARKRIDER, D. G., STEVENS, J. L., and ARCHAMBEAU, C. B. (1994), *Theoretical Rayleigh and Love Waves from an Explosion in Prestressed Source Regions*, Bull. Seismol. Soc. Am. *84*, 1410–1442.

HELLE, H. B., and RYGG, E. (1984), *Determination of Tectonic Release from Surface Waves Generated by Nuclear Explosions in Eastern Kazakhstan*, Bull. Seismol. Soc. Am. *74*, 1883–1898.

LANDER, A. V., *Frequency-time analysis*. In *Seismic Surface Waves in a Laterally Inhomogeneous Earth*. (V.I. Keilis-Borok, ed.), (Kluwer Acad. Publ., Dordrecht, 1989) pp. 153–163.

LASSERRE, C., BUKCHIN, B., BERNARD, P., TAPPONIER, P., GAUDEMER, Y., MOSTINSKY, A., and RONG DAILU (2001), *Source Parameters and Tectonic Origin of the June 1, 1996 Tianzhu (Mw = 5.2 and July 21, 1995 Yongden (Mw = 5.6) Earthquakes, near Haiyuan Fault (Gansu, China)*, Geophys. J. Intl. *144*, 206–220.

LEVSHIN, A. L. (1985), *Effects of Lateral Inhomogeneity on Surface Wave Amplitude Measurements*, Annales Geophysicae *3*(4), 511–518.

LEVSHIN, A. L., YANOVSKAYA, T. B., LANDER, A. V., BUKCHIN, B. G., BARMIN, M. P., RATNIKOVA, L. I., and ITS, E. N. *Seismic Surface Waves in Laterally Inhomogeneous Earth* (ed. V. I. Keilis-Borok) (Kluwer. Publ., Dordrecht, 1989)

LEVSHIN, A. L., RATNIKOVA, L. I., and BERGER, J. (1992), *Peculiarities of Surface Wave Propagation across the Central Eurasia*, Bull. Seismol. Soc. Am. *82*, 2464–2493.

LEVSHIN, A., RITZWOLLER M., and RATNIKOVA, L. (1994), *The Nature and Cause of Polarization Anomalies of Surface Waves Crossing Northern and Central Eurasia*, Geophys. J. Int. *117*, 577–591.

LEVSHIN, A. L., and RITZWOLLER, M. H. (1995), *Charactersistics of Surface Wave Generated by Events on and near the Chinese Nuclear Test Site*, Geophys. J. Intl. *123*, 131–148.

LI, Y., TOKSÖZ, M. N., and RODI, W. (1995), *Source Time Functions of Nuclear Explosions and Earthquakes in Central Asia Determined Using Empirical Green's Functions*, J. Geophys. Res. *100*, 659–674.

MASSE, R. P. (1981), *Review of Seismic Source Models for Underground Nuclear Explosions*, Bull. Seismol. Soc. Am. *71*, 1249–1268.

MUELLER, R. A., and MURPHY, J. R. (1971), *Seismic Characteristics of Underground Nuclear Detonations. Part I: Seismic Spectrum Scaling*, Bull. Seismol. Soc. Am. *61*, 1675–1692.

PATTON, H. J. *Seismic moment estimation and the scaling of the long-period explosion source spectrum*. In *Explosion Source Phenomenology, Geophysical Monograph 65, AGU* (eds. S. R. TAYLOR, H. J. PATTON, and P. G. RICHARDS) pp. 171–184, 1991.

PAULSSEN, H., LEVSHIN, A. L., LANDER, A. V., and SNIEDER, R. (1990), *Time and Frequency Dependent Polarization Analysis: Anomalous Surface Wave Observations in Iberia*, Geophys. J. Int. *103*, 483–496.

PRESS, F., and ARCHAMBEAU, C. B., (1962) *Release of Tectonic Strain by Underground Explosions*, J. Geophys. Res. *67*, 337–343.

STEVENS, J. L. (1986), *Estimation of Scalar Moments from Explosion-generated Surface Waves*, Bull. Seismol Soc. Am. *76*, 123–151.

STEVENS, J. L., BARKER, T. G., DAY, S. M., MCLAUGHLIN, K. L., RIMER N., and SHKOLLER, B. *Simulation of teleseismic body waves, regional seismograms, and Rayleigh wave phase shifts using two-dimensional nonlinear models of explosion sources*. In *Explosion Source Phenomenology, Geophysical Monograph 65, AGU* (eds. S. R. TAYLOR, H. J. PATTON, and P. G. RICHARDS) pp. 239–252, 1991.

TOKSÖZ, M. N., and KEHRER, H. H. (1972), *Tectonic Strain Release by Underground Nuclear Explosions and its Effect on Seismic Discrimination*, Geophys. J. R. Astr. Soc. *31*, 141–161.

WALLACE, T. C., HELMBERGER, D. V., and ENGEN, G. R. (1985), *Evidence of Tectonic Release from Underground Nuclear Explosions in Long-period S Waves*, Bull. Seismol. Soc. Am. *76*, 123–151.

WALLACE, T. C., *Body wave observations of tectonic release*. In *Explosion Source Phenomenology, Geophysical Monograph 65, AGU* (eds. S. R. TAYLOR, H. J. PATTON, and P. G. RICHARDS) pp. 161–170, 1991.

WALLACE, T. C., and TINKER, M. A. (1996), *The Last Nuclear Weapons Test? A Brief Review of the Chinese Nuclear Weapons Program*, IRIS Newsletter *15* (3).

WALTER, W. R., and PATTON, H. J., (1990) *Tectonic Release from SJVE*, Geophys. Res. Lett. *17*, 1517–1520.

WOODHOUSE, J. H. (1974), *Surface Waves in the Laterally Varying Structure*, Geophys. J. R. Astr. Soc. *90*, 713–728.

ZHANG, J. *Polarization characteristics of seismic waves from the May 21, 1992 Lop Nor nuclear explosion using IRIS/GSN broadband data*. In *Proc. of the 16th Annual Seismic Research Symposium*, 7–9 Sept. 1994, 393–397, Phillips Lab., Directorate of Geophysics, 1994.

(Received June 25, 1999, revised March 20, 2000, accepted May 15, 2000)

Pure appl. geophys. 158 (2001) 1517–1530
0033–4553/01/081517–14 $ 1.50 + 0.20/0

⌷ Pure and Applied Geophysics

Theoretical and Observed Depth Correction for M_s

MARIJAN HERAK,[1] GIULIANO F. PANZA,[2,3] and GIOVANNI COSTA[2]

Abstract — Modal summation technique is used to generate 5000, three-component theoretical seismograms of Love and Rayleigh waves, assuming modified PREM (PREM-C) and AK135F global earth models. The focal depth h and the geometrical fault parameters are randomly chosen so as to uniformly cover possible source mechanisms and obtain uniform distribution of log h in the interval $1 < h < 600$ km. The amplitudes of 20 ± 2-s waves measured on each of the synthetic seismograms yield curves of amplitude vs. depth, and consequently the theoretical surface waves magnitude depth correction for all three components of ground motion. Predicted surface wave amplitudes are practically constant for $h < 20$ km, then decrease with the focal depth. This decrease is not uniform, and depends on the excitation level of higher modes. For PREM-C model and for shallow sources, computed Love wave amplitudes are nearly an order of magnitude larger than those of Rayleigh waves, which is why the AK135F model is given preference over PREM-C. The theoretical depth correction ranges between zero, for shallow sources, to about $+1$ magnitude unit, for the deepest ones. The theoretical results are compared with 74,480 individual station measurements of 20-s surface wave amplitudes reduced to the same distance, period and seismic moment. It is found that empirical data closely match the predictions made by using the AK135F model. Considering both theoretical modeling and observational data, we propose the introduction of a piecewise log-linear depth correction for M_s of the form:

$$\begin{aligned}
\Delta M_s(h) &= 0 & &\text{for } h < 20\,\text{km,} \\
\Delta M_s(h) &= 0.314 \log(h) - 0.409 & &\text{for } 20 \leq h < 60\,\text{km,} \\
\Delta M_s(h) &= 1.351 \log(h) - 2.253 & &\text{for } 60 \leq h < 100\,\text{km,} \\
\Delta M_s(h) &= 0.400 \log(h) - 0.350 & &\text{for } 100 \leq h < 600\,\text{km .}
\end{aligned}$$

After applying the above correction, the relationship between the surface wave magnitude and the scalar seismic moment for the observational data set significantly improves, and becomes independent of the source depth. In relation to CTBT, no depth correction is needed for M_S when the $m_b - M_S$ discriminant is computed, because the proposed correction is zero for earthquakes with foci above 20 km.

Key words: Earthquake magnitude, depth correction, M_S, surface waves, earth models, focal depth, $m_b - M_S$ discriminant.

---

[1] Department of Geophysics, Faculty of Science, University of Zagreb, Horvatovac bb, 10000 Zagreb, Croatia. E-mail: herak@rudjer.irb.hr
[2] Department of Earth Sciences, Via Weiss 1, 34127 Trieste, Italy.
[3] Abdus Salam International Center for Theoretical Physics, SAND Group, 34100 Trieste, Italy.

1. Introduction

The surface wave magnitude, M_S, is one of the commonly used measures for the quantification of earthquakes. The M_S scale, as originally defined by GUTENBERG (1945), is based on the measurements of 20-s surface waves from shallow events. The standard formula for the computation of M_S – the Moscow-Prague formula (KARNIK *et al.*, 1962; VANEK *et al.*, 1962):

$$M_s = \log(A/T)_{\max} + \sigma(D) = \log(A/T)_{\max} + 1.66\log(D) + 3.3 \qquad (1)$$

is used for M_S estimation by worldwide agencies like ISC or USGS. In the above expression $(A/T)_{\max}$ is the maximum of the ground motion [amplitude (μm)/period (s)] ratio for the waves with periods near 20 s, and D is the epicentral distance, in degrees. It has been shown, however, both theoretically (PANZA *et al.*, 1989) and empirically (e.g., HERAK and HERAK, 1993; REZAPOUR and PEARCE, 1998) that the calibrating function $\sigma(D)$ in (1) yields distance biased M_S.

The fact that M_S is computed only for shallow events (with focal depths h less than 60 and 50 km in the case of ISC and USGS agencies, respectively) severely limits the use of M_S as the standard for comparing the strengths of earthquakes worldwide. This shortcoming may be overcome by introducing a depth correction for M_S, replacing $\sigma(D)$ in (1) by $\sigma(D, h)$, similarly to the m_b calibrating function.

The problem of defining the M_S depth correction was theoretically discussed for Love waves by PANZA *et al.* (1989), and by e.g. AMBRASEYS and FREE (1997) who studied empirical data on European earthquakes. PANZA *et al.* (1989) examined theoretical amplitudes of the fundamental mode 20-s Love waves, assuming the validity of the PREM global model (DZIEWONSKI and ANDERSON, 1981). The source was represented by a pure strike-slip on a fault dipping at 45°, and synthetic seismograms were computed with azimuth increments of 45°. The epicentral distance ranged from 1° to 180°, with assumed focal depths between the surface and 80 km. Their main conclusions are: a) The theoretical attenuation of amplitudes with distance differs from the standard calibrating functions; b) depth-dependent excitation for 20-s Love waves necessitates the introduction of the M_S depth correction ranging from −0.4 to +0.7 for focal depths between 5 and 80 km; and c) introduction of the depth correction for M_S contributes to better separation of nuclear explosions and earthquakes in the m_b vs. M_S diagrams.

Here we aim to use the same approach to re-examine and extend some of these conclusions by:

1) considering another global structural model (AK135F) in addition to PREM,
2) synthesizing Rayleigh waves as well as Love waves, with all relevant higher modes,
3) analyzing a large number of focal mechanisms,
4) extending the focal depth range to 600 km, and
5) comparing theoretical results with observations.

2. Seismogram Synthesis and Results

The Preliminary Reference Earth Model (PREM) (DZIEWONSKI and ANDERSON, 1981) is compiled along the guidelines established by the Standard Earth Model Committee in 1975, by considering a large data set of normal mode periods, travel-time observations, Q values and data on the earth's mass, moment of inertia, density distribution, etc. In order to satisfy the observational data – in particular those related to body shear waves and short-period ($T < 200$ s) surface waves – the authors had to drop the assumptions of isotropy. Instead, they proposed the upper 200 km of the mantle to be transversely isotropic (with the vertical symmetry axis), which removed the Love wave-Rayleigh wave discrepancy, and improved the overall fit to the data. The required anisotropy is about 2–4% for both P and S waves. In order to enable both Love and Rayleigh wave synthesis at the free surface we assumed a continental crust instead of the oceanic type crust given in the original formulation (Fig. 1). We shall therefore further refer to the model as PREM-C.

The AK135 velocity model (KENNETT et al., 1995) was designed to provide a good fit to a wide variety of seismic phases, and represents an improvement on the IASP91 model (KENNETT and ENGDAHL, 1991). The AK135F variant of this model is supplemented with a density and Q model by combining the study of travel times with those of free oscillations (MONTAGNER and KENNETT, 1995). This is an isotropic model with Q values needed to bring the 1 Hz travel times into a match with free oscillations, and gives good fit to the observed Q values for normal modes.

The two models are presented in Figure 1. It is seen that, unlike PREM-C, there is no upper mantle low-velocity zone in AK135F. This is reflected in the complex density structure in the upper mantle of AK135F with a density inversion. The attenuation is larger in AK135F above 400 km, and in PREM-C below this depth.

In order to theoretically examine the effect of focal depth on the 20-s surface waves generation, we use the modal summation technique (PANZA, 1985; FLORSCH et al., 1991) to generate synthetic seismograms consisting of the first 45 modes of Love and Rayleigh waves (down to the period of 10 s) assuming the PREM-C and AK135F models. For each of the models 5000 three-component synthetic seismo-grams were computed. The source depth for each of them was randomly chosen so as to obtain a uniform distribution of $\log(h)$ in the range 1–600 km. The strike, dip and rake of the causative fault were also randomly selected to uniformly represent possible double-couple sources. Since relative depth dependence of surface wave amplitudes is independent of epicentral distance (PANZA et al., 1989) all computa-tions have been made, with no loss of generality, assuming the constant distance of 2000 km. For such epicentral distance it is, in fact, possible to use the very fast codes for flat geometry (PANZA, 1985; FLORSCH et al., 1991), while keeping, in the period range of interest, an absolute accuracy of at least two digits with respect to the signals computed in the spherical geometry. The irrelevance of the distance chosen has been verified by considering a smaller set of synthetics generated for larger distances. All

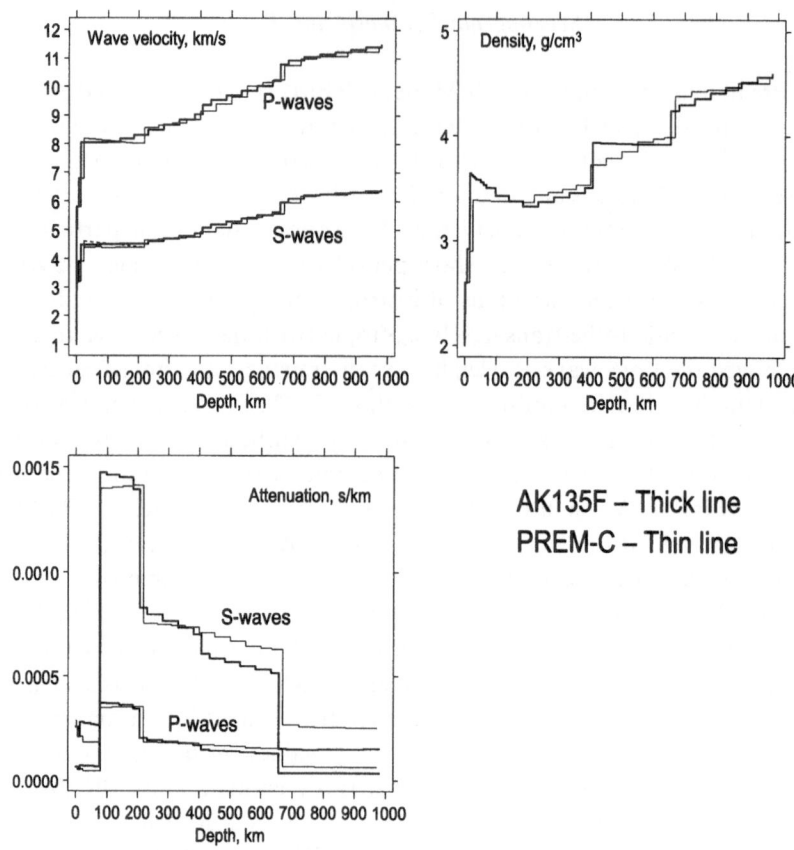

Figure 1

Depth profiles (for $h < 1000$ km) of body-wave velocities, density and attenuation coefficient (equal to $1/(2cQ)$, $c = P$- or S-wave phase velocity, $Q =$ respective quality factor) for AK135F (thick line) and PREM-C (thin line). For PREM-C, solid and dashed lines indicate horizontal and vertical components of S-wave velocities in the upper mantle, respectively. PREM-C is the same as PREM (DZIEWONSKI and ANDERSON, 1981), with the exception of the continental crust instead of the oceanic one.

seismograms are computed for a point source with the constant scalar seismic moment. The largest amplitudes of 20 ± 2 s waves are measured on each of the synthesized seismograms, and plotted against h as is shown in Figure 2. In the figure the amplitudes, A, are scaled so as to obtain the mean log $A = 0$ for the vertical and radial components of motion in case of shallow sources. In this way the average curves shown in the figure can be interpreted as the negative theoretical depth corrections for M_S for the respective components. Comparison of the results obtained for AK135F and PREM-C models leads to the following conclusions:

1) The amplitudes of 20 ± 2 s waves are practically independent of source depth for shallow sources ($h < 30$ km for Rayleigh waves and $h < 15$ km for Love waves), for both models.

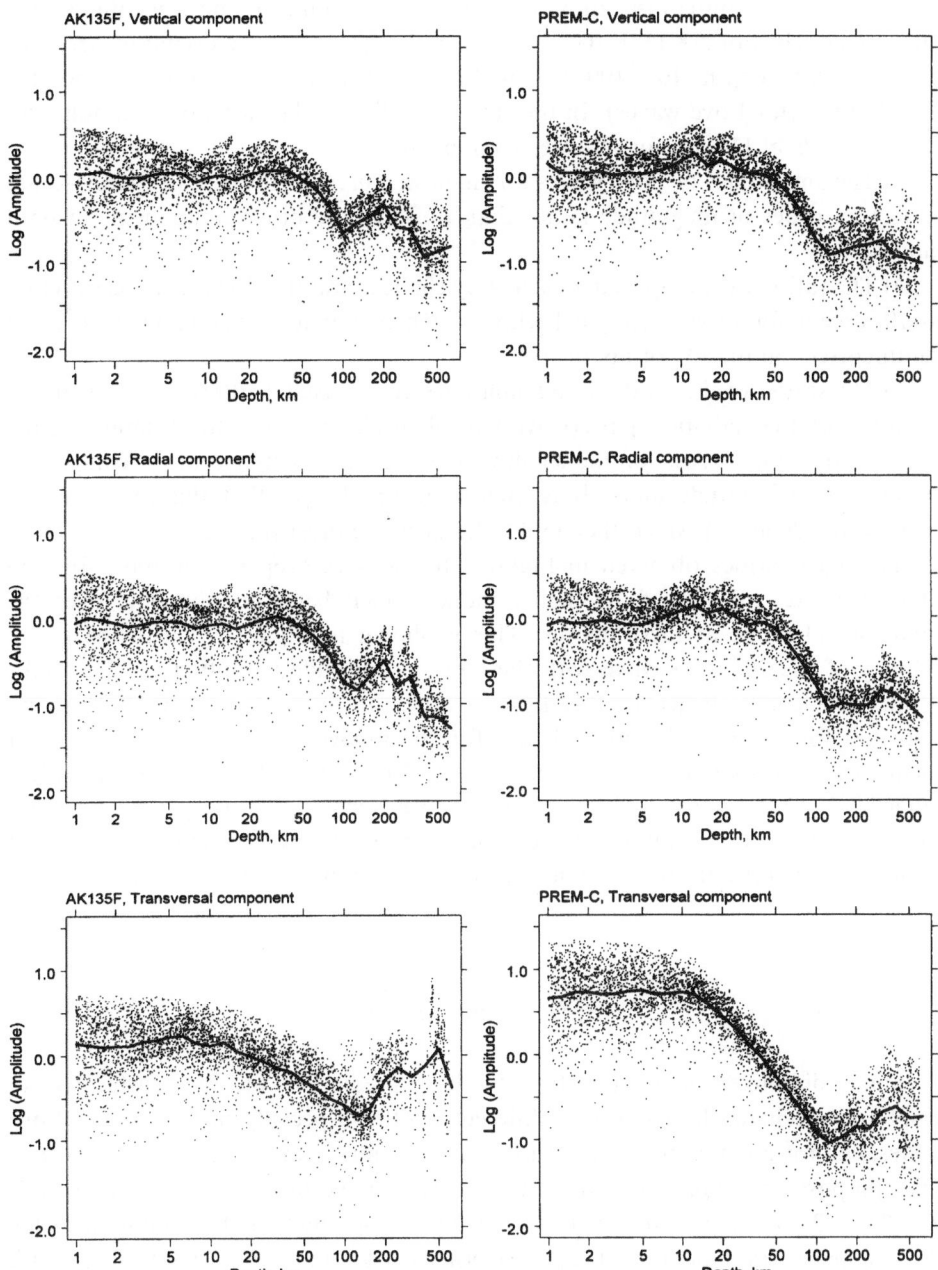

Figure 2

Theoretical depth dependence of 20-s Rayleigh and Love wave amplitudes for AK135F and PREM-C models for 5000 sources with randomly chosen fault mechanisms. Thick line is the average measured log(amplitude). See text for details.

2) For deeper sources the amplitude decay is not uniform, and it is different for the two models. For AK135F the efficiency of 20-s surface wave excitation seems to increase with depth for $100 < h < 200$ km (for Rayleigh waves) and for $h > 120$ km (for Love waves). In the case of PREM-C the variation of amplitudes is small for $h > 110$ km for all three components.

3) The amplitudes of Love waves for shallow sources in the case of PREM-C are almost an order of magnitude larger than those of the other components in both models.

4) The ratio of the amplitudes excited by shallow and the deepest sources is close to 10, except for Love waves in PREM-C, which implies a maximum theoretical depth correction for M_S of about $+1$.

5) The scatter of the synthesized amplitudes at any fixed depth is solely due to the variation of the radiation pattern with focal mechanism and the azimuth of the receiver. It is the largest for the shallow and the deepest hypocenters, comprising about ± 0.5 magnitude units. It is interesting to observe that this scatter is the smallest for depths at which the amplitudes decrease most rapidly.

The irregularities observed in Figure 2 for sources deeper than about 100 km (item 2 above) are caused by higher modes which begin to dominate over the fundamental mode at these depths. This is shown in Figure 3, where the group velocity of the 20 ± 2 s wave with the largest amplitude is plotted vs. the source depth. The empty boxes in this figure define the interval of theoretical group velocities for the period range 18–22 s for the mode(s) indicated beside each of them. It is seen that the fundamental mode (1) dominates for $h < 100$ km, while modes 2–12 are responsible for the largest waves generated by deeper earthquakes. Modes higher than 12, although used when synthesizing the seismograms, do not significantly contribute to the signal in the period range considered, regardless of the focal depth.

3. Observational Data

Some disagreement in the two models' prediction of 20-s surface wave amplitudes radiated by sources at different depths, especially in the case of Love waves, call for a comparison of the theoretical results with observational data. For this purpose we take the amplitudes, A_T, of surface waves with periods $T = 20 \pm 2$ s for 2752 earthquakes from the period 1990–1994 as reported in the ISC database (INTERNATIONAL SEISMOLOGICAL CENTRE, 1996, 1997), along with the data on the focal depth h, epicentral distance, D, of reporting stations, and the Harvard CMT scalar seismic moment, M_0. A total of 74,480 amplitude readings on all three components have been collected, and reduced to the common period and reference distance using the 20-s surface wave calibrating function proposed by HERAK and HERAK (1993):

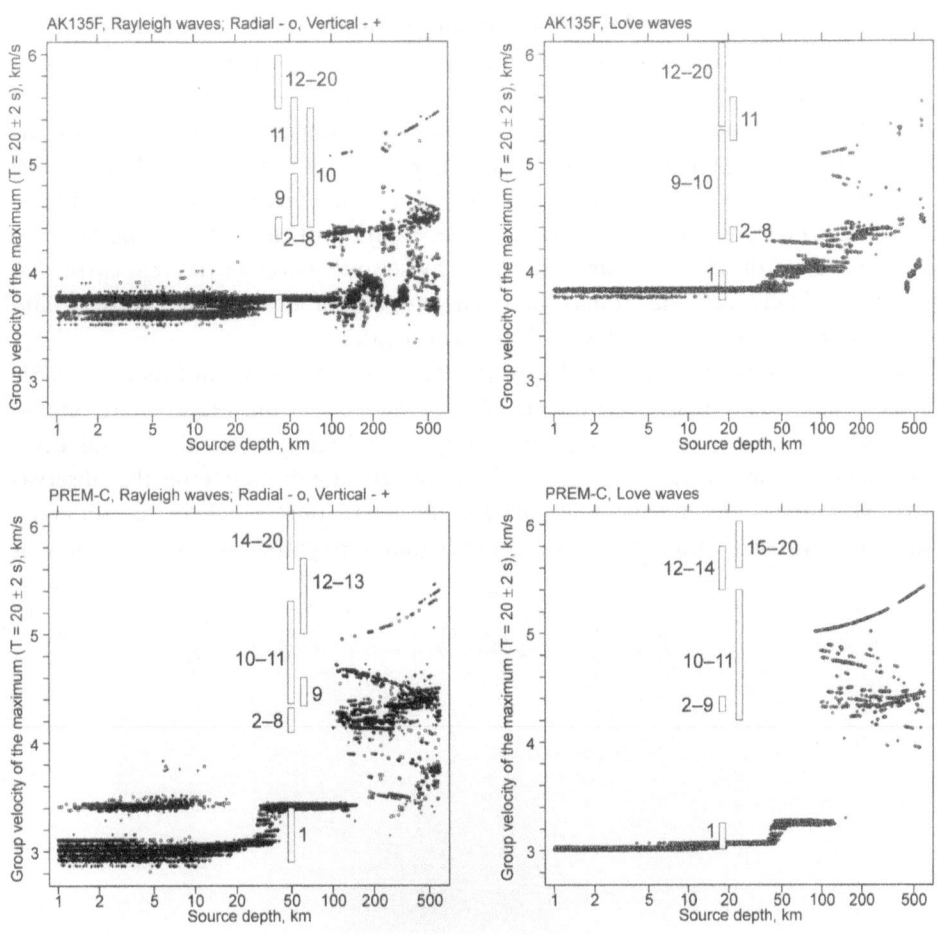

Figure 3
Group velocities of the largest wave with the period between 18 and 22 s for 5000 three-component
synthetic seismograms, plotted against the assumed source depth. The empty boxes denote the theoretical
group velocity interval for this period range and the mode number(s) indicated beside each of them, for the
AK135 (top) and PREM-C (bottom) models. The fundamental mode is marked as 1.

$$A_{20,100} = A_T \frac{20}{T} \left(\frac{D}{100} \right)^{1.094},$$

where $A_{20,100}$ is the amplitude reduced to $T = 20$ s and $D = 100°$. This formula is
chosen instead of the standard Moscow-Prague calibrating function because the
later yields distance-biased results for the 20 s waves (HERAK and HERAK, 1993;
AMBRASEYS and FREE, 1997; REZAPOUR and PEARCE, 1998). The amplitudes are
then scaled to the same seismic moment by assuming the validity of the ω^2 model
of AKI (1967).

Figure 4 shows the set of reduced amplitudes shifted along the ordinate so as to obtain the mean logarithm of the amplitude equal to zero for the shallowest sources. The bulk of data is concentrated in the depth range 10–60 km, as the consequence of inefficient generation of surface waves by deep earthquakes, and because of the practice that 20-s wave amplitudes are reported routinely only for shallow events. The scatter of observations around the mean is somewhat larger than in the synthetic cases (Fig. 2) because here, in addition to the variation of the radiation pattern, we see the effect of diverse structures traversed by seismic waves in the real earth. No systematic differences are found between the amplitudes of the vertical and horizontal components, regardless of the focal depth.

The observations are compared with the theoretical modeling in Figure 5, where the mean curve as well as 1 σ confidence limit boundaries from Figure 4 are plotted over the measured synthetic amplitudes shown in Figure 2 (jointly for all three components). The AK135F model produces results closely matching the observations. The largest disagreement is seen for the deep sources, where the observed amplitudes decrease with depth more regularly than the synthetic ones. The match of

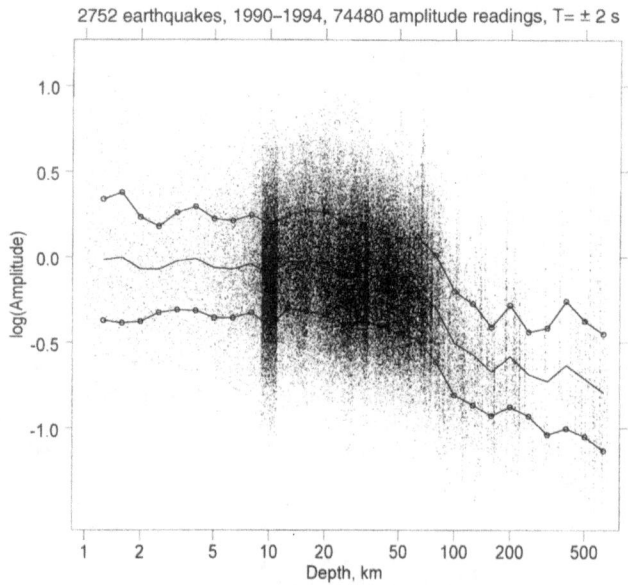

Figure 4

74,480 observational data from individual stations of the (20 ± 2) s surface wave amplitudes reduced to the same period, epicentral distance and seismic moment (for details see text) plotted against the depth of the hypocenter as reported by the INTERNATIONAL SEISMOLOGICAL CENTRE (1996, 1997). The strong concentration of events around 10 and 33 km is the consequence of depths being fixed by ISC. It is seen that no systematic mislocation of depth occurred in these cases. The artificial random jitter of the depth coordinate of ±0.5 km is introduced to prevent overplotting of symbols at integer depths. Superimposed (thick solid lines with circles) are the 1 σ confidence limits and the mean of log(amplitude) (solid line).

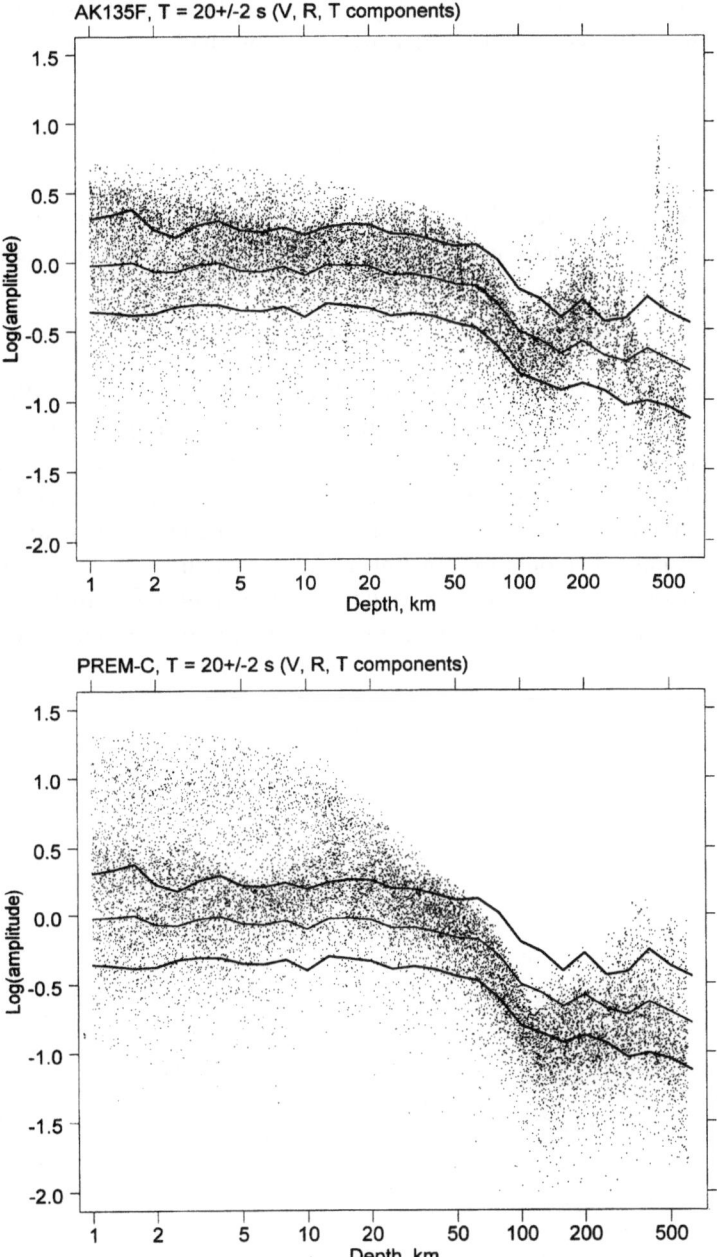

Figure 5

Mean and $\pm 1\ \sigma$ confidence bounds for observed amplitudes (from Fig. 4) plotted over the theoretical amplitudes (dots, jointly for all three components, from Fig. 2) for the AK135F (top) and PREM-C (bottom) models.

theory and observations is considerably poorer in the case of PREM-C, mostly because of the too large excitation of Love waves by shallow sources.

4. The M_S Depth Correction

The M_S depth correction, ΔM_S, is defined as the logarithm of the ratio (A_0/A_h) of the amplitudes of 20 s waves generated by the same source at the surface and at the depth h:

$$\Delta M_s(h) = \log(A_0/A_h) = \log(A_0) - \log(A_h) \ .$$

Obviously $\Delta M_S(h)$ is equal to the mean curves in Figures 2 and 4 with their sign reversed. They are plotted together in Figure 6. For depths larger than about 50 km the curve representative of PREM-C systematically deviates from observations by more than 1 σ and is therefore considered unacceptable. The AK135F mean theoretical curve closely follows the mean observed one everywhere within the 1 σ confidence bounds. The depth correction is zero for sources shallower than about 20 km, and then increases to approximately $+0.75$ magnitude units for the deepest foci. This increase seems to be the steepest for depths in the interval 60–100 km.

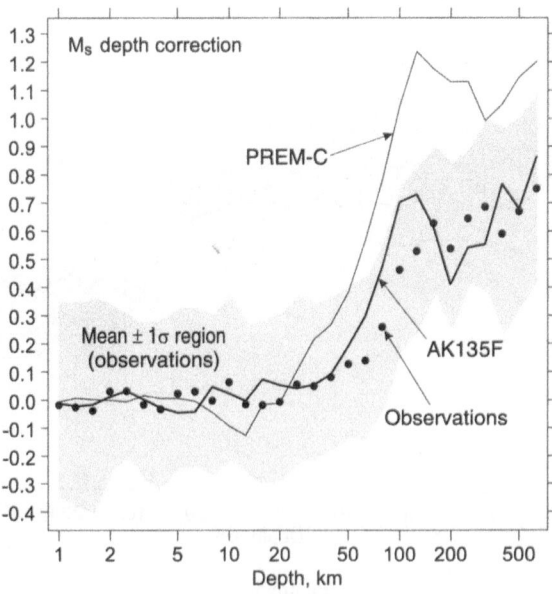

Figure 6

Mean (of all three components) depth corrections based on theoretical modeling assuming the AK135F (thick line) and PREM-C (thin line) models, and observations (solid dots). Shaded is the $\pm 1 \sigma$ confidence region for observational data.

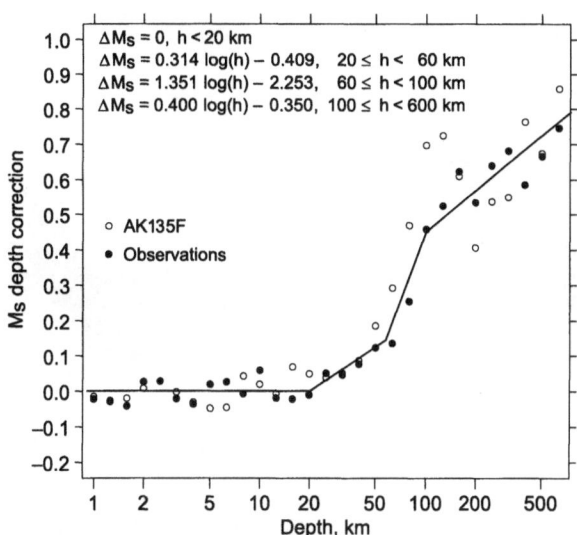

Figure 7

M_s depth correction based on theoretical modeling (AK135F model) and on observations. Solid line represents the proposed M_s depth correction.

Given the rather wide confidence regions for the mean observed as well as mean theoretical curves, for practical use we propose to describe the depth correction by a three-fold log-linear relationship for $h \geq 20$ km, i.e.,

$$\begin{aligned}
\Delta M_s(h) &= 0 &&\text{for } h < 20\,\text{km,}\\
\Delta M_s(h) &= 0.314\log(h) - 0.409 &&\text{for } 20 \leq h < 60\,\text{km,}\\
\Delta M_s(h) &= 1.351\log(h) - 2.253 &&\text{for } 60 \leq h < 100\,\text{km,}\\
\Delta M_s(h) &= 0.400\log(h) - 0.350 &&\text{for } 100 \leq h < 600\,\text{km ,}
\end{aligned}$$

as shown in Figure 7. This correction is applied to the ISC data set described above. Figure 8 shows that, with the depth correction applied, the scatter of observations in the relation between M_S and the logarithm of the seismic moment (M_0) is significantly reduced if compared to the uncorrected data. Regressing M_S against $\log(M_0)$ and $\log(h)$ we obtain results presented in Table 1.

The coefficient b, as well as the respective correlation coefficient, $r(M_S, \log h)$, are both very small regardless of the M_S interval considered, which indicates that the proposed correction effectively removed the influence of focal depth on M_S. Correlation coefficients between M_S and $\log M_0$ with the depth correction applied are all significantly larger than for the uncorrected M_S. These results represent improvement of the early finding of ROMANELLI and PANZA (1995). Coefficients a and c systematically decrease with an increasing lower M_S threshold, which means that, at least for our data set, a linear relationship between M_S and $\log(M_0)$ is not appropriate. For M_S in the range from about 5.5 to 8.1 the coefficients a and c are

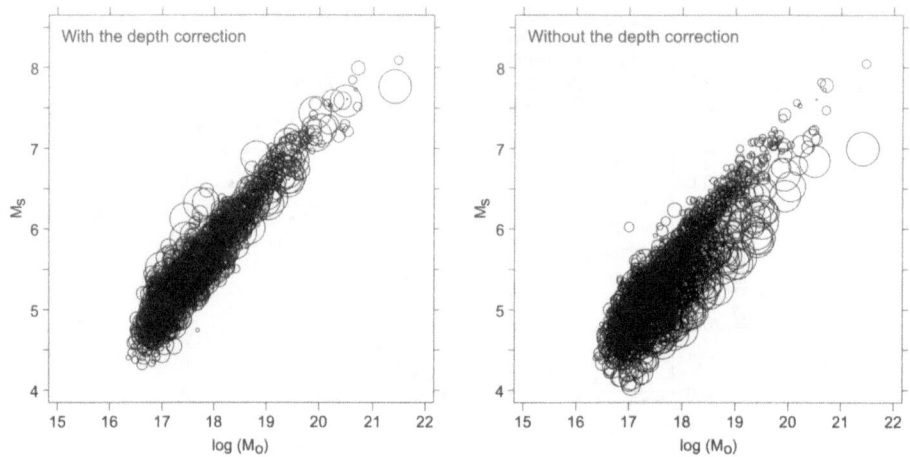

Figure 8

Comparison of the surface wave magnitude, M_s, and the logarithm of the seismic moment, $\log(M_0)$, before (right) and after (left) application of the depth correction from Figure 7. A total of 2752 events from the period 1990–1994 are considered (see text). The symbol size logarithmically increases with the focal depth.

Table 1

Regression coefficients in the formula $M_s = a \log M_0 + b \log h - c$. $r(M_s, \log M_0)$ and $r(M_s, \log h)$ are correlation coefficients. Regressions are made for depth-corrected M_s in the interval given in the first column. N is the number of data for each regression. The values in parentheses correspond to uncorrected M_s

Interval of M_s	a	b	c	N	$r(M_s, \log M_0)$	$r(M_s, \log h)$
4.5–8.1	0.745	+0.03	7.69	2734	0.94 (0.89)	0.09
5.0–8.1	0.699	+0.02	6.83	2198	0.93 (0.87)	0.06
5.5–8.1	0.649	+0.01	5.84	1059	0.92 (0.83)	0.01
6.0–8.1	0.580	+0.01	4.45	429	0.90 (0.76)	0.04
6.5–8.1	0.550	−0.02	3.75	160	0.90 (0.65)	0.04
7.0–8.1	0.427	−0.01	1.21	52	0.80 (0.51)	0.02

similar to those in the definition of the moment magnitude, $M_w = 0.667$ $\log(M_0) - 6.0$ (KANAMORI, 1977).

5. Discussion and Conclusions

The proposed depth correction for M_S may be compared to the one estimated from observational data for the European region by AMBRASEYS and FREE (1997). Their correction factor, $\Delta M_S(h) = 0.0036(h - 30)P$ ($P = 0$ for $h < 30$ km, $P = 1$ for $h \geq 30$ km) is linear with h and gives ΔM_S between 0 and +0.98 in the depth range 30–300 km. This is higher than $\Delta M_S = +0.64$ obtained here for $h = 300$ km.

Comparing the amplitude depth dependence as reported here with the earlier theoretical results for the PREM-C model and Love waves (PANZA *et al.*, 1989) one finds that this study points to much less pronounced influence of the focal depth (in the range 5–80 km considered in PANZA et al., 1989) even for the PREM-C model. This is due to the choice of the focal mechanisms made by PANZA et al. (1989), which is shown here not to be representative of the average behavior.

In relation to the Comprehensive-Test-Ban Treaty (CTBT), our findings show that no depth correction is needed for M_S when the $m_b - M_S$ discriminant is computed because the proposed correction is zero for earthquakes with foci above 20 km. Furthermore, station correction for M_S should be strictly based on shallow events ($h < 20$ km), unless the depth correction is applied.

The comparison of experimental data with the theoretical amplitude-focal depth relationships clearly gives preference to the AK135F global model over PREM-C. In view of the considerably different performance of these two earth models, the ability to enable realistic simulation of the surface wave excitation for sources at various depths could be introduced as one of the criteria of the model's validity.

In theoretical modeling it is assumed that all types of focal mechanisms are equally probable, regardless of the source depth. LEVSHIN et al. (1999), however, demonstrate that thrust mechanisms prevail for earthquakes with hypocenters below 20 km. If M_S depends on the type of faulting (see e.g., LEVSHIN and GRUDEVA, 1974), this may have produced bias in our results. Good agreement of theory and observations presented here indicates that, if such a bias exists, it is smaller than resolution of empirical data. Possible influence of predominant focal mechanism and tectonic setting on the M_S depth correction will be the topic of a forthcoming study.

The introduction of depth correction for M_S enables computation of surface wave magnitude for all earthquakes, regardless of their focal depth. This is especially important for the quantification of deep historical earthquakes, for which the seismic moment may be difficult to estimate from recordings of early mechanical seismographs. We therefore propose adoption of the new M_S calibrating function, based on this and previously cited studies, which would yield both distance- and depth-independent magnitude estimates.

Acknowledgments

We wish to thank Professors Anatoli Levshin, Lowell Whiteside and an anonymous reviewer for their most helpful comments on the first version of the manuscript. This study was supported by the Ministry of Science and Technology of the Republic of Croatia (project No. 119298), by the GNDT grants CNR 96.00318.05, CNR 98.0244.PF05, CNR 96.02968.PF54, and by the EU contract ENV4-CT96-0296. All support is gratefully acknowledged.

REFERENCES

AKI, K. (1967), *Scaling Law of Seismic Spectrum*, J. Geophys. Res. *65*, 729–740.

AMBRASEYS, N. N., and FREE, M. W. (1997), *Surface-wave Magnitude Calibrating Function for European Region Earthquakes*, J. Earthq. Engin. *1*, 1–22.

DZIEWONSKI, A. M., and ANDERSON, D. L. (1981), *Preliminary Reference Earth Model*, Phys. Earth and Planet. Inter. *25*, 297–356.

FLORSCH, N., FÄH, D., SUHADOLC, P., and PANZA, G. F. (1991), *Complete Synthetic Seismograms for High-frequency Multimode SH Waves*, Pure appl. geophys. *136*, 529–560.

GUTENBERG, B. (1945), *Amplitudes of Surface Waves and Magnitude of Shallow Earthquakes*, Bull. Seismol. Soc. Am. *35*, 3–12.

HERAK, M., and HERAK, D. (1993), *Distance Dependence of M_s and Calibrating Function for 20 Second Rayleigh Waves*, Bull. Seismol. Soc. Am. *83*, 1881–1892.

INTERNATIONAL SEISMOLOGICAL CENTRE (1996), *ISC Bulletins 1986–1991, 1992–1993*, CD-ROM.

INTERNATIONAL SEISMOLOGICAL CENTRE (1997), *ISC Bulletins 1964–1977, 1978–1985, 1994*, CD-ROM.

KANAMORI, H. (1977), *The Energy Release in Great Earthquakes*, J. Geophys. Res. *8*, 2981–2987.

KARNIK, V., KONDORSKAYA, N. V., RIZNITCHENKO, Y. V., SAVARENSKY, E. F., SOLOV'EV, S. L., SHEBALIN, N. V., VANEK, J., and ZATOPEK, A. (1962), *Standardization of the Earthquake Magnitude Scale*, Studia Geophysica et Geodaetica *6*, 41–47.

KENNETT, B. L. N., and ENGDAHL, E. R. (1991), *Traveltimes for Global Earthquake Location and Phase Identification*, Geophys. J. Internat. *105*, 429–465.

KENNETT, B. L. N., ENGDAHL, E. R., and BULLAND, R. (1995), *Constraints on Seismic Velocities in the Earth from Traveltimes*, Geophys. J. Internat. *122*, 108–124.

LEVSHIN, A. L., and GRUDEVA, N. P., *Some problems of the magnitude theory*. In *Magnitude and Energy of Earthquakes, I* (Nauka, Moscow, 1974), pp. 172–180 (in Russian).

LEVSHIN, A. L., RITZWOLLER, M. H., and RESOVSKY, J. S. (1999), *Source Effects on Surface Wave Group Travel Times and Group Velocity Maps*, Phys. Earth and Planet. Inter. in press.

MONTAGNER, J. P., and KENNETT, B. L. N. (1995), *How to Reconcile Body-wave and Normal-mode Reference Earth Models?* Geophys. J. Internat. *125*, 229–248.

PANZA, G. F. (1985), *Synthetic Seismograms: The Rayleigh Waves Modal Summation*, J. Geophys. *58*, 125–145.

PANZA, G. F., DUDA, S. J., CERNOBORI, L., and HERAK, M. (1989), *Gutenberg's Surface-wave Magnitude Calibrating Function: Theoretical Basis from Synthetic Seismograms*, Tectonophys. *166*, 35–43.

REZAPOUR, M., and PEARCE, R. G. (1998), *Bias in Surface Magnitude M_s Due to Inadequate Distance Corrections*, Bull. Seismol. Soc. Am. *88* (1), 43–61.

ROMANELLI, F., and PANZA, G. F. (1995), *Effect of the Source Depth Correction on the Estimation of Earthquake Size*, Geophys. Res. Lett. *22* (9), 1017–1019.

VANEK, J., ZATOPEK, A., KARNIK, V., KONDORSKAYA, N. V., RIZNICHENKO, Y. V., SAVARENSKY, E. F., SOLOV'EV, S. L., and SHEBALIN, N. V. (1962), *Standardization of Magnitude Scales*, Bull. Acad. Sci. USSR, Geophys. Ser. (English translation), *No. 2*, 108–111.

(Received April 26, 1999, revised September 22, 1999, accepted January 15, 2000)

To access this journal online:
http://www.birkhauser.ch

Pure appl. geophys. 158 (2001) 1531–1545
0033–4553/01/081531–15 $ 1.50 + 0.20/0

┃Pure and Applied Geophysics

Automated Detection, Extraction, and Measurement of Regional Surface Waves

A. L. Levshin[1] and M. H. Ritzwoller[1]

Abstract — Our goal is to develop and test an effective method to detect, identify, extract, and quantify surface wave signals for weak events observed at regional stations. We describe an automated surface wave detector and extractor designed to work on weak surface wave signals across Eurasia at intermediate periods (8 s–40 s). The method is based on phase-matched filters defined by the Rayleigh wave group travel-time predictions from the broadband group velocity maps presented by Ritzwoller and Levshin (1998) and Ritzwoller *et al.* (1998) and proceeds in three steps: Signal compression, signal extraction or cleaning, and measurement. First, the dispersed surface wave signals are compressed in time by applying an anti-dispersion or phase-matched filter defined from the group velocity maps. We refer to this as the 'compressed signal.' Second, the surface wave is then extracted by filtering 'noise' temporally isolated from the time-compressed signal. This filtered signal is then redispersed by applying the inverse of the phase-matched filter. Finally, we adaptively estimate spectral amplitude as well as group and phase velocity on the filtered signal. The method is naturally used as a detector by allowing origin time to slide along the time axis. We describe preliminary results of the application of this method to a set of nuclear explosions and earthquakes that occurred on or near the Chinese Lop Nor test site from 1992 through 1996 and one explosion on the Indian Rajasthan test site that occurred in May of 1998.

Key words: Surface waves, Rayleigh waves, matched filters, group velocity, nuclear monitoring, $M_s : m_b$ discriminant.

1. Introduction

The $M_s : m_b$ discriminant and its regional variants are the most reliable transportable means of discriminating earthquakes from explosions. To measure surface wave amplitudes accurately in order to estimate M_s is very challenging for small events in which surface waves may not be readily identifiable in raw seismograms. To provide these amplitude measurements, it is crucial to be able to reliably detect small amplitude surface wave-packets and extract all and only the desired wave-packets reliably so that spectral amplitude measurements can be obtained.

[1] Department of Physics, University of Colorado at Boulder, Boulder, CO, 80309-0390, USA.
E-mail: levshin@ciei.colorado.edu

We describe a surface wave detector and extractor designed to work on weak surface wave signals across Eurasia at intermediate periods (8 s–40 s). It is founded on a long history of surface wave frequency-time analysis (e.g., DZIEWONSKI et al., 1969; LEVSHIN et al., 1972, 1989, 1992; CARA, 1973; RUSSELL et al., 1988). However, successful detection and wave-packet extraction are both dependent on the ability to make accurate predictions of surface wave arrival times at intermediate periods. Our method is based on the Rayleigh wave group travel-time predictions from the recent broadband group velocity maps of RITZWOLLER and LEVSHIN (1998) and RITZWOL-LER et al. (1998) and proceeds in three steps: Signal compression, signal extraction or cleaning, and measurement.

First, the dispersed surface wave signals are compressed in time by applying an anti-dispersion or phase-matched filter defined from our group velocity maps. We refer to this as the 'compressed signal.' Second, the surface wave is then extracted by filtering 'noise' temporally isolated from the time-compressed signal. This filtered signal is then redispersed by applying the inverse of the phase-matched filter. We refer to this wave-form as the 'filtered' or 'cleaned signal.' Finally, we adaptably estimate spectral amplitude as well as group and phase velocity on the filtered signal. After amplitudes are measured, M_s is estimated using an empirical relation such as that recently presented by REZAPOUR and PEARCE (1998).

The general methodology of matched filtering was developed previously by a number of other researchers (e.g., HERRIN and GOFORTH, 1977; STEVENS, 1986; RUSSELL et al., 1988; STEVENS and MCLAUGHLIN, 1997). We introduce three innovations here: (1) the use of recent group velocity maps to define the matched-filters, (2) automation of the procedure, and (3) the use of the method as a detector.

We describe preliminary results of the application of this method to a set of nuclear explosions and earthquakes that occurred on or near the Chinese Lop Nor test site from 1992 through 1996 and one explosion on the Indian Rajasthan test site that occurred on May 11, 1998.

2. Group Velocity Maps and Correction Surfaces

Elsewhere we have described the construction of intermediate period group velocity maps across Eurasia (e.g., RITZWOLLER and LEVSHIN, 1998; RITZWOLLER et al., 1998), the Arctic (LEVSHIN et al., 2001), South America (e.g., VDOVIN et al., 1999), and Antarctica (e.g., VDOVIN, 1999). The method of tomography and the construction of group velocity correction surfaces is described by BARMIN et al. (2001) in this volume. In this paper we will use the somewhat dated group velocity maps presented by RITZWOLLER et al. (1998).

Figure 1 displays group velocity correction surfaces computed from the 20 s group velocity map of RITZWOLLER et al. (1998) for four stations: AAK (Ala-Archa,

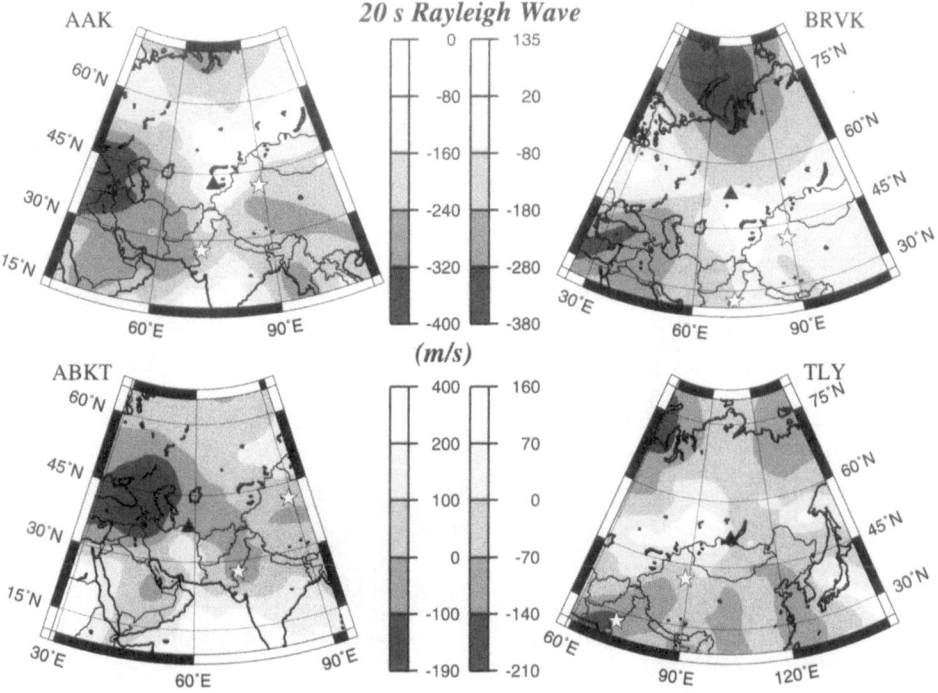

Figure 1

Group velocity correction surfaces for four stations in Central and Southern Asia for the 20 s Rayleigh wave. For each geographical point, the maps define the group velocity perturbation that should be applied to a 20 s Rayleigh wave observed at a station if an event were located at the chosen point. Perturbations are relative to the group velocity at the station. Units are m/s. The locations of the Chinese and Indian test sites are indicated with stars.

Kirghizstan), ABKT (Alibek, Turkmenistan), BRVK (Borovoye, Kazakhstan), and TLY (Talaya, Russia). For a given period, the value at each point on these maps represents the group velocity perturbation that a surface wave that originated at the point would experience if recorded on the specified station. The perturbations are relative to the group velocity at the station. In this form, group velocity maps can be used efficiently to predict dispersion curves for any event:station pair. BARMIN et al. (2001) present examples of group velocity correction surfaces for Rayleigh waves at 40 s period. We note two circumstances regarding the correction surfaces in Figure 1 and those shown by BARMIN et al. (2001). First, the corrections can be very large. For example, the 20 s Rayleigh wave from an event in the Caucasus 2500 km to AAK would experience a total group velocity perturbation of almost 400 m/s relative to the group velocity at AAK or a perturbation in arrival time of more than 2.5 minutes. Second, the correction surfaces at 20 s and 40 s are very different. This is because sedimentary basins control the 20 s map and crustal thickness controls the 40 s map.

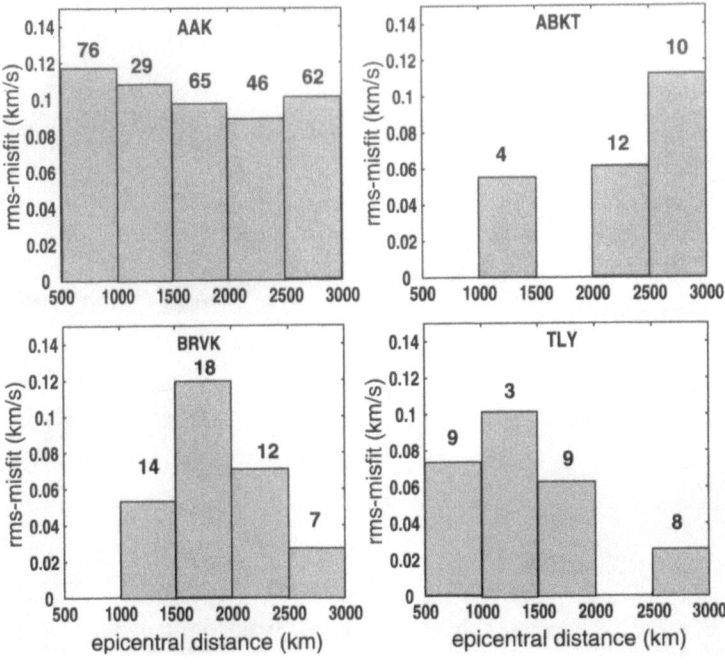

Figure 2

RMS-misfit between observed and predicted group velocities at four stations, plotted as a function of epicentral distance. Predicted group velocities are from RITZWOLLER *et al.* (1998). The measurements are from a database that includes earthquakes throughout Central and Southern Asia. The number of measurements per epicentral distance bin (summed over periods of 15 s, 20 s, and 25 s) is indicated above each histogram bar.

Figures 2–4 provide information about how well the correction surfaces constructed from the group velocity maps of RITZWOLLER *et al.* (1998) predict group velocity observations. Figure 2 shows the overall rms-misfit for the Rayleigh waves from 15 s–25 s period segregated by epicentral distance for observations made at four different stations. There is some indication of a diminishment of misfit with epicentral distance, and misfit is highly variable between stations. Misfit also varies strongly as a function of wave path. More or less homogeneous paths are fit better than paths through complicated structures. Figures 3 and 4 exemplify this. Figure 3 presents the observed group velocities between about 8 s and 35 s period obtained for the set of explosions and earthquakes on or near the Chinese Lop Nor test site, identified in Table 1. Because of the strong structural variability in Central Asia, there are very different dispersion curves observed at the different stations. The trends of the observed dispersion curves agree fairly well with those predicted from our group velocity maps. In the case of ABKT, however, the agreement is very poor, presumably because the path from Lop Nor to Turkmenistan is along structural gradients which complicate the wavefield in ways not represented by our group

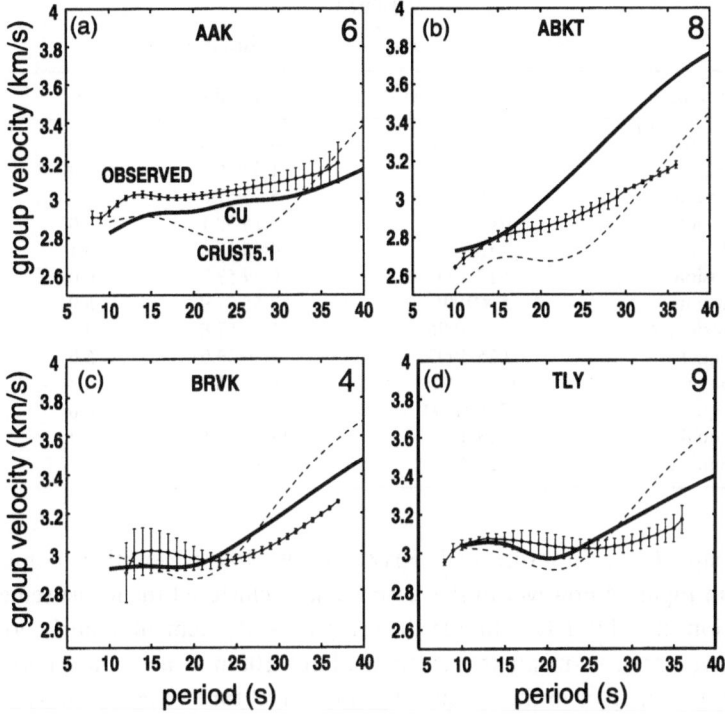

Figure 3
Comparison between observed and predicted group velocity curves. Group velocity measurements (error bars) at four stations in Central and Southern Asia following several nuclear explosions and earthquakes (Table 1) that occurred on or near the Lop Nor test site. The error bars represent the standard deviations of the measurements obtained at a given station following several Lop Nor events. The number of events used for each station is listed in the upper right-hand corner of each panel. Comparison is made with predictions RITZWOLLER *et al.* (1998) (solid line) and predictions from the hybrid crustal and mantle model CRUST5.1/S16B30 (dashed line).

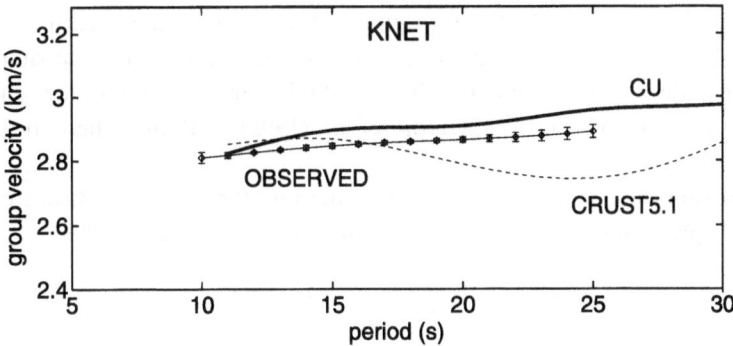

Figure 4
Like Figure 3, but for one of the Indian nuclear tests on May 11, 1998 observed at 5 stations of KNET.

Table 1

Events on or near the Lop Nor test site used in Figure 3

Event type	Date (mm/dd/yyyy)	Time (Z)	m_b	M_s
explosion	5/21/1992	4:59:57.5	6.5	5.0
explosion	10/5/1993	1:59:56.6	5.9	4.7
explosion	6/10/1994	6:25:57.8	5.8	–
explosion	10/7/1994	3:25:58.1	6.0	–
explosion	5/15/1995	4:05:57.8	6.1	5.0
explosion	8/17/1995	0:59:57.7	6.0	–
explosion	6/8/1996	2:55:57.9	5.9	4.3
explosion	7/29/1996	1:48:57.8	4.9	–
earthquake	9/25/1992	7:59:59.9	5.0	–
earthquake	11/27/1992	16:09:09.1	5.3	4.8
earthquake	12/26/1994	16:58:46.1	4.6	–
earthquake	3/18/1995	18:02:36.6	5.2	–
earthquake	3/20/1996	2:11:21.9	4.8	

velocity maps (e.g., Levshin and Ritzwoller, 1995). Another comparison is presented in Figure 4 however in this case we have clustered the measurements made at five stations in KNET to estimate the error bars. Agreement is fairly good in this case. On average, rms-misfit is less than about 100 m/s near 20 s period, which corresponds to an error in the predicted group travel time of less than 3%. The rms-misfit for the hybrid crustal and mantle model CRUST5.1/S16B30 (Masters et al., 1996; Mooney et al., 1998) is about twice this value.

3. Phase Matched Filter for Automated Detection, Extraction, and Measurement

We describe here the method for detecting, extracting, and measuring surface waves. The method is entirely automated. We will describe the extraction and measurement first, assuming that the detection has been made and then will discuss the detector. To extract surface wave signals we use a phase-matched filter based on our group velocity correction surfaces, and to detect weak signals we simply allow the filter to slide along the time axis. The method of extraction and measurement is similar to that described by Levshin et al. (1992), although here it has been automated.

Assume that the coordinates of the epicenter and the epicentral distance Δ to the station are approximately known. Let the surface wave signal $s(t)$ be

$$s(t) = \pi^{-1} Re \int_{\omega_0}^{\omega_1} |S(\omega)| e^{i(\omega t - \Psi(\omega))} \, d\omega \ , \tag{1}$$

where the origin time is assumed to be at zero time, $\Psi(\omega) = k(\omega)\Delta = \omega\Delta/C(\omega)$ is the negative of the phase spectrum relative to the origin time, $C(\omega)$ is the phase velocity curve, k is wavenumber, and $|S(\omega)|$ is the amplitude spectrum of $s(t)$. The effects of dispersion on $s(t)$ are contained within the phase spectrum.

We wish to compress the signal by undispersing it in order to maximize the signal-to-noise ratio. To do this we would like to apply the correction $\psi(\omega) = \Psi(\omega)$, but typically we do not know the phase velocity or wavenumber curves reliably at periods below about 40 s. Thus we must estimate the phase correction either from a 3-D model or from group velocity maps. The 3-D model approach is taken by STEVENS and MCLAUGHLIN (2001) in this volume. The advantage of this approach is discussed further below. Models, however, are not as well resolved as group velocity maps and the signal-to-noise enhancement of the matched filter will depend on the accuracy and resolution of the dispersion maps used to undisperse the surface waves. In order to optimize resolution, we use recent group velocity maps to build the matched filter. To do so, we must utilize the relation between group and phase velocities, $U = d\omega/dk, C = \omega/k$, from which we see that:

$$k(\omega) = k_0 + \int_{\omega_0}^{\omega} \frac{d\omega'}{U(\omega')} \, , \tag{2}$$

where

$$k_0 = \int_{0}^{\omega_0} \frac{d\omega'}{U(\omega')} \, . \tag{3}$$

$U(\omega)$ is the group velocity curve, and the phase correction is

$$\psi(\omega) = k(\omega)\Delta \, . \tag{4}$$

In practice, the continuous curve $U(\omega)$ is obtained by spline interpolating the discrete curve $U(T_i)$ from the group velocity maps for frequencies $\omega \in (\omega_0, \omega_1)$, the low and high cut-off frequencies of a bandpass filter applied to the observed seismogram. We typically construct the discrete curve $U(T_i)$ for each source-receiver pair from at a set of periods $T_i = 10, 15, 20, 25, 30, 35, 40$ s, set $2\pi/\omega_0 = 10$ s and $2\pi/\omega_1 = 40$ s, and taper quickly at the lower and higher frequencies. Using equations (2) and (4) we compute the envelope function $E(t)$ of the compressed or undispersed signal as follows:

$$E(t) = \pi^{-1} \left| \int_{\omega_0}^{\omega_1} |S(\omega)| \exp[i\omega t - i\Psi(\omega) + i\psi(\omega)] \, d\omega \right| \, . \tag{5}$$

If the source time is known we apply the phase correction given by equation (5) described above to the spectrum of the observed seismogram and return to the time domain. If the group velocity curve is accurate this should have effectively undispersed the surface wave. We demonstrate this with applications to synthetic and real seismograms. Synthetics are computed by fundamental mode summation (e.g., LEVSHIN et al., 1989) using a spherically symmetric model EUS. The velocities and Q values that define EUS are chosen to simulate stable, Eurasian structures. Compressed signals produced from the synthetic seismograms in Figure 5a are shown

in Figure 5b and from the real data following a large magnitude event shown in
Figure 6a are contained in Figure 6b. Perfect compression would result in a sharply
peaked function, as shown in the synthetic example. The example on real data is
peaked, but not as sharply as the synthetic. Figure 5b shows that as the wave
disperses, the amplitude of the compressed signal increases relative to the dispersed
waveform. Thus, the signal-to-noise enhancement of the compressed signal grows
with distance. The signal-to-noise enhancement at 1000 km in the synthetic example
is about 20–30%, which is consistent with the ~20% enhancement observed on real
data at 1200 km in Figure 6b.

To filter noise and unwanted signals we extract the compressed signal using a
temporal window of fixed width centered on the peak of the compressed signal. The
choice of the window width is *ad hoc*, however important. It should be broad enough
to encompass the broadening of the compressed signal caused by the difference
between the predicted and real dispersion curves, but narrow enough to filter out
unwanted signals. Figure 2 demonstrates that for several stations in Central Asia
using our entire surface wave database, the rms-misfit is less than about 100 m/s and
is roughly independent of range below 3000 km. We find that a time window
centered at the peak of the compressed signal with a full width in time corresponding
to a group velocity range of about 200 m/s is broad enough to work in most cases.
More work is still required to calibrate this width, perhaps increasing it in
structurally complex regions and decreasing it elsewhere. Finally, we redisperse the
extracted waveform by applying the inverse of the matched filter. Spectral amplitude
as well as group and phase velocity are measured on the extracted waveform exactly
as described by LEVSHIN *et al.* (1992) and RITZWOLLER and LEVSHIN (1998). In these
papers, however, applications involved human interaction to control the waveform

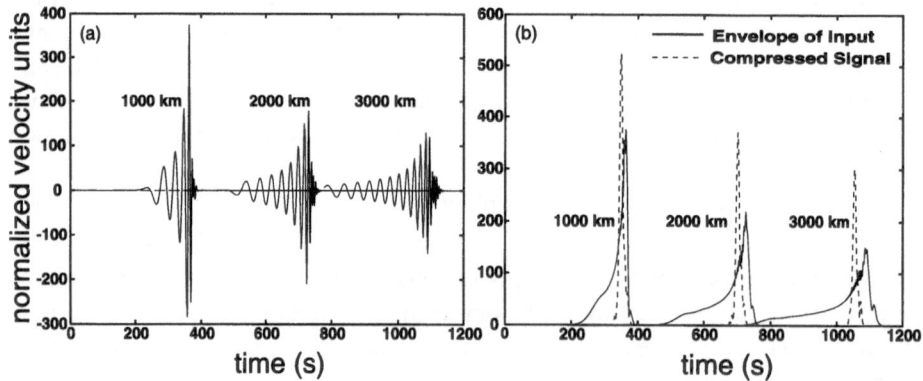

Figure 5
(a) Three synthetic Rayleigh waves computed for an explosive source at epicentral distances of 1000 km,
2000 km, and 3000 km. (b) The envelopes of the synthetic seismograms (solid lines) are compared with the
envelopes of the compressed signals (dashed lines), $E(t)$, that result from the application of the phase-
matched filter.

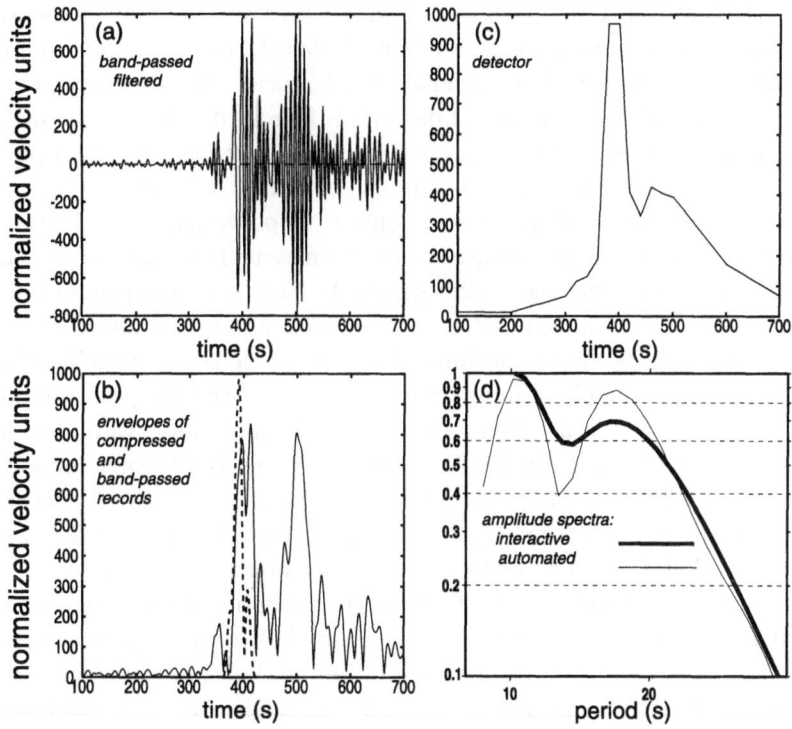

Figure 6

Demonstration of waveform compression, detection, and spectral amplitude estimation of strong surface wave signals. Nuclear explosion at the Lop Nor test site on June 8, 1996, $m_b = 5.9$, $\Delta \sim 1200$ km to AAK. (a) Band-passed (10 s^{-1}–40 s) vertical seismogram. The strong coda results from surface wave channeling by the Tarim Basin (LEVSHIN and RITZWOLLER, 1995). (b) Comparison of the envelopes of the band-passed filtered seismogram (solid line) and the compressed signal (dashed line). The compressed signal, $E(t)$, results from the application of the phase-matched filter. (c) Peak amplitude of the envelope of the compressed signal, $D(t - t_0)$, plotted for origin times from several minutes before to several minutes after the PDE origin time. (d) Rayleigh wave spectrum found by automated frequency-time analysis compared with the spectrum obtained with human interaction during the frequency-time analysis.

extraction. This method is adaptive in that it finds the dispersion ridge and measures amplitudes and velocities on the ridge. Figure 6d presents an example of an automated spectral amplitude compared with the spectral amplitude estimated with human interaction to define the extraction filter. Discrepancies of 20% may exist below 20 s period, particularly near spectral holes.

In the description above we have glossed over an important subtlety. Knowledge of the phase correction, $\psi(\omega)$, requires us to know k_0, which in turn depends on an estimate of $U(\omega)$ to zero frequency. In fact, we estimate $U(\omega)$ only above some minimum frequency, ω_0. The result is that we only know $\psi(\omega)$ reaching a constant of unknown value equal to $k_0\Delta$. Thus the resulting envelope function will appear shifted in time by about $k_0\Delta$ from where it would appear if we knew the phase correction perfectly. The shift will always be toward shorter times relative to the location with the unknown constant included. Notice in Figure 5b that the location of the compressed signal marches to earlier times relative to the peak of the envelope of the uncompressed signal as the epicentral distance is increased. This uncertainty in the location of the envelope of the compressed signal is no obstacle for most purposes, because we make no measurements on the compressed signal. Rather in our applications the compressed signal exists only to improve the detectability and to facilitate the extraction of the surface waves. All measurements are made when the signals have been uncompressed or redispersed. In the process of uncompression the time uncertainty that is introduced by compression is reversed and the extracted waveform suffers no distortions in time. However, if the compressed signal would have resulted from a highly accurate phase correction, then the peak times of the compressed signal could be used to locate the event. Event location with weak surface wave signals probably is not a major desirable at this time. If it becomes important in the future and if 3-D models improve substantially in order to ensure the accuracy of the predicted phase velocity curves, then the model based method of STEVENS and MCLAUGHLIN (2001) would be preferable to the method we discuss here.

If the source time is not known, the method we describe above may be naturally used to detect weak surface wave signals. We simply vary the source time, t_0, in a systematic way and plot the peak amplitude of the envelope of the compressed signal for each source time. We call this the detector time series, $D(t - t_0)$, which is a time series of amplitudes that will peak near the travel time of the compressed signal from the source. Signals with phase-content similar to the phase-matched filter are amplified and other signals are reduced. Figure 6c shows an example of a strong event. The signal-to-noise ratio (SNR) of this detection is about 75:1. Therefore, the detector clearly has identified a signal. Strong detections need not be delta-like. The detector time series in Figure 6c displays a hump following the main detection caused by surface wave energy scattered into the Tarim Basin (LEVSHIN and RITZWOLLER, 1995). Figure 7 demonstrates that the detector based on our group velocity maps works considerably better than one based on the spherically symmetric model EUS. The signal-to-noise ratio of the detector time series is usually about 1.5 times higher using the group velocity maps than the model EUS.

4. Preliminary Application to Weak Events

The detection of strong events, such as the example in Figure 6, is straightforward. Detection and spectral estimation of weak events is the motivation of this study, however. We have applied the detector/extractor/spectral estimator to broadband records following a number of weak events at the Lop Nor, China test site and one of the nuclear explosions at the Indian Rajasthan test site that occurred on May 11, 1998.

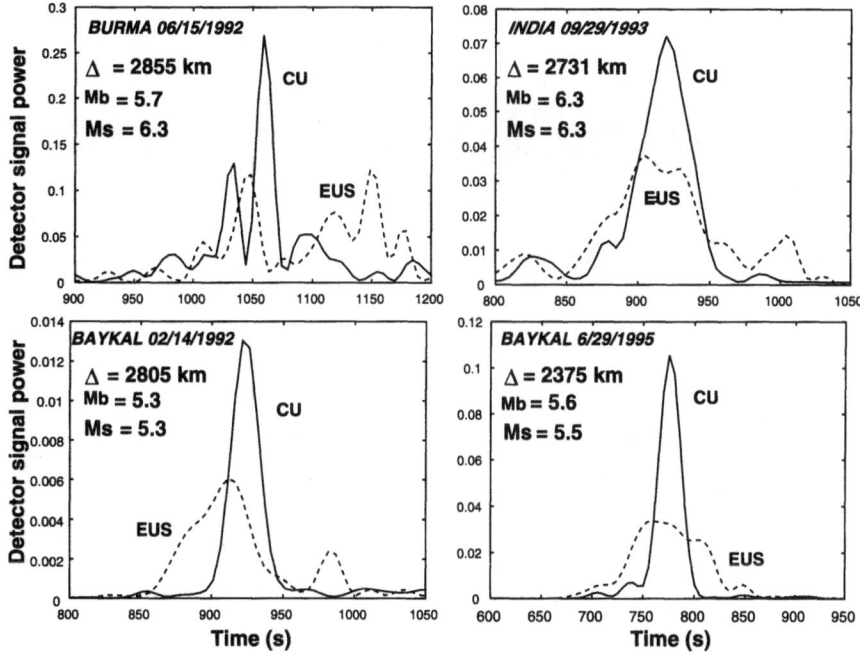

Figure 7

Comparison of the effectiveness of the automated detector based on two different group velocity correction surfaces: the group velocity maps of RITZWOLLER *et al.* (1998) denoted as CU (solid lines) and the group velocity curve predicted by an average model of the crust and uppermost mantle for Eurasia denoted as EUS (dashed lines). All measurements are from the station AAK and the event location is indicated in each panel. Signal compression and hence detection is greatly improved with the group velocity maps.

Figure 8 shows the results for a nuclear test at Lop Nor with $m_b = 4.9$ that occurred on July 29, 1996. The M_s of this event is probably in the middle to high 3's, because, as Table 1 shows, the M_s for nuclear explosions at Lop Nor is usually more than a magnitude unit below m_b. As can be seen in Figure 8, the surface wave can barely be discerned on raw or band-passed records. However, the detection algorithm described above demonstrates a clear detection with a SNR of about 4:1, and the automated amplitude estimate approximates that which is obtained with direct human interaction.

Figure 9 presents similar results for the Indian test. However, for the Indian test we have applied the method to five stations of KNET. The stations are located in structurally very different areas, and this apparently manifests itself as substantial differences in the observed amplitude spectra below about 18 s period. These differences are borne out when the measurements are obtained with human interaction. This highlights the difficulty in reducing the period at which M_s is measured below its current value of 20 s. Near-receiver structural variations with spatial scales well below the resolution of group velocity maps or 3-D models can result in amplitude effects as large as 50% at periods below 15 s.

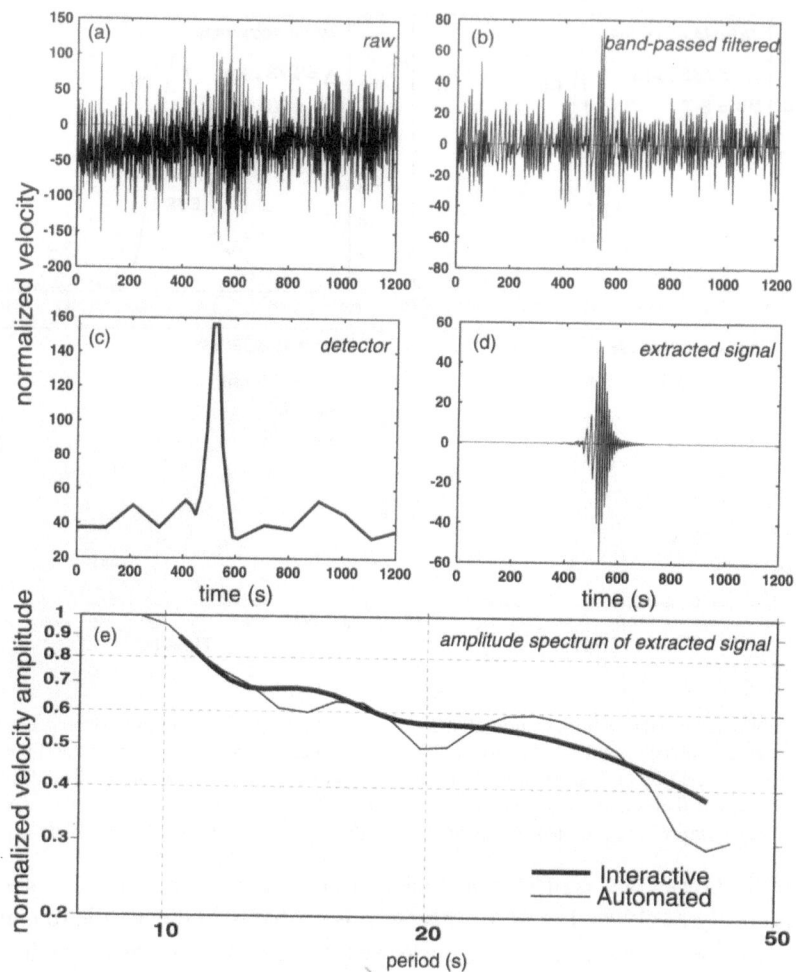

Figure 8

Demonstration of the detection, extraction, and spectral amplitude estimation of weak surface wave signals. Nuclear explosion at the Lop Nor test site on July 29, 1996, $m_b = 4.9$, $\Delta \sim 1600$ km to TLY. (a) Raw vertical seismogram observed at TLY following a nuclear explosion at Lop Nor. (b) Band-passed (10 s–40 s) seismogram. (c) Same as Figure 6c. (d) Extracted waveform on which the spectral amplitude measurement is obtained. (e) Rayleigh wave spectrum found by the automated frequency-time analysis compared with the spectrum obtained with human interaction during the frequency-time analysis.

5. Conclusions

We describe an automated surface wave detector and extractor designed to work on weak surface wave signals across Eurasia at intermediate periods (8 s–40 s). The method is based on phase-matched filters defined by the Rayleigh wave group travel-time predictions from the broadband group velocity maps presented by RITZWOLLER

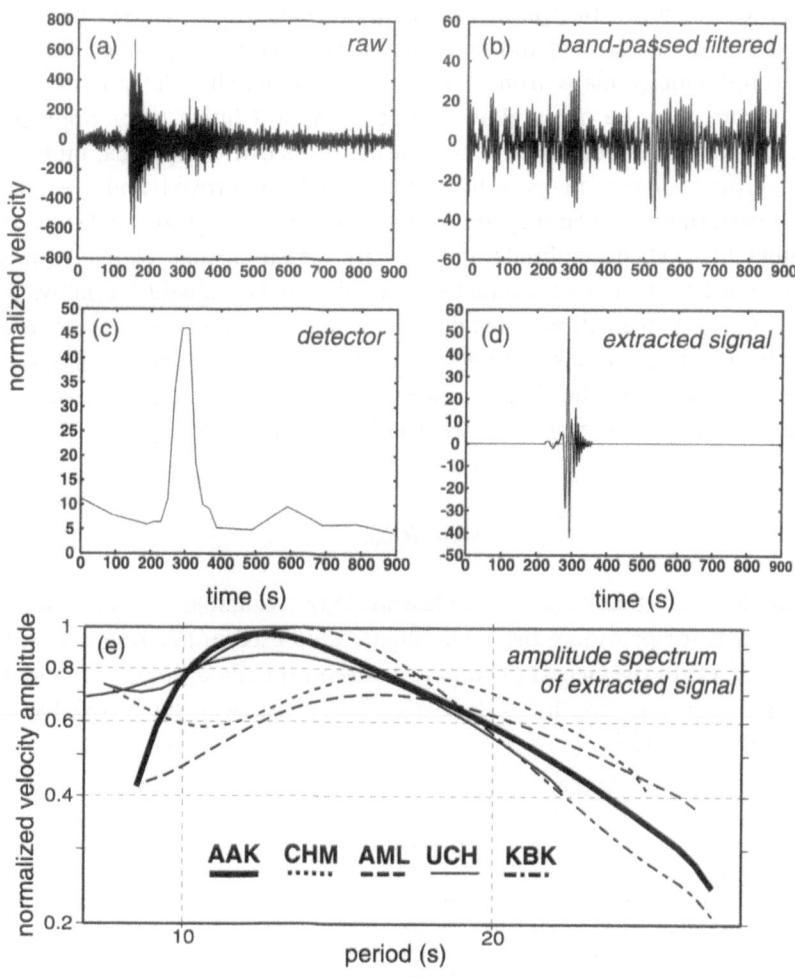

Figure 9

Same as Figure 8, but for a nuclear test in India. Nuclear explosion at the Rajasthan test site on May 11, 1998, $m_b = 5.1$, $\Delta \sim 1700$ km to KNET. The spectral amplitude measurements are at five KNET stations. Panels (a)–(d) are for AAK.

and LEVSHIN (1998) and RITZWOLLER et al. (1998). We describe preliminary results of the application of this method to a set of nuclear explosions and earthquakes in Central Asia. These and other applications lead us to conclude that this method shows considerable promise as a surface wave detector and of yielding high quality surface wave measurements automatically. It appears to be feasible to obtain automated spectral amplitude measurements for events in Central and Southern Asia with M_s down to as low as 3.5–4.0.

The method, however, requires further tuning and a more complete statistical evaluation. As discussed above, the extraction algorithm requires further work in that we need to calibrate the width of the temporal extraction window. The quality of the automated measurements and the SNR of the detection depend on the accuracy of the group velocity maps from which the phase-matched filter is defined. For example, more accurate maps allow sharper temporal filters to be applied which reduces bias in amplitude and velocity measurements caused by coda, multipathing, etc. The group velocity maps will continue to be improved and we anticipate improved performance when they are used. A more complete statistical evaluation of the automated spectral amplitude and velocity measurements relative to measurements obtained with human interaction would also be valuable. Finally, a more complete statistical evaluation of the detector is in order. In particular, the frequency and character of false-alarms and missed events should be characterized and a more complete comparison of the signal-to-noise characteristics of the detector based on our dispersion maps and earlier models should be performed.

Acknowledgments

We would like to thank Antonio Villaseñor for his valuable review and the staff at the IRIS DMC for providing the broad digital data used in this research. All maps were generated with the Generic Mapping Tools (GMT) data processing and display package (WESSEL and Smith, 1991, 1995). This work was supported by DSWA contract 001-97-C-0157.

REFERENCES

BARMIN, M. P., RITZWOLLER, M. H., and LEVSHIN, A. L. (2001), *A Fast and Reliable Method for Surface Wave tomography*, Pure appl. geophys., this volume.

CARA, M. (1973), *Filtering of Dispersed Wavetrains*, Geophys. J. R. astr. Soc. *33*, 65–80.

DZIEWONSKI, A., BLOCH, S., and LANDISMAN, N. (1969), *A Technique for the Analysis of Transient Seismic Signals*, Bull. Seismol. Soc. Am. *59*, 427–444.

HERRIN, E., and GOFORTH, T. (1977), *Phase-matched Filters: Application to the Study of Rayleigh Waves*, Bull. Seismol. Soc. Am. *67*, 1259–1275.

LEVSHIN, A. L., and RITZWOLLER, M. H. (1995), *Characteristics of Surface Waves Generated by Events on and near the Chinese Nuclear Test Site*, Geophys. J. Int. *123*, 131–149.

LEVSHIN, A. L., PISARENKO, V. F., and POGREBINSKY, G. A. (1972), *On a Frequency-time Analysis of Oscillations*, Ann. Geophys. *28*, 211–218.

LEVSHIN, A. L., RATNIKOVA, L., and BERGER, J. (1992), *Peculiarities of Surface Wave Propagation across Central Eurasia*, Bull. Seismol. Soc. Am. *82*, 2464–2493.

LEVSHIN, A. L., YANOVSKAYA, T. B., LANDER, A. V., BUKCHIN, B. G., BARMIN, M. P., RATNIKOVA, L. I., and ITS, E. N. *Seismic Surface Waves in a Laterally Inhomogeneous Earth* (V. I. Keilis-Borok, ed.) (Kluwer Publ., Dordrecht, 1989).

LEVSHIN, A. L., RITZWOLLER, M. H., BARMIN, M. P., VILLASEÑOR, A., and PADGETT, C. A. (2001), *New Constraints on the Arctic Crust and Uppermost Mantle: Surface Wave Group Velocities, P_n, and S_n*, Phys. Earth Planet. Int. *123*, 185–204.

MASTERS, G., JOHNSON, S., LASKE, G., and BOLTON, H. (1996), *A Shear-velocity Model of the Mantle*, Phil. Trans. R. Soc. Lond. *A 354*, 1385–1411.

MOONEY, W. D., LASKE, G., and MASTERS, G. (1998), *CRUST 5.1: A Global Crustal Model at 5 Degrees by 5 Degrees*, J. Geophys. Res. *103*, 727–748.

REZAPOUR, M., and PEARCE, R. G. (1998), *Bias in Surface Wave Magnitude M_s due to Inadequate Distance Corrections*, Bull. Seismol. Soc. Am. *88*, 43–61.

RITZWOLLER, M. H., and LEVSHIN, A. L. (1998), *Eurasian Surface Wave Tomography: Group Velocities*, J. Geophys. Res. *103*, 4839–4878.

RITZWOLLER, M. H., LEVSHIN, A. L., RATNIKOVA, L. I., and EGORKIN, A. A. (1998), *Intermediate Period Group Velocity Maps across Central Asia, Western China, and Parts of the Middle East*, Geophys. J. Int. *134*, 315–328.

RUSSELL, D. W., HERRMANN, R. B., and HWANG, H. (1988), *Application of Frequency-variable Filters to Surface Wave Amplitude Analysis*, Bull. Seismol. Soc. Am. *78*, 339–354.

STEVENS, J. L. (1986), *Estimation of Scalar Moments from Explosion-generated Surface Waves*, Bull. Seismol. Soc. Am. *76*, 123–151.

STEVENS, J. L., and MCLAUGHLIN, K. L. (1997), *Improved methods for regionalized surface wave analysis.* In Proceed. 19th Seismic Res. Symp. on Monitoring a CTBT, pp. 171–180.

STEVENS, J. L., and MCLAUGHLIN, K. L. (2001), *Optimization of Surface Wave Identification and Measurement*, this volume.

VDOVIN, O. Y. (1999), *Surface Wave Tomography of South America and Antarctica*, Ph.D. Thesis, Department of Physics, University of Colorado at Boulder.

VDOVIN, O. Y., RIAL, J. A., LEVSHIN, A. L., and RITZWOLLER, M. H. (1999), *Group Velocity Tomography of South America and the Surrounding Oceans*, J. Geophys. Res. *136*, 324–330.

WESSEL, P., and SMITH, W. H. F. (1991), *Free Software Helps Map and Display Data*, EOS Trans. AGU *72*, 441.

WESSEL, P., and SMITH, W. H. F. (1995), *New Version of the Generic Mapping Tools Released*, EOS Trans. AGU *76*, 329.

(Received February 20, 2000, revised May 25, 2000, accepted June 15, 2000)

 To access this journal online:
http://www.birkhauser.ch

Pure appl. geophys. 158 (2001) 1547–1582
0033–4553/01/081547–36 $ 1.50 + 0.20/0

Optimization of Surface Wave Identification and Measurement

JEFFRY L. STEVENS,[1] and KEITH L. MCLAUGHLIN[2]

Abstract — Accurate and reliable measurement of surface waves is important to Comprehensive Nuclear-Test-Ban Treaty (CTBT) monitoring because the M_s:m_b discriminant and its regional variants can in many cases unambiguously identify events as earthquakes or explosions. Surface wave processing at the International Data Center (IDC) is designed to be completely automated and is performed using the program Maxsurf. Maxsurf searches for surface wave characteristics in the expected surface wave arrival time window for all continuous long-period and broadband data in the IDC processing stream. The Prototype IDC GSETT3 Reviewed Event Bulletin (REB) now contains a very large and growing data set of surface wave measurements. Users of this data set need to be aware of processing changes and calibration errors in the GSETT3 experimental bulletin. The prototype International Monitoring System (IMS) surface wave detection threshold is approximately one magnitude unit lower than the detection threshold of other global networks that use visual identification of surface waves. Surface wave identification and measurement can be improved through development of regionalized earth models, phase-matched filtering and the use of path corrected spectral magnitudes in place of M_s. Regionalized earth models are developed through tomographic inversion of a very large data set of phase and group velocity dispersion measurements. Discrimination capability can be improved through the use of maximum likelihood magnitudes and maximum likelihood upper bounds.

Key words: Surface wave, dispersion curve, phase matched filter, regionalization, moment, earthquake/explosion discrimination.

Introduction

Surface waves are of primary importance for CTBT monitoring because the M_s:m_b discriminant and its regional variants are among the most reliable means of determining whether an event is an earthquake or an explosion. With the International Data Center detecting and locating approximately 20,000 events per year, it is particularly important to be able to unambiguously "screen" as many events as possible (see FISK *et al.*, 1999). In this paper, we describe the surface wave processing procedures that are now in place at the Prototype IDC. We then discuss methods for optimizing surface wave identification and measurement in order to reduce the magnitude threshold for which surface waves can be reliably measured,

[1] Science Applications International Corporation, 10260 Campus Point Drive, MS X1145 San Diego, CA 92121, USA. E-mail: Jeffry.L.Stevens@saic.com
[2] Center for Monitoring Research, 1300 N. 17th St., Suite 1450, Arlington, VA 22209-2308, USA.

minimize the number of events that require more detailed analysis, and decrease the number of unidentified events.

Because of the large volume of data acquired daily from the IMS, it is necessary for the IDC to be as automated as possible. An automatic processor initially performs *P*-wave phase identification and event location, however considerable analyst review and correction is required to ensure the quality of the results before the REB is published. Surface wave processing in contrast is completely automatic, with no operator review of the results, and is applied to all continuous data for each event in the REB. Consequently it is very important that the surface wave processing procedures be reliable and robust.

Surface wave processing at the IDC is performed using the automatic processing program Maxsurf, which has been developed and maintained by the Maxwell Technologies Geophysics Group (now part of SAIC). Maxsurf 1.0 was first implemented at the Prototype International Data Center (PIDC) in May, 1995. Several revisions have been made to the code and to the procedures used since that time, and it was included in the first delivery of software to the permanent IDC in Vienna. Improvements in surface wave processing have been tested and implemented first at the PIDC, prior to installation in the IDC, consequently the IDC software and procedures lag behind those in place at the PIDC by about a year. All of the processing described in this paper was performed at the PIDC, and all of the dates when changes were made and errors occurred or were corrected refer to the PIDC and will in general be different for the IDC. Some of the analysis described in this paper was performed as part of the testing of Maxsurf, and has been described in reports to the PIDC Configuration Control Board, which approves all software changes, in problem reports to the PIDC and in technical reports (STEVENS and MCLAUGHLIN, 1997).

In this paper we discuss the surface wave processing procedures that are currently in the operational system at the PIDC, as well as procedures that have been tried and then removed as better procedures were found. We will then discuss ways to optimize the procedures, and ways to improve surface wave identification and measurement, and show how improved procedures can be used to improve discrimination capability. The improved procedures include development of path corrected spectral magnitudes in place of M_s, phase-matched filtering, and implementation of maximum likelihood magnitudes and maximum likelihood upper bounds.

Automatic Surface Wave Identification and Measurement at the IDC

Surface wave processing is applied to all continuous long-period and broadband data received by the IDC. The existing IMS stations that produce this data are shown in Figure 1. There is a total of 32 stations – 5 long-period arrays and 27 three-component stations. 21 of the three-component stations and all five arrays are currently operational and transmitting data to the IDC. Surface waves are only

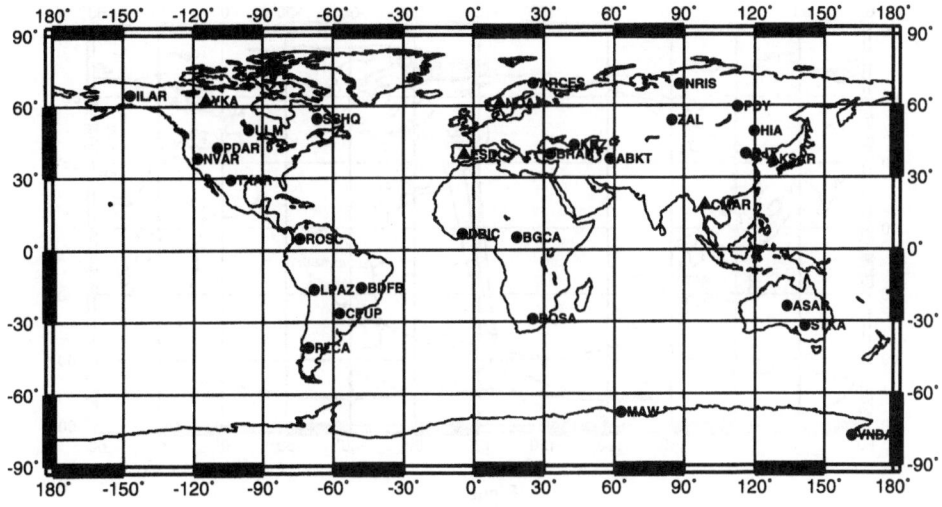

Figure 1

Map showing locations of current IMS stations that produce continuous long-period or broadband data. Circles indicate three-component stations, triangles arrays.

measured using primary stations; for economic reasons, the IDC does not currently request auxiliary station data in the surface wave arrival time window. Processing of auxiliary long-period data may be added in the future. Surface waves are processed for all stations within 100 degrees of each event.

Maxsurf runs in the processing stream after events have been identified and located. Maxsurf then examines the arrival window in which a surface wave would be expected and applies a dispersion test to see if a surface wave can be identified. If so, then the amplitude is measured and stored in the IDC database. We start with an example that illustrates this procedure. Figure 2 shows the location of an m_b 3.9 South Pacific earthquake that occurred on June 15, 1997, together with the 12 IMS stations within 100 degrees that recorded the event, and the great circle paths between the earthquake and the stations. Figure 3 shows the data recorded at these stations after conversion to a common (KS36000) instrument. The surface wave is visible at most of the stations, however it is obscured by noise and difficult to see at some of them. Surface waves are identified in the following way: a set of narrow band filters is applied to the data over a set of 8 frequencies from 0.02 to 0.06 Hz. A long-period or broadband beam is formed at arrays with the expected azimuth and slowness. The arrival times at each frequency are then compared with predicted arrival times generated from the regionalized group velocity model which will be described later in this paper. Figure 4 delineates the bounds on the allowed dispersion and the measured group velocities at the 8 frequencies for several stations. The dispersion test requires that 6 of the 8 measured data points lie within the predicted

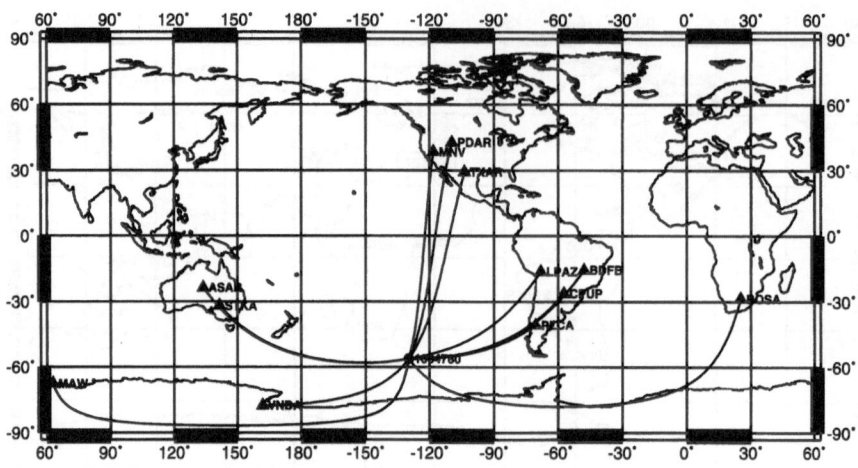

Figure 2

Map showing location of a 1997 South Pacific earthquake and the 12 IMS stations within 100 degrees that recorded the event. Lines are great-circle paths.

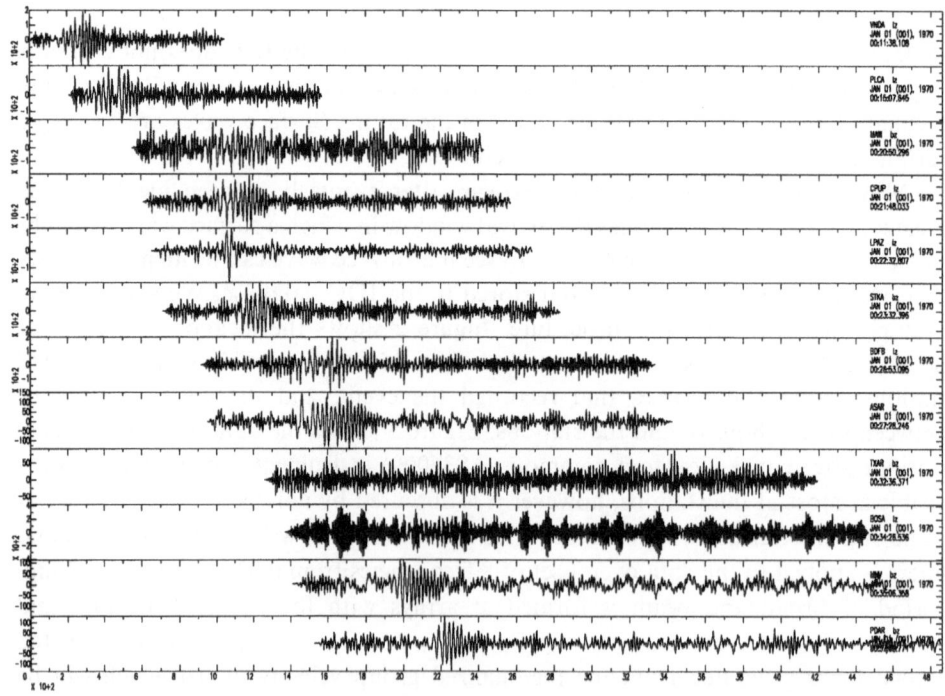

Figure 3

Long-period data from the 1997 South Pacific earthquake. Seismograms are ordered by distance from the earthquake. All seismograms have been transformed to a common KS36000 instrument.

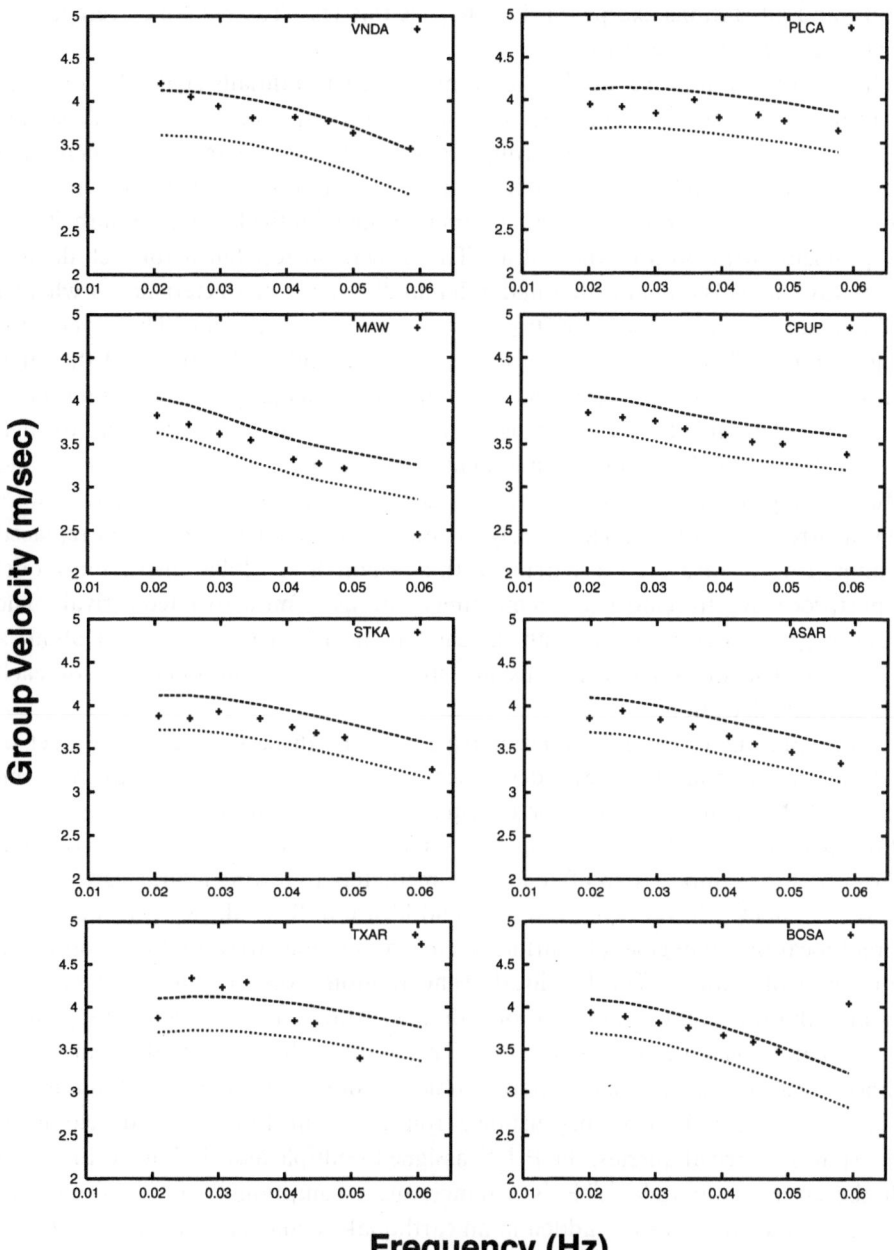

Figure 4

Measurements of group velocity made by narrow band filtering of eight of the seismograms, and the predicted dispersion window. Plus signs indicate measured data. Lines indicate the predicted dispersion curve offset above and below to determine the allowable time window as discussed in the text.

bounds, and all stations except for TXAR pass this test. The network averaged M_s measured for this event is 4.16.[3]

The identification procedure has been improved substantially since Maxsurf was first installed in 1995. The first version of Maxsurf applied two tests: A simple dispersion test that looked for group velocity that decreased with increasing frequency, and an azimuth test that compared the azimuth obtained from either an array beam or three-component polarization filtering with the known azimuth. There were problems with both of these tests. The dispersion test failed for well-defined surface wave arrivals on paths that had different dispersion characteristics or where a small amount of noise caused variations in the measured dispersion. There were also a large number of misassociations of arrivals and multiple associations of the same arrival with more than one event. The azimuth test, particularly for three-component stations, failed frequently for otherwise well-defined surface waves because of high noise levels on the horizontal components, and because of polarity errors and malfunctioning channels. Consequently a large fraction of surface wave arrivals were being incorrectly dismissed. These early identification tests have now been replaced by comparison of measured dispersion with a regionalized global dispersion model and postprocessing to remove any remaining spurious or misassociated arrivals. The change in procedures resulted in an increase in the LR detection rate of about a factor of 3. The measured azimuths are still recorded in the database for each associated LR detection.

A few spurious or misassociated arrivals pass the dispersion test, and a set of postprocessing queries has been developed to identify and remove them. Such misassociated arrivals may occur for several reasons. First, two events closely spaced in time may have surface waves in the same arrival time window. A common occurrence, for example, is an aftershock immediately following a large earthquake, or two aftershocks closely spaced in time and location. Second, two earthquakes in different locations may generate surface waves in the same arrival time window at a small subset of stations. Third, a local event, random noise, or an isolated arrival from an unknown event or an event outside the surface wave processing range may just by chance pass the dispersion test. This is most likely to happen at larger distances because the predicted arrival time window, which is based on group velocity, increases with increasing distance from the event. Prior to the development of the spurious arrival queries, the PIDC assigned multiple associations to the event with the larger magnitude. This is a dangerous assumption in a CTBT context, however, where an explosion hidden in an earthquake coda is a serious concern and it is very important not to incorrectly associate a surface wave from an earthquake with an explosion. A better assumption is to assign the arrival to the event with more arrivals at other stations.

---

[3] The results presented here for this event differ from the Reviewed Event Bulletin (REB) because of instrument correction problems (discussed later) that existed at the PIDC at the time of this event.

A set of four queries is used to remove spurious and misassociated arrivals. First, a query is performed to identify all events with four or more arrivals and an initial assessment is made that these arrivals are properly associated. Second, association of any of these arrivals with other events is assumed to be incorrect and the redundant arrivals are removed. Third, isolated arrivals, which are defined as events with fewer than three arrivals all at distances greater than 60 degrees, are removed. The parameters for this test were empirically determined and will depend to some extent on the number of stations and distance ranges used in the processing. Fourth, arrivals associated with more than one well-recorded event are removed. A side effect of this is that two large events closely spaced in time and location may have no surface wave arrivals in the REB because it is not possible to determine which event generated the surface waves. This is the correct approach for a CTBT monitoring system where it is more important to correctly associate arrivals than to identify every arrival.

Surface wave measurement procedures have also evolved over time. The first step in measurement of a well-dispersed wavetrain is to decide which part of the wavetrain to measure. Prior to February 15, 1997, the measurement algorithm picked a peak close to 20 seconds using a weighting scheme based on the difference between the measured peak and 20 seconds. This caused a small bias to occur between PIDC measurements and USGS measurements, therefore the procedures were revised and are now identical to those used by the USGS, where the largest peak in the 18–22 second period range is used. To ensure consistency in measurement, seismograms are first transformed to a common (KS36000 long-period) instrument. The largest peak to peak instrument corrected amplitude in the 18–22 second period range within the predicted arrival time window is then identified and measured and the LR amplitude, which is defined as half the peak to peak measurement, is stored in the IDC database. At three-component stations, the measurement is made on the vertical component, and at arrays it is made on a beam of the vertical components where the beam is constructed using the known azimuth to the event.

Prior to September, 1998, the surface wave magnitude M_s was calculated using the standard IASPEI formula $M_s = \log A/T + 1.66 \ \log \Delta + 0.3$ where A is the zero to peak amplitude in nanometers, Δ is the distance in degrees and T is the measured period in seconds, in order to maintain consistency with USGS M_s. However, M_s based on the IASPEI formula becomes anomalously small at distances less than about 25°, and M_s values from stations at close range are inconsistent with M_s measured at more distant stations. This is a serious problem because we seek to use surface waves recorded at short distances where the signal-to-noise ratio is best. The appropriate distance correction for M_s has been discussed by a number of authors (VON SEGGERN, 1975a, 1975b; THOMAS et al., 1978; MARSHALL and BASHAM, 1972; HERAK and HERAK, 1993; REZAPOUR and PEARCE, 1998). REZAPOUR and PEARCE (1998) used the entire ISC data set to derive the following relation for a distance independent M_s:

$$M_s = \log\frac{A}{T} + \frac{1}{3}\log(\Delta) + \frac{1}{2}\log(\sin(\Delta)) + 0.0046\Delta + 2.370 . \tag{1}$$

This new definition was adopted by the PIDC in September, 1998. After implementation of this formula, the PIDC also relaxed its previous restriction to measure surface waves only in the 20–100 degree distance range, and surfaces waves are now processed for all stations less than 100 degrees.

Parameters for Surface Wave Processing

Narrow band filtering (ARCHAMBEAU et al., 1966; DZIEWONSKI et al., 1969) and the dispersion test require a few processing parameters that must be defined and optimized. Following are a list of these parameters and their optimum values as determined by processing a large body of GSETT3 data and reviewing the results. These parameters are:

1. The frequencies used for narrow band filtering to compare with predicted group velocity arrival times. The following eight frequencies are used: 0.02, 0.025, 0.03, 0.035, 0.04, 0.045, 0.05, 0.06 Hz. These are good average global values, however they could be adjusted and performance possibly improved by using higher frequencies at shorter distances. Notice in Figure 4 that the measured frequency is slightly different from the input frequency. This occurs because the finite width of the narrow band filter combined with amplitude variation of the spectra causes a shift in the instantaneous frequency of the output time series.

2. The allowable error in the group velocity arrivals, and the fraction of group velocities that are required to match the predicted arrival times. We require that a minimum of 70% of the group velocity points, 6 out of 8 in this case, fall within the predicted arrival time window. This is sufficient to remove most "accidental" arrivals where the arrival peaks of noise by chance fall into the arrival window. Frequently one or two arrival times will be out of range because of low signal/noise, interference, or other factors (see Fig. 4, for example), and this requirement allows such arrivals to be identified. The group arrival time t is required to be within the time window given by

$$\frac{r}{v_p + v_0} - p_0 T - t_0 < t < \frac{r}{v_p - v_0} + p_0 T + t_0 , \tag{2}$$

where r is the source to receiver distance, v_p is the predicted group arrival time, T is the period, and v_0, p_0, and t_0 are user-definable constants. The parameters currently used are $v_0 = 0.2$, $p_0 = 1.0$, and $t_0 = 0$. This has the effect of changing the allowed group velocity window from about 0.2 km/s at vast distances to about 0.3 km/s at regional distances. This is necessary because a fixed group velocity window corresponds to a very large time window at extended distances, increasing the chance for spurious arrivals. The LR amplitude is also required to be measured within the time limits defined by Equation (2).

3. The narrow band filter Q. The narrow band filter $F(f)$ is defined by $F(f) = \exp(-\alpha(f - f_c)^2)$ where $\alpha = \ln 2/2(Q/f_c)^2$, and f_c is the center frequency. In general Q should be smaller for closer distances and larger for long distances, with a reasonable range being from about 8 to 20. A lower Q value gives better time resolution, while a higher Q smoothes the time series for more distant seismograms that have traveled on complex paths. A narrow band filter Q of 15 is a good average value to use over a wide distance range.

These and other parameters are specified in a parameter file in the operational system and may be adjusted (or tuned) in the future to improve detection. Identification at regional distances, for example, may be improved by using higher frequencies for the dispersion test, and a narrower time window at very short distances.

Performance of PIDC Surface Wave Processing

The GSETT3 Reviewed Event Bulletin (REB) of the PIDC is an excellent resource for seismic studies. As of this June 1999 it contained over 70,000 surface wave arrivals from over 16,000 events. However, users of these data must be aware of the changes in processing that have occurred over the life of the PIDC, as well as calibration errors and other instrumentation problems which have occurred frequently. Except as noted below, surface wave amplitudes can be considered consistent and reliable for dates after February 15, 1997. Prior to that date measurements were made by a slightly different method and identification was less reliable as discussed above. The major changes that have been made since that time have been implementation of improved dispersion curves in May, 1998 and the change in magnitude definition and distance range in September, 1998.

A serious and persistent problem with PIDC magnitudes is incorrect instrument corrections. This has led to errors of as much as a magnitude unit, or more in some cases. These incorrect values must be removed before the database can be used for research. Following is a list of all major instrumentation errors that have been identified to date since 1997.

1. A change in the PIDC database was made on April 3, 1997 that caused Maxsurf to use unnormalized instrument responses. This was corrected on June 24, 1997. This produced errors in the amplitudes at 9 of the IMS stations and caused no arrivals to be reported from 10 other stations. All LR amplitude data in the database during this time period should be considered unreliable.

2. Station ILAR had an incorrect calibration that caused amplitudes to be low by about 0.6 magnitude units from installation until October 22, 1997.

3. Beginning in early 1998, a series of errors occurred as instruments were changed from KS36000 to KS54000-M1-01 without a corresponding change in the database at several stations. This led to errors of about 0.94 magnitude units at the following stations: ASAR, beginning 1997/12/01; BRAR, starting with station installation; CMAR, measurements made on the broadband channel, all dates;

CMAR measurements made on LP channel, starting 1998/10/01; ESDC, starting 1998/05/15; KSAR, starting 1998/10/15. These errors were all corrected as of May 25, 1999 although ASAR is still not being processed pending verification of the response by the Australian NDC.

4. Station TXAR lost its calibration factor on 1997/07/15. This was also corrected May 25, 1999.

5. Station PDAR registered a calibration factor error from August 1, 1998 through October 1, 1998.

6. Station BGCA had a series of calibration errors following a lightning strike on March 2, 1999 that were corrected as of June 8, 1999.

7. Station ARCES had a large calibration error causing amplitudes to be high by about one magnitude unit beginning October 20, 1999. This has not been corrected as of this writing (December, 1999).

Error number 3 is particularly unfortunate because it includes 3 of the 5 long-period arrays (CMAR, ESDC, and KSAR; the others are YKA and NAO) and a very large fraction of all surface wave arrivals in the database. In an automatic processing system it is very important to implement a calibration checking system that regularly checks magnitude residuals and looks for consistent outliers in order to identify and correct this sort of problem. Such a system does not currently exist at the PIDC and problems like these can easily go unnoticed for extended periods of time. Users of the REB are advised to check station residuals for calibration errors before drawing conclusions from the data.

The US Geological Survey (USGS) publishes reports of surface wave arrivals in the Earthquake Data Reports (EDR). It is instructive to compare the arrivals in the EDRs with the arrivals in the GSETT3 Reviewed Event Bulletin (REB) produced by the PIDC. There are significant differences in the way the events are processed. All PIDC arrivals are generated automatically as described above, while USGS M_s values are measured manually. The network reporting to the USGS is substantially larger than the IMS network. The average number of stations reporting surface wave arrivals in the EDRs is 24, compared with 5 in the REB. The networks are mostly disjoint. There are only 9 stations mutual to the two networks. All surface wave seismograms measured by the IDC are first transformed to a KS36000 instrument response, while data reported to NEIC is recorded on a wide variety of long-period and broadband systems. The IDC processes surface wave arrivals only at distances less than 100° while the EDR includes more distant stations. Comparison of the two bulletins therefore can be used to test the consistency of surface wave measurements made mostly independently.

The following analysis uses data from the EDR and REB for the time period February 15, 1997 through July 29, 1997 excluding the time period from April 3, 1997 through June 23, 1997. During this time period there were 187 events with M_s reported in both bulletins. There were 234 arrivals measured by stations that are common to both networks. There was a total of 6600 surface wave arrivals from 269 events reported in the EDR and 5414 arrivals from 1221 events reported in the REB, therefore although there were 20% more arrivals reported in the EDR, surface waves were reported from 4.5 times as many events in the REB. During this time period, the REB used the same M_s definition as the EDR, and used only stations at distances between 20 and 100 degrees.

Figure 5 shows a cumulative distribution (fraction of arrivals reported greater than a given amplitude) of arrival amplitudes in the EDR and REB for the time period studied, with the REB bulletin also separated into arrivals reported by three-component stations and the four long-period arrays (KSAR, CMAR, ESDC, and YKA were operating during this time period). The 90% detection thresholds calculated from this data are 55 nm, 95 nm, and 645 nm for IMS arrays, IMS three-component stations, and the EDR bulletin, respectively, hence the detection threshold for the REB is about an order of magnitude lower than for the EDR. This clearly demonstrates both the utility of automated processing and the utility of using long-period arrays for reducing the routine detection threshold.

There is also a small bias between the REB and the EDR. The average difference in M_s between the two bulletins is $M_s(\text{EDR}) - M_s(\text{REB}) = 0.12 \pm 0.23$. However, as noted above, station ILAR had a calibration error and was consistently low during this time period. Removing this station reduces the bias to $M_s(\text{EDR}) - M_s(\text{REB}) = 0.08 \pm 0.23$. Figure 6 shows $M_s(\text{REB})$ plotted as a function of $M_s(\text{EDR})$ with ILAR removed from the REB.

We investigated several possible causes for the bias. First, we looked at measurements at the stations that are common to the two networks. For these

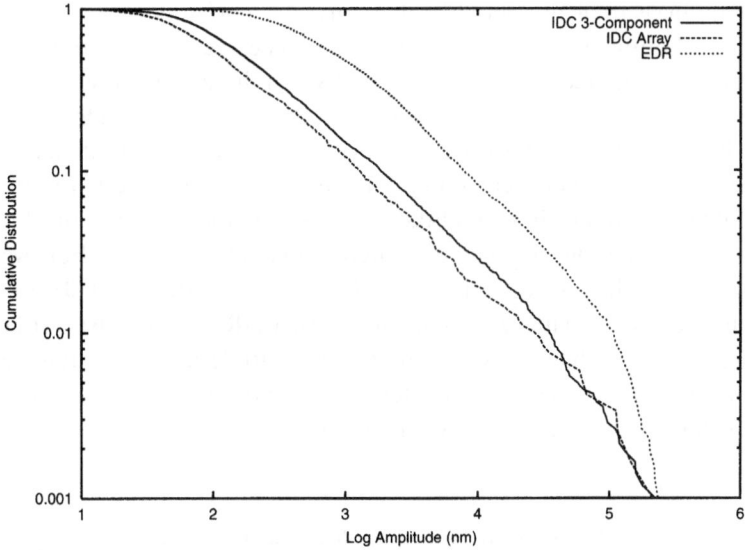

Figure 5

Cumulative distribution of IDC and EDR amplitudes. This figure shows that the IDC detection threshold is approximately one magnitude unit lower than the EDR detection threshold.

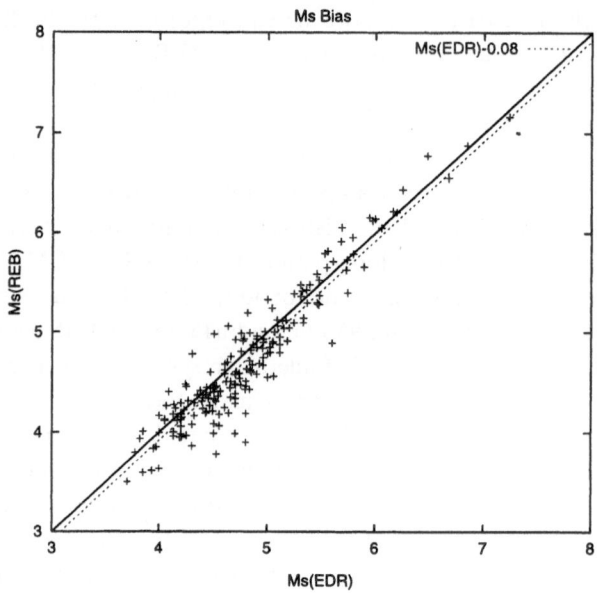

Figure 6

IDC vs. NEIS M_s shows a bias that increases with decreasing magnitude.

arrivals, the bias is $M_s(\text{EDR}) - M_s(\text{REB}) = -0.02 \pm 0.10$, therefore the station bias is very small and in the opposite direction from the network bias. The effects of the difference in measurement methods therefore do not appear to be significant. Another contributing factor is the distance range. The average reporting distance for the REB is 63° compared with 79° in the EDR. As discussed earlier, the $1.66 \log \Delta$ distance correction used in the M_s formula overcorrects the amplitude, causing M_s to increase with distance. This results in a bias of about 0.03 magnitude units between the two networks. The main remaining difference appears to be due to censoring in the EDR data, the loss of low amplitude signals near the detection threshold. This can be seen by examining the bias for different magnitude ranges. For events with $M_s > 5$ in both bulletins, the bias is $M_s(\text{EDR}) - M_s(\text{REB}) = -0.03 \pm 0.016$ (48 events), while for events with $M_s < 5$ the bias is $M_s(\text{EDR}) - M_s(\text{REB}) = 0.11 \pm 0.023$ (124 events). Thus the bias increases by 0.14 magnitude units over this magnitude range. Censoring also occurs in the REB data; however, because the detection threshold is lower, it occurs at a lower magnitude.

Development of Regionalized Earth Models

In order to predict dispersion curves, as well as the other quantities needed for optimization of surface wave measurements discussed in the next section, we need a set of regionalized earth models. Using these models, we can calculate all of the quantities needed to predict surface waves for any source and receiver point. In addition, we can use the phase velocity for the path to construct phase matched filters, and use the predicted group velocity arrival times as part of an existence test for surface waves in automatic processing as discussed earlier.

Observed surface waves provide strong constraints on earth structure, so development of regionalized earth models can be a self-correcting process. That is, surface wave dispersion and amplitudes can be used to infer earth structure, and earth structure can be used to calculate surface wave dispersion and amplitudes. Therefore with a data center such as the IDC which collects surface wave data on a continuous basis, it is possible to implement a program of continuous improvements in regionalization and surface wave processing using this extensive data set. In the following, we have used dispersion data measured from GSETT3 PIDC data, merged with dispersion data from other studies to develop regionalized global earth models.

Previous Earth Structure and Dispersion Models

Many studies have been performed to analyze surface wave dispersion and infer earth structure in specific regions, however few studies have attempted to regionalize the entire world, particularly at shorter periods. MOONEY et al. (1998) developed the Crust 5.1 earth model which consists of 139 distinct earth models regionalized on a 5° by 5° grid, each using a seven-layer crustal model over a single-layer mantle model. Each model is characterized by compressional velocity, shear velocity, and density for each layer. STEVENS and MCLAUGHLIN (1996) developed a set of regionalized group velocity models on a 10° by 10° grid which was used at the PIDC prior to May, 1998, and STEVENS and MCLAUGHLIN (1997) developed the set of regionalized earth models on a 5° by 5° grid that was used to calculate the dispersion curves that are currently installed at the PIDC and the IDC. In the following, we discuss the procedure for developing the 5° resolution model, and improvements to the model that have been made since that time.

Tomographic Inversion to Improve Crust 5.1 Models

The Crust 5.1 model provides an excellent starting point for development of regionalized surface wave parameters. Comparisons with surface wave dispersion data revealed that although group velocities predicted by Crust 5.1 were quite good on continental paths at periods close to 20 seconds, they were not as good as the 10° gridded model that was installed at the time at the PIDC for oceanic paths, particularly at periods <25 seconds, or for continental paths at periods >30 seconds. In order to make the models more useful for surface wave analysis, we improved the models using the following procedure.

We assumed that the 139 models in Crust 5.1 are an adequate classification of the earth's structure in continental regions, nonetheless that the constraints on the shear velocity in each model are weak and that a first-order improvement can be made by varying the shear velocity in each model. In oceanic regions we created additional models corresponding to different ocean ages, using the Crust 5.1 model as a starting model, but then allowing the new models in these distinct regions to vary independently. We also removed five models with very thick low velocity sediments that led to unrealistic dispersion curves. The final model has 149 distinct earth structures.

The following data sets were used in the inversion:

1. Global surface wave group velocities from earthquakes derived using PIDC GSETT3 data (STEVENS and MCLAUGHLIN, 1996), augmented with more recent measurements derived from PIDC data, for a total of 1500 paths at 6 frequencies from 0.02–0.06 Hz.

2. Surface wave phase and group velocity dispersion curves from underground nuclear test sites (STEVENS, 1986; STEVENS and MCLAUGHLIN, 1988), calculated from earth models for 270 paths (test site – station combinations) at 10 frequencies between 0.015 and 0.06 Hz.

3. Phase and group velocity measurements for western Asia and Saudi Arabia from MITCHELL et al. (1996) for 12 paths at 17 frequencies between 0.012 and 0.14 Hz.

4. Global phase velocity model of EKSTRÖM et al. (1996) for 9 periods between 35 and 150 seconds calculated for each 5° grid block from a spherical harmonic expansion of order $l = 40$.

5. Group velocity measurements for Eurasia from RITZWOLLER et al. (1996) and LEVSHIN et al. (1996) for 20 frequencies between 0.004 and 0.1 Hz with 500 to 5000 paths per frequency.

The complete data set of approximately 90,000 data points was used to invert for shear velocity structure in each model. This was accomplished using tomographic inversion of the entire data set for all models simultaneously. That is, we solved the equation $\mathbf{Am} = \mathbf{d}$ where $\mathbf{m}$ is the change in the shear slowness of each model layer with dimension equal to the total number of layers being varied in the 149 models, $\mathbf{d}$ is the difference between the observed and predicted slowness of each data point with dimension equal to the number of data points, and $\mathbf{A}$ is a matrix constructed from the partial derivatives of phase or group velocity with respect to shear velocity in each layer, and the fraction of each model crossed by each data point. This can be written explicitly as:

$$\sum_j \Delta m_j \sum_g \frac{\partial S_g}{\partial m_j} \Delta X_{ig} = \Delta T_i \ , \tag{3}$$

where Δm_j is the change in model shear slowness and the index j runs over all layers used in the inversion and all models; ΔT_i is the difference between the observed and calculated travel times where the index i runs over all data points, where each travel time corresponds to a particular observed frequency and may correspond to either phase or group velocity depending on the observation; S_g is the slowness calculated for each grid cell crossed by the raypath calculated for either phase or group velocity at the frequency of the observed data point; and ΔX_{ig} are the distances crossed by each ray i in each grid cell g. The partial derivatives $\partial S_g/\partial m_j$ are calculated using the method of TAKEUCHI and SAITO (1972). A minor variation of Equation (3) allows model data, such as the Harvard global phase velocity models, to be included in the same inversion. To stabilize the inversion, we write following NOLET (1987):

$$\begin{pmatrix} C_d^{-1/2}A \\ C_m^{-1/2} \end{pmatrix} \Delta m = \begin{pmatrix} C_d^{-1/2}\Delta T \\ 0 \end{pmatrix} , \tag{4}$$

where A and ΔT are the matrix and data misfit vector defined in Equation (4), $C_d^{-1/2}$ is a diagonal weighting matrix, and $C_m^{-1/2}$ is a smoothing matrix constructed from the model. The two sets of weighting factors $C_d^{-1/2}$ and $C_m^{-1/2}$ can be adjusted to set the optimum tradeoff between the data fit and the initial model, smoothness of the model, or other smoothness criterion.

The advantage of inverting for earth structure rather than the more common tomographic inversion of travel times on a frequency by frequency basis is that we can use the entire data set at once. Each frequency and each data type resolves different parts of the model, so for example higher frequency data will affect shallow structure, while lower frequency data affect deeper structure. Small errors in the data are smoothed because they must match a continuous (in frequency) dispersion curve calculated from the model. After the inversion is complete, dispersion curves and the other surface wave excitation and propagation factors discussed earlier can be calculated from the models. It is also easy to make continuous improvements in the inversion by adding new dispersion data sets.

The inversion was limited to depths between 3 and 200 km, with fixed water layers, and with Crust 5.1 extended to greater depths using PREM (DZIEWONSKI and ANDERSON, 1981). A smoothness condition that minimizes the change in adjacent layer velocities was also applied. The calculations were performed on a DEC Alpha 2100 dual-processor computer using the LSQR algorithm as described by NOLET (1987). The group velocities calculated from the models at 50 seconds are contoured in Figure 7. Phase and group velocities calculated for all models are shown in Figure 8.

The results indicate significant improvements over both the 10° gridded model and the Crust 5.1 model. The following tables show group velocity residuals (and standard deviations) in percent for the three models for data from nuclear test sites and for the global travel times derived from PIDC data. The three group velocity models below are the 10° PIDC model (STEVENS and McLAUGHLIN, 1996), as modified by Harkrider (personal communication) to a 5° grid with a better match to ocean/continent boundaries; the original Crust 5.1 model, and the 5° inversion results described above which have now been incorporated into the operational system at the IDC.

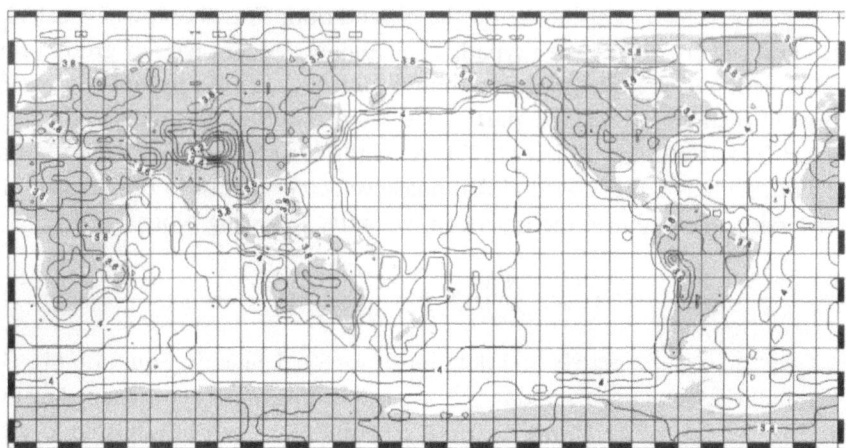

Figure 7
Group velocity contours of the inversion model at 50 seconds period.

The average shear velocity change was −0.16 (2.57)%, with extreme values of −11% and 15%. Additional improvements could be made in these models, particularly in the western Pacific, where many rays follow grazing paths along continental boundaries, and in areas such as the South Pacific where ray coverage is limited. There is more variation and more error at periods less than 25 seconds than

Table 1

40 second group velocity % average residuals (standard deviations)

Source	IDC 10°	Crust 5.1	IDC 5° (Inversion)
NTS (59)	0.93 (1.80)	2.08 (2.46)	0.15 (1.40)
East Kazakh (40)	1.20 (3.53)	5.00 (3.47)	0.80 (2.21)
Mururoa (13)	−0.80 (1.74)	−2.45 (1.73)	−0.78 (1.43)
Novaya Zemlya (99)	2.01 (2.39)	2.00 (2.28)	−0.06 (2.06)
Amchitka (55)	1.68 (1.58)	0.51 (2.54)	0.25 (1.39)
Earthquakes (1572)	1.44 (4.98)	1.68 (5.32)	0.46 (3.48)

Table 2

20 second group velocity % average residuals (standard deviations)

Source	IDC 10°	Crust 5.1	IDC 5° (Inversion)
NTS (58)	−1.83 (4.85)	−1.17 (2.48)	−0.47 (2.25)
East Kazakh (40)	−2.61 (3.23)	−0.31 (2.46)	−0.75 (1.96)
Mururoa (11)	−1.43 (2.30)	−1.47 (1.84)	0.41 (1.43)
Novaya Zemlya (99)	2.44 (6.15)	0.35 (4.68)	−0.26 (3.61)
Amchitka (54)	−1.23 (4.43)	1.42 (4.33)	0.83 (3.76)
Earthquakes (1673)	0.31 (5.82)	2.17 (6.31)	2.00 (5.46)

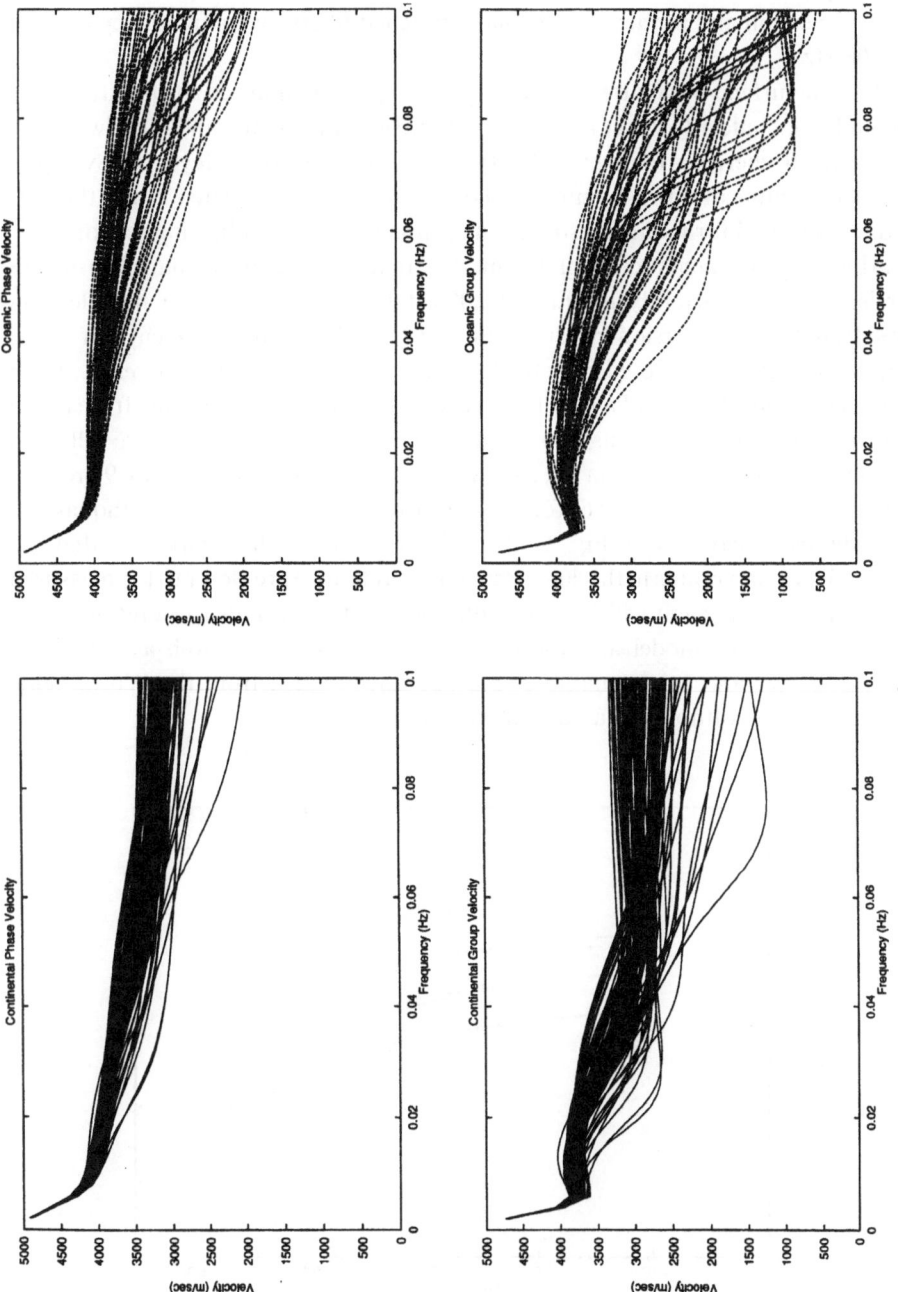

Figure 8

Phase (top) and group (bottom) velocities calculated from continental (left) and oceanic (right) models.

at longer periods. Nevertheless, the models work quite well for predicting group velocity dispersion and for generating phase-matched filters over a frequency band of 0.01–0.06 Hz.

Although surface wave attenuation was not explicitly modeled in this study, the earth models include a Q structure based on PREM in the mantle and on "Swanger's law" $Q = \beta/10$ where β is the shear velocity in m/sec in each crustal layer (H. Swanger, personal communication). Attenuation coefficients were calculated from these Q structures and used for moment estimation as discussed in the following section.

Further improvements to the global model are now under development, using an expanded data set currently totaling 166,000 data points. New data include data derived from PIDC dispersion measurements, group velocity measurements for South America and Antarctica provided by the University of Colorado (VDOVIN *et al.*, 1999; RITZWOLLER *et al.*, 1999) and a very large set of dispersion measurements from Saudi Arabia provided by St. Louis University. These additional data have filled in important gaps in the earlier data set, such as in the Antarctic region. Figure 9 presents the old and new dispersion curves for the path from the South Pacific earthquake to VNDA discussed earlier (see Fig. 4). The dispersion curve has shifted so that the predicted dispersion points in the 20–25 second period range are closer to the measured values. Further refinement will require additional models with finer resolution.

The current IDC model and the model discussed above as well as calculated dispersion curves and other quantities may be downloaded from the World Wide Web. Contact the authors for a current web address.

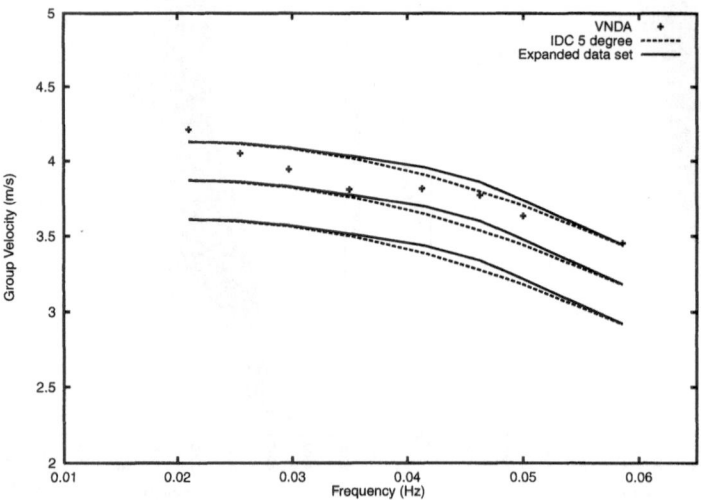

Figure 9
Change in dispersion curve for the path from the South Pacific earthquake to VNDA (see Figs. 2–4) with new inversion model. The three lines shown for each model correspond to the calculated dispersion curve (middle) and upper and lower bounds for the dispersion test.

Optimization of Surface Wave Processing and Analysis

Surface wave processing can be optimized by defining a regionalized path corrected spectral magnitude. We want to find a measure of surface wave excitation that can be used in the same manner as M_s for M_s:m_b discrimination, but which can be regionalized to allow for geographic variations in source excitation and attenuation, is free of amplitude variations caused by dispersion, and can also be measured over different frequency bands in order to optimize signal-to-noise ratio. To do this, we base the magnitude on the equation for surface waves in a plane-layered structure. This equation can be factored into functions that depend on the source and receiver earth structure and the phase velocity and attenuation integrated over the path. The theory for generation of surface waves from an explosion or other source and surface wave transmission through a regionalized earth model are summarized in Appendices A and B. The notation used in this section follows the notation of these appendices. The displacement spectrum for a Rayleigh wave at distance r from an *explosion* is given by (from Equations (B.8)–(B.10)):

$$U(\omega, r) = M_0' \frac{S_1^x(\omega, h_x) S_2(\omega) \exp[-\gamma_p(\omega)r + i(\varphi_0 - \omega r/c_p(\omega))]}{\sqrt{a_e \sin(r/a_e)}} . \tag{5}$$

S_1^x depends on the source region elastic structure and the explosion source depth. S_2 depends on the receiver region elastic structure, γ_p is the attenuation coefficient that depends on the attenuation integrated over the path between the source and receiver. c_p is the phase velocity integrated over the source to receiver path. φ_0 is the initial phase of the source. a_e is the radius of the earth. $M_0' = 3\beta^2/\alpha^2 M_0$ where M_0 is the explosion isotropic moment and α and β are the P and S velocities at the source. This definition is introduced so that the function S_1^x does not depend explicitly on the material properties at the source depth.

We can use Equation (5) to define a spectral magnitude corrected for distance and spectral shape. We define, for any event, earthquake or explosion, the scalar moment (STEVENS, 1986):

$$M_0' = \left| U(\omega, r, \theta) \frac{S_1^x(\omega, h_x) S_2(\omega) \exp[-\gamma_p(\omega)r + i(\varphi_0 - \omega r/c_p(\omega))]}{\sqrt{a_e \sin(r/a_e)}} \right| . \tag{6}$$

log M_0' is then a path corrected spectral magnitude that can be evaluated over any desired frequency band. This is similar to the approach taken by OKAL and TALANDIER (1987) in defining a mantle magnitude M_m, except that they used an averaged earthquake source spectrum which they referred to as the Rayleigh Wave "Excitability" instead of the explosion excitation function S_1^x at the reference depth h_x. Although the imaginary part of the exponential is removed by the absolute value, it is shown here explicitly because in practice the phase is used to generate a phase-matched filter (HERRIN and GOFORTH, 1977) to compress the signal and improve signal/noise ratio prior to taking the spectrum. The spectrum is then averaged over a frequency band to smooth the spectrum and obtain a stable measurement.

For an isotropic explosion source at depth h_x, M_0' is independent of frequency. Equation (6) therefore corrects completely for all frequency dependent and distance dependent factors in the observed spectrum. In general, M_0' for an earthquake is not completely frequency independent, although it is partially corrected for frequency dependence by removal of the path attenuation and receiver structure and similarities in the explosion and earthquake excitation functions in the same source region. The remaining differences mean that the earthquake magnitude will vary somewhat when measured over different frequency bands while the explosion will not. In particular, the spectra of deeper earthquakes will decline more rapidly with increasing frequency, while the path corrected spectra of shallow earthquakes is approximately flat over the frequency band of about 0.01–0.08 Hz. Figure 10 shows the scalar moment calculated with Equation (6) of an explosion and a typical earthquake with strike 0, dip 80, rake 15, observed at an azimuth of 45 degrees for several depths in a Eurasian earth structure. The explosion has unit scalar moment, and the earthquake has unit double couple moment.

By defining the path corrected spectral magnitude as the logarithm of Equation (6), we obtain a measure of surface wave magnitude that is independent of range, nearly independent of frequency, and regionalizeable. The functions S_1^x and S_2

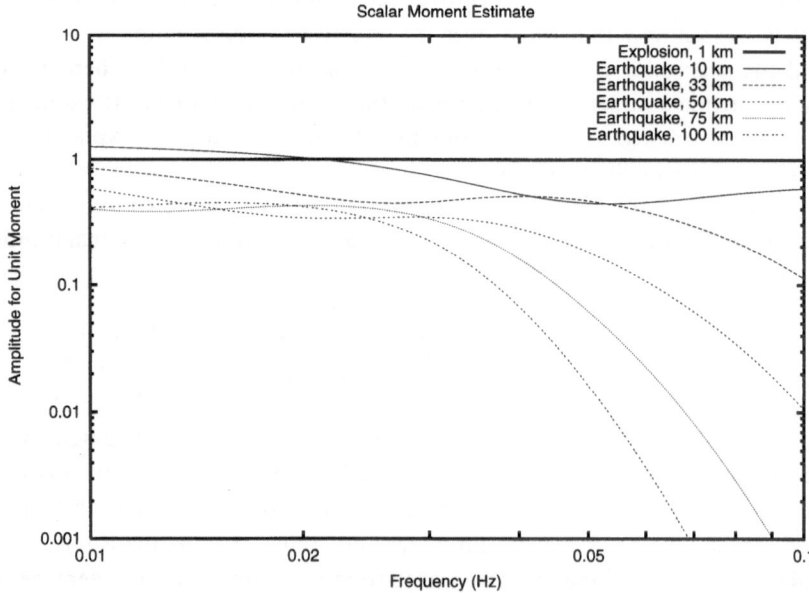

Figure 10
Scalar moment for an explosion and earthquake (Eq. 6) calculated for several earthquake depths. The spectral amplitude decays more rapidly with increasing frequency for deeper sources.

depend only on the source and receiver points and can be stored in a simple lookup table. The functions γ_p and c_p depend on the source to receiver path and can be found by integrating along a great circle path between the source and receiver in a regionalized earth model.

Although it has units of moment, the scalar moment M_0' defined in equation (6) is actually a measure of the strength of the surface wave. It is related to the double-couple moment through a distribution function. Figure 11 shows the frequency distribution of the difference between the predicted scalar moment and the double-couple moment for a large set of earthquakes. One way to interpret this figure is as the relative excitation of surface waves by a double-couple observed at a particular receiver point compared to a surface wave excited by an explosion with the same moment. The very small values of scalar moment correspond to nodal or near-nodal observation points. These are rarely if ever observed since even a small amount of lateral heterogeneity will cause surface waves from a slightly different azimuth with significantly larger amplitudes to be measured. For a few orientations, notably shallow thrust faults, the earthquake scalar moment is larger than the double-couple moment because of strong surface wave excitation. For most orientations and observation points, however, the scalar moment is less than the double-couple moment, with the average value about half of the double-couple moment. The observation that earthquakes with the same moment in general produce smaller surface waves than explosions may seem counterintuitive given the success of the $M_s:m_b$ discriminant, however there are several other factors that account for this including spectral differences and the body wave radiation pattern (STEVENS and DAY, 1985).

Path Corrected Spectral Magnitudes and Phase-matched Filtering

A second automatic surface wave processing program, Maxpmf, has been developed which is similar to Maxsurf except that it applies phase-matched filtering to seismograms and calculates path-corrected spectral magnitudes in addition to M_s. Maxpmf integrates a regionalized phase velocity model to generate a phase-matched filter and applies amplitude corrections to generate a path-corrected spectral magnitude (log M_0'). An advantage of the frequency domain processing is that a spectral magnitude of either signal or noise can always be measured over the specified frequency band, while it may not be possible to measure data in a specified frequency band in the time domain. Maxsurf, for example, will reject a seismogram if it cannot find a 20-second arrival within the predicted arrival time window. This often occurs at regional distances and there is no reason for such a restriction for spectral magnitudes. Consequently, moments can be measured for regional seismograms in cases where standard M_s measurements cannot be made.

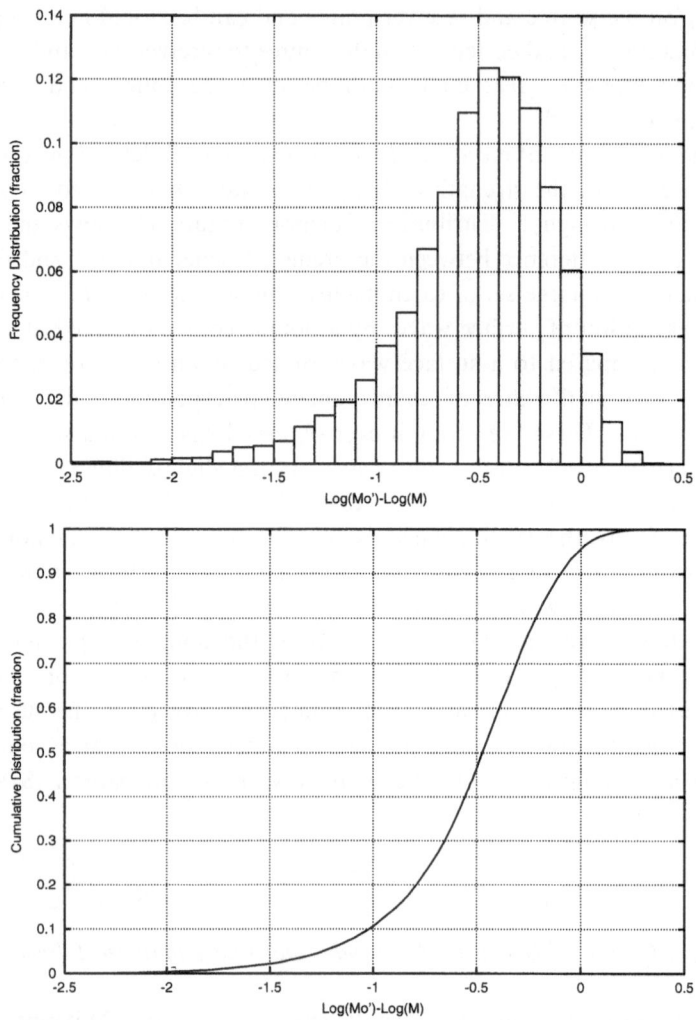

Figure 11

Frequency distribution (top) and cumulative distribution (bottom) of $\log(M_0') - \log(M)$ where M_0' is the scalar moment measured as described in the text, and M is the earthquake double-couple moment. Calculations were performed for 100 random fault orientations and depths between 0 and 50 km for all 304 earth models for a total of over 30,000 realizations. These distributions give the relationship between the scalar moment measured from surface waves and the double-couple moment, or, equivalently, as the distribution of surface wave amplitudes for a random set of earthquakes compared to an explosion with the same moment.

Figure 12 illustrates an example of phase-matched filtering derived from the regional phase velocity model applied to the data set discussed earlier and shown in Figure 3. The surface waves are compressed into a narrow time window near zero and amplified relative to the noise.

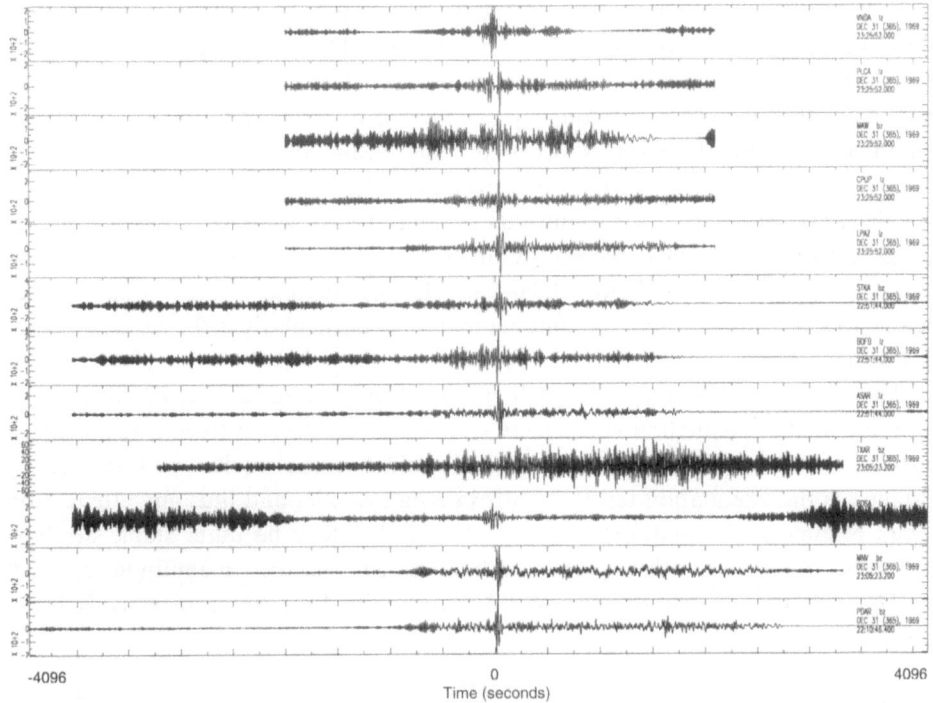

Figure 12

Waveforms from Figure 3 after phase-matched filtering. The surface wave is compressed into a narrow time window near zero.

There are additional parameters that must be set in order to calculate spectra and path-corrected spectral magnitudes. These are:

1. The phase-matched filtering time window. Phase-matched filtering (e.g., HERRIN and GOFORTH, 1977) compresses the waveform into a narrow time window centered near zero time, allowing noise to be windowed out by taking the spectrum of this narrow window rather than the full seismogram time window. However, the amount of compression depends on how well the phase-matched filter matches the actual phase of the seismogram. With accurate phase-matched filters and event origin time and locations, it is possible to use a time window as small as ±50 seconds. However, if the time window selected is too small, part of the signal will be windowed out, resulting in inaccurate, low amplitude estimates. Our review of the compressed spectra showed that while compression along most paths is very good, complex paths such as paths grazing the Pacific rim, and certain other paths not well constrained by data in the inversion, such as paths near and across Antarctica, either do not compress as well, or compress to an arrival time different from zero. We found that a time window of ±150 seconds is sufficient to capture the complete surface wave arrival for nearly all paths while still providing significant noise reduction.

2. The frequency band used to average the spectrum to estimate the moment. For our test cases we used a frequency band of 0.02–0.05 Hz for all data, and found that this worked quite well, giving a nearly flat spectrum over this frequency band for most data. In principle, any frequency band could be used, and a higher frequency band may be required for very short paths. For oceanic paths, however, the surface wave spectra and dispersion curves become quite variable above frequencies of about 0.06–0.08 Hz. In our tests we found that oceanic spectra were flat approaching this range and then dropped precipitously at higher frequencies. We therefore recommend using frequencies below 0.06–0.08 Hz for oceanic paths, while higher frequencies should be usable for continental paths, particularly if the earth structure and dispersion are well defined.

The distance correction of the path-corrected spectral magnitude depends on the accuracy of the attenuation coefficients (Equation (6)) derived from a regionalized earth model. However, errors in log M_0' should become smaller with decreasing distance as the attenuation term $-\gamma_p(\omega)r$ becomes smaller and thus any attenuation errors become less important. Log M_0' should therefore be quite stable at short distances. Figure 13 shows the path-corrected spectral magnitude residual $\log M_0 - \overline{\log M_0}$ at each station for all events with four or more arrivals for a

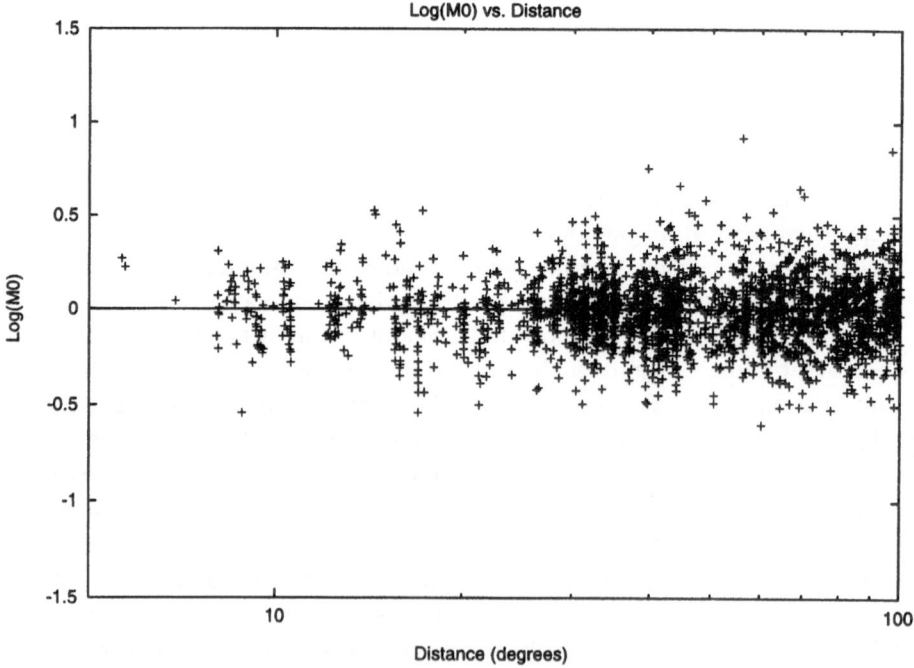

Figure 13
Path-corrected spectral magnitude residuals vs. distance derived using regionalized earth models. The residuals exhibit very little distance dependence.

10-day continuous PIDC data set (see next section). The log M_0 residuals are remarkably distance independent, particularly considering the relatively simple Q models used in the earth structures. This indicates that the attenuation coefficients derived from the Q models are approximately correct. The average attenuation coefficient for the earth models at a period of 20 seconds is 0.0114 ± 0.0022/degree.

Maximum Likelihood Moment and M_s

Maximum likelihood magnitudes were originally developed to correct for censoring. As discussed earlier, for small events it is common for larger arrivals to be measurable while smaller arrivals are lost in noise, causing the average magnitude of the observed arrivals to be biased high by the large amplitude tails of the magnitude distribution. Similarly, for very large events, the largest arrivals may be clipped and therefore discarded, causing the average magnitude of the remaining arrivals to be biased low. Maximum likelihood magnitudes correct for censoring by including the measured noise level as an upper bound on the observed amplitude at a station. This correction, together with station corrections that are derived as part of the processing, lead to more consistent and reliable network magnitudes.

To evaluate the effect of using maximum likelihood time domain and spectral magnitudes, we ran the automatic surface wave processing program Maxpmf on a data set of 10 days of continuous PIDC data from June, 1997, with data from 517 earthquakes, together with a data set of historical explosion seismograms from 253 underground nuclear tests from several test sites. All data were processed and measurements from seismograms for which a signal was not found were used as noise measurements in the maximum likelihood processing. Maximum likelihood magnitudes and moments were calculated using the method of MCLAUGHLIN (1988). Network magnitudes for all events and station corrections for all stations were calculated simultaneously using a maximum likelihood general linear model (GLM). One complication is that the networks used for recording nuclear explosions and the current IMS network have no stations in common, which causes nonuniqueness in the GLM calculation. The nonuniqueness exists despite the standard constraint that the sum of all station corrections is zero. To address this problem we added a fictitious station with a magnitude equal to the average magnitude of each event to all of the events. This has the effect of linking the old and new data and stabilizing the inversion. The expectation maximization algorithm used in the GLM minimizes the offset for the fictitious "station" and therefore we find the solution "closest" to the network average.

Figures 14 and 15 show the effect of the censoring correction on M_s and log M_0' for this data set. Note the resemblance between Figure 14, M_s with and without the censoring correction, and Figure 6 showing REB M_s plotted vs. EDR M_s.

In a CTBT context, the most important consequence of using maximum likelihood magnitudes is the ability to determine an upper bound on a magnitude when there is no

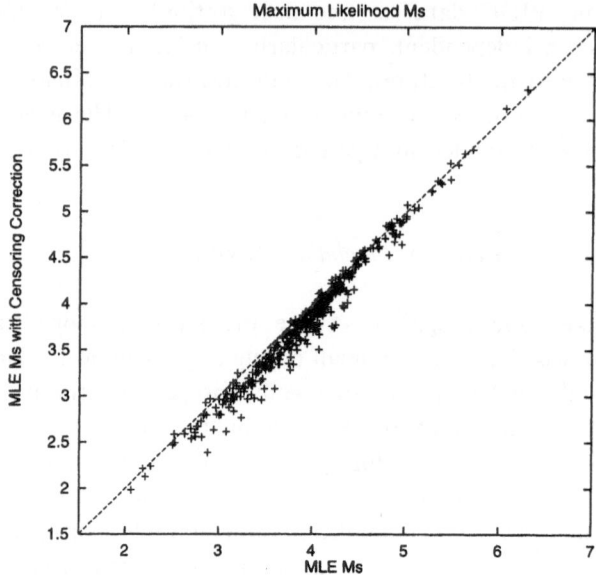

Figure 14
Maximum likelihood GLM M_s with and without censoring correction.

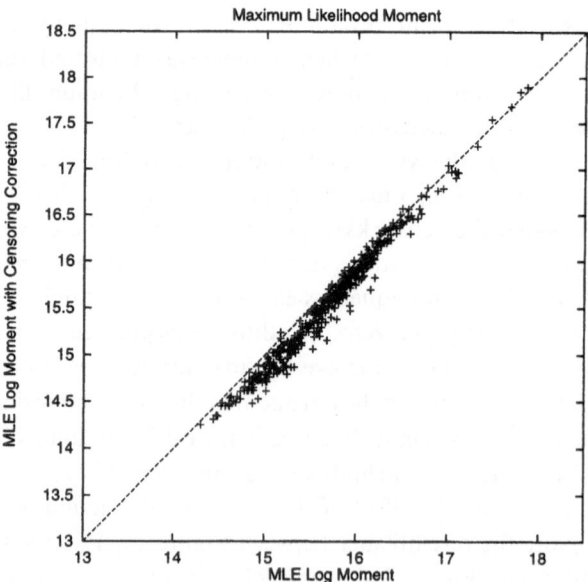

Figure 15
Maximum likelihood GLM log M_0 with and without censoring correction.

measurable LR data. Because $m_b - M_s$ may be as large as two magnitude units for an underground nuclear test, surface waves will rarely be observable for explosions with m_b less than 4, and may be difficult to observe up to m_b 5. A large earthquake may obscure surface waves from even larger explosions. However, it is possible to determine an upper bound on M_s for these events and in many cases this will be sufficient to identify the event as an explosion. The definition of "upper bound" used here is the magnitude that has a 50% probability of having no detections at the stations recording noise in the arrival time window. This definition was selected because it is consistent with the definition of maximum likelihood magnitudes. The "upper bound" magnitude of an event with only noise measurements is approximately the same as the magnitude of an event with the same noise measurements except for a single signal with the amplitude of the smallest noise measurement. Both magnitudes and upper bounds are calculated using the maximum likelihood station corrections described above. For the events studied here, we found upper bounds on M_s and log M_0' for 298 earthquakes and 46 explosions. Figure 16 shows maximum likelihood M_s plotted vs. maximum likelihood moment, including both values and upper bounds. Maximum likelihood M_s and log M_0' are related by log $M_0' = M_s + 11.74 \pm 0.21$. IDC software routinely computes the network average M_s, a network maximum likelihood M_s, and in the cases of no LR detections a network M_s upper bound. In the future, the network M_s estimates will include station corrections.

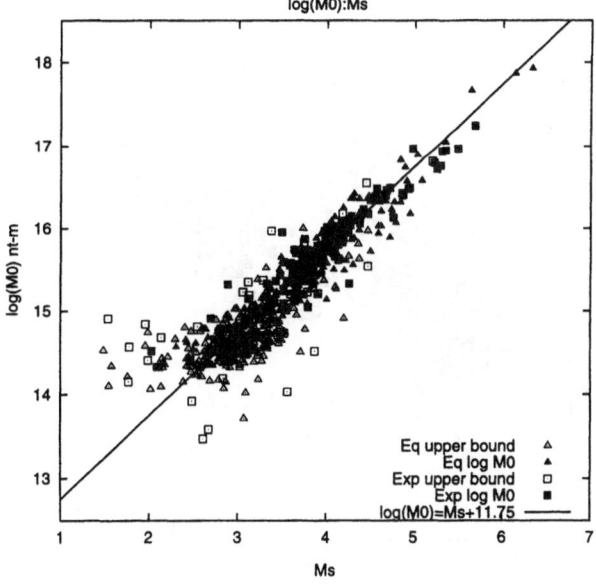

Figure 16

Maximum likelihood log moment plotted vs. maximum likelihood M_s. Note that while the M_s values have all been reevaluated using current procedures, the m_b values have not. For an unbiased distribution, the historical m_b values should be reevaluated using current procedures.

Following are some factors that must be kept in mind when using maximum likelihood magnitudes.

1. Noise measurements must be made as accurately as signal measurements. In particular, it is extremely important to avoid including bad data from a malfunctioning instrument. In a maximum likelihood calculation a low but meaningless value will strongly bias the results, causing an unrealistically low magnitude. This is true both for maximum likelihood magnitudes and maximum likelihood upper bounds.

2. For time domain magnitudes, it may not be possible to obtain a reliable noise measurement in the 18–22-second period range, particularly when broadband instruments are used. Replacing the instrument with a standard long-period instrument helps significantly, but will not eliminate the problem. Spectral magnitudes including $\log M_0'$ are not affected by this problem.

3. Very small events may still be biased by censoring. This occurs, for example, when a small event is recorded only at one close station and all noise measurements are significantly higher than the signal. In that case, the maximum likelihood magnitude will be equal to the signal measured magnitude.

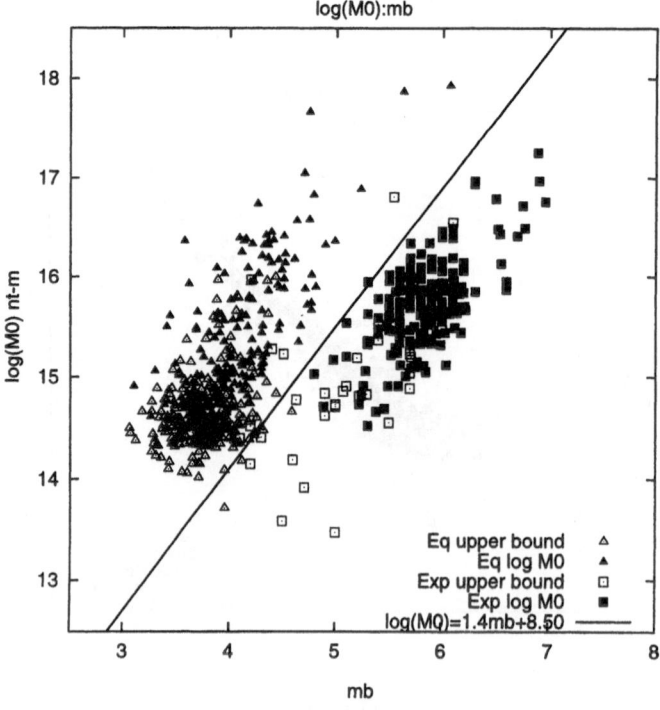

Figure 17

Maximum likelihood GLM station corrected $\log M_0$ and $\log M_0$ upper bounds plotted vs. m_b for PIDC earthquakes and historical explosion data.

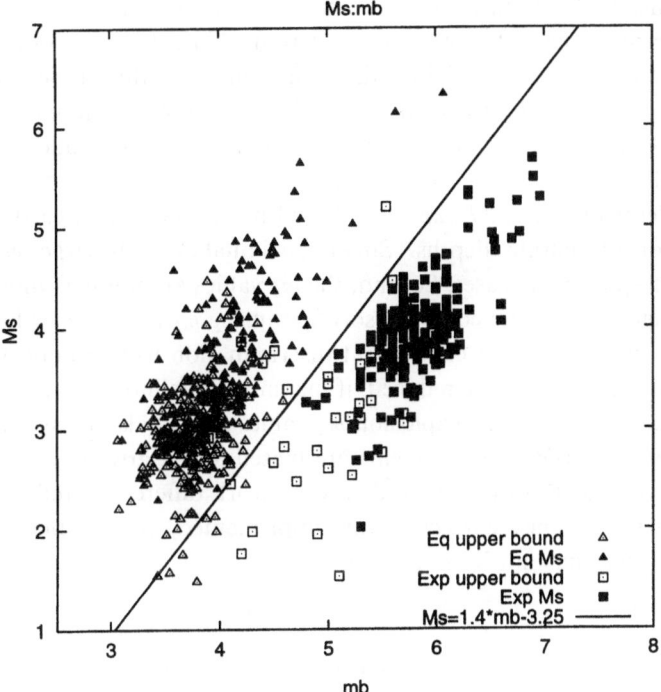

Figure 18

Maximum likelihood GLM station corrected M_s and M_s upper bounds plotted vs. m_b for PIDC earthquakes and historical explosion data.

4. Maximum likelihood procedures and upper bounds reduce but do not eliminate the problem of masking by coda of large events. As can be seen in the figures in the following section, it is sometimes possible to identify an event as an explosion if surface waves from the explosion are hidden in earthquake coda; however, if the coda is large enough the explosion will still be obscured. This was the case with the May 30, 1998 Pakistan nuclear test, which was obscured by a large earthquake in Afghanistan that occurred shortly before the test.

Earthquake/Explosion Discrimination Using the $M_s{:}m_b$ and $log(M_0')\colon m_b$ Methods

Figures 17 and 18 show maximum likelihood GLM log M_0' and M_s plotted vs. m_b for the PIDC data set and historical explosion data described above. m_b values for earthquakes are from the PIDC database, and for explosions are a mixture of values from AWRE, NEIS and ISC. Open symbols indicate upper bounds on moment and M_s, while solid symbols indicate events with at least one measured surface wave.

Also delineated in Figures 17 and 18 is an approximate discrimination line between the earthquake and explosion populations. Note that most of the explosions with no observations clearly fall into the explosion population based on the upper bound of either moment or M_s for the event. The best separation line has a slope of 1.4, which differs from the 1.0 slope that is expected at low magnitudes (e.g., STEVENS and DAY, 1985). There are several possible reasons for this. First, although only events with depths less than 75 km are plotted here, many small earthquakes have constrained or inaccurate depths. Small earthquakes with large $m_b - M_s$ may therefore be deep in many cases. Second, the m_b values are not maximum likelihood values, and therefore are subject to censoring, causing m_b to be biased high for small events. Consequently, we can expect that discrimination will be improved, and the slope of the discrimination line reduced, if maximum likelihood m_b is used and deep earthquakes are removed. The explosion m_b values should also be systematically re-evaluated since the PIDC measurements are biased low relative to current NEIS m_b values (MURPHY and BARKER, 1996) due to network differences and the manner in which instrument response corrections are implemented, and the bias compared to historical m_b values has not been assessed.

Conclusions

In this paper we have described procedures for optimization of surface wave processing under a Comprehensive Nuclear-Test-Ban Treaty for the purpose of earthquake/explosion discrimination. Because the number of events increases rapidly at small magnitudes, a decrease in the threshold of reliable surface wave detection and measurement can greatly reduce the number of unidentified events. Improved surface wave analysis methods can reduce the surface wave magnitude threshold, improve screening capability, and reduce the likelihood of unnecessary on-site inspections under a CTBT. Recommended techniques include the use of: regionalized earth models and surface wave parameters, phase-matched filtering to improve signal/noise ratio, path-corrected spectral magnitudes in place of and in addition to M_s, maximum likelihood magnitudes and maximum likelihood upper bounds on magnitudes.

The regionalized models can be used to calculate surface wave parameters, generate phase-matched filters, and predict dispersive arrival times. In addition, the techniques used here to develop these models can be used on a continuing basis to improve the models. With the large amount of data now coming continuously into the PIDC and the IDC, it is possible to maintain a rapidly increasing database of dispersion curves, adding to the database used in this study and improving the results by filling in regions with poor coverage and extending the frequency range.

In addition to the improvements discussed here, some changes in operational procedure would improve the reliability of surface wave identification and measure-ment. Clearly, routine identification and correction of instrument calibration errors is necessary if the database is to be used reliably. The addition of data from auxiliary stations would also increase surface wave detections and improve reliability. Finally, although Maxsurf and Maxpmf were designed to be completely automatic, some operator review would be beneficial, particularly for removal of bad data and for phase identification of surface waves in cases where the automatic processing results are marginal.

Acknowledgements

We would like to thank Gabi Laske of the University of California at San Diego for providing the Crust 5.1 earth models, Mike Ritzwoller of the University of Colorado, Brian Mitchell and Bob Herrmann of St. Louis University, Goran Ekstrom of Harvard University, and their coworkers for the use of their data and models in this project, and David Harkrider for many useful discussions. We would also like to thank Mariana Eneva for helping to identify and track down the time periods of instrument correction errors, and David Adams for continuing the tomographic inversions with larger data sets. Mike Skov and Hans Israelsson significantly contributed to the implementation and testing of Maxsurf at the PIDC. This work was supported by U. S. Air Force contract F19628-95-C-0110 and Defense Threat Reduction Agency contract DSWA01-98-C-0154.

Appendix A
Surface Wave Excitation by Arbitrary Sources and Explosions

The vertical component of a Rayleigh wave from a source located at the radial origin and depth h and measured at distance r and depth z in a plane layered medium has the following form[4]

$$u_z(\omega, r, z, \varphi) = A_R \sqrt{2/\pi\omega cr} \, \exp[i(\pi/4 - \omega r/c)] F_s(\omega, \varphi, h) \, y_1(\omega, z) \qquad (A.1)$$

and the radial component is given by

$$u_r(\omega, r, z, \varphi) = -iA_R \sqrt{2/\pi\omega cr} \, \exp[i(\pi/4 - \omega r/c)] F_s(\omega, \varphi, h) \, y_3(\omega, z) \ , \qquad (A.2)$$

---

[4] Equations (A.1–4) appear in a variety of forms in the literature with slightly different notations. The definition of I_1 used here follows TAKEUCHI and SAITO (1972) and HARKRIDER et al. (1994). KENNETT (1983) uses the same definition of I_1. AKI and RICHARDS (1980) define an energy integral I_1 equal to 1/2 the I_1 defined here. McGARR and ALSOP (1967) define an integral I_7 which is equivalent to I_1. The sign of the Fourier transform used by Aki and Richards and Kennett is opposite the sign used in the other references mentioned here, therefore the equations are the complex conjugate of the equations given here.

where A_R is the Rayleigh wave amplitude, c is the phase velocity, ω is the angular frequency, F_s is a function that depends on the source type and source depth, y_1 and y_3 are the vertical and radial Rayleigh wave eigenfunctions as defined by TAKEUCHI and SAITO (1972), respectively, evaluated at receiver depth z, and φ is the source to receiver azimuth. F_s can represent any source, and can be expressed as an expansion in cylindrical harmonics which can be derived from a source represented by an expansion in spherical harmonics (HARKRIDER et al., 1994). Equations (A.1) and (A.2) are equivalent to Equation (46) of HARKRIDER et al. (1994) except that the following changes were made to make it easier to compare to observations: the Hankel functions have been replaced by their asymptotic expansions, the right-hand side of the equation was divided by $i\omega$ so that a source function with constant moment would have a step rather than delta time function, u_z is oriented with vertical up, and eigenfunctions at the receiver are included so that the receiver depth may be non-zero. A_R is related to the kinetic energy in the mode through the relation

$$A_R = \frac{1}{2cUI_1} \; , \tag{A.3}$$

where U is the group velocity and the kinetic energy I_1 is given by

$$I_1 = \int_0^\infty \rho(z)[y_1(z)^2 + y_3(z)^2]\, dz \; , \tag{A.4}$$

where $\rho(z)$ is the density at depth z.

For the specific case of an explosion source

$$F_s = M_0 \frac{\beta^2}{\alpha^2}\left(y_3 - \frac{y_2}{2\mu k}\right) \; , \tag{A.5}$$

where α, β, and μ are the compressional speed, shear speed, and shear modulus at the source, respectively, k is the wavenumber ω/c and y_2 is the normal stress eigenfunction at the source depth. Since y_2 vanishes at the free surface and is small at typical explosion depths, F_s is sensitive to Poisson's ratio at the source. As discussed by STEVENS (1986), we can define a normalized moment

$$M_0' = 3\frac{\beta^2}{\alpha^2}M_0 \tag{A.6}$$

so that when M_0 is replaced by M_0' in Equation (A.5), F_s becomes

$$F_s = \frac{1}{3}M_0'\left(y_3 - \frac{y_2}{2\mu k}\right) \; . \tag{A.7}$$

For a fixed M_0' at shallow depths, F_s depends on the eigenfunction y_3 which is a function of the average earth structure and only weakly dependent on material properties at the source.

Appendix B
Surface Wave Transmission Through Laterally Heterogeneous Media

We are interested in cases where a surface wave is generated in one structure, passes through any number of different structures, and is recorded at a final structure. We can modify Equations (A.1) and (A.2) to be applicable to this case using the approximation that energy is conserved and that there is no mode conversion (MCGARR, 1969; BACHE et al., 1978). The kinetic energy in the mode is defined by

$$K = \tfrac{1}{2}\omega^2 \int_0^\infty \rho\left[|u_z|^2 + |u_r|^2\right] dz .$$ (B.1)

Energy conservation requires that the energy flux of the surface wave through a cylindrical surface be constant. This condition can be written:

$$K(r_1)U(r_1)r_1 = K(r_2)U(r_2)r_2 ,$$ (B.2)

where U is the group velocity. The displacement at point r_2 can be written in terms of the displacement at r_1 and a transmission coefficient $T(r_1, r_2)$.

$$u_z(r_2) = T(r_1, r_2)u_z(r_1) .$$ (B.3)

Substituting (A.1) and (A.2) into (B.1), $K(r) \sim I_1(r)/r$, so $I_1(r_1)U(r_1) = T^2 I_1(r_2)U(r_2)$ and

$$T = \sqrt{\frac{I_1(r_1)U(r_1)}{I_1(r_2)U(r_2)}}$$ (B.4)

or using (A.3)

$$T = \sqrt{\frac{A_R(r_2)c(r_2)}{A_R(r_1)c(r_1)}} .$$ (B.5)

We can then rewrite Equation (A.1) in the form

$$u_z(\omega, r, z, \varphi) = \sqrt{2A_{R_1}/\pi\omega c_1^2 r}\sqrt{c_2 A_{R_2}} \exp\left[i(\pi/4 - \omega r/c_p)\right] F_s(\omega, \varphi, h)(y_1(\omega, z))_2 ,$$ (B.6)

where the 1 and 2 subscripts refer to the source and receiver location, respectively, and the "p" subscript refers to the path averaged value of the phase velocity (the phase slowness is averaged over the path). $(y_1(\omega, z))_2$ refers to the y_1 eigenfunction for the receiver structure evaluated at the receiver depth z. In an attenuating, spherical earth, Equation (B.6) must also be multiplied by

$$\exp(-\gamma r_p)\left(\frac{r}{a_e \sin(r/a_e)}\right)$$

where a_e is the radius of the earth and γ is the frequency dependent attenuation coefficient, as a result (B.6) becomes

$$u_z(\omega,r,z,\varphi) = \frac{1}{\sqrt{a_e \sin(r/a_e)}} \sqrt{\frac{2A_{R_1}}{\pi\omega c_1^2}} \sqrt{c_2 A_{R_2}} \exp\left[i\left(\pi/4 - \omega r/c_p - \gamma_p r\right)\right]$$
$$\times F_s(\omega,\varphi,h)\, y_1(\omega,z) \ . \tag{B.7}$$

Each factor in Equation (B.7) depends only on the source location, receiver location, or path. For an explosion source we can write (B.7) in the simple form:

$$u_z(\omega,h_x,r,z) = M_0' \frac{S_1^x(\omega,h_x)S_2(\omega)\exp[-\gamma_p(\omega)r + i(\varphi_0 - \omega r/c_p(\omega))]}{\sqrt{a_e \sin(r/a_e)}} y_1(\omega,z) \ , \tag{B.8}$$

where φ_0 is the initial phase equal to $-3\pi/4$,

$$S_1^x(\omega,h_x) = \sqrt{\frac{2A_{R_1}}{9\pi\omega c_1^2}}\left(\frac{1}{2\mu k}y_2(h_x) - y_3(h_x)\right) \ , \tag{B.9}$$

$$S_2(\omega) = \sqrt{c_2 A_{R_2}} \ . \tag{B.10}$$

S_1 and S_2 as defined here correct an extra factor of $1/\sqrt{c_1}$ in S_1 and $\sqrt{c_2}$ in S_2 in the corresponding definitions of STEVENS (1986). The equations in that paper as well as the corresponding equations in BACHE et al. (1978) from which they were derived incorrectly placed a phase velocity factor resulting from the asymptotic expansion of the Hankel function into the receiver term S_2 leading to the difference noted above.

REFERENCES

AKI, K., and RICHARDS, P. G., Quantitative Seismology: Theory and Methods (W. H. Freeman, San Francisco, 1980).

ARCHAMBEAU, C. B., FLINN, E. A., and LAMBERT, D. G. (1966), Detection, Analysis, and Interpretation of Seismic Energy in the Upper Mantle, J. Geophys. Res. 71, 3483–3501.

BACHE, T. C., RODI, W. L., and HARKRIDER, D. G. (1978), Crustal Structures Inferred from Rayleigh-wave Signatures of NTS Explosions, Bull. Seismol. Soc. Am. 68, 1399–1413.

DZIEWONSKI, A. M., and ANDERSON, D. L. (1981), Preliminary Reference Earth Model, J. Phys. Earth Planet. Inter. 25, 297–356.

DZIEWONSKI, A. M., BLOCH, J., and LANDISMAN, M. (1969), A New Technique for the Analysis of Transient Seismic Signals, Bull. Seismol. Soc. Am. 59, 427–444.

EKSTRÖM, G., DZIEWONSKI, A. M., SMITH, G. P., and SU, W. (1996), Elastic and inelastic structure beneath Eurasia. In Proceed. 18th Annual Seismic Res. Symp. on Monitoring a Comprehensive Test Ban Treaty, 4–6 September, 1996, Phillips Laboratory Report PL-TR-96-2153, July, pp. 309–318, ADA313692.

FISK, M. D., JEPSEN, D., and MURPHY, J. R. (1999), Experimental Event-Screening Criteria at the Prototype International Data Center, Pure appl. geophy., submitted.

HARKRIDER, D. G., STEVENS, J. L., and ARCHAMBEAU C. B. (1994), *Theoretical Rayleigh and Love Waves from an Explosion in Prestressed Source Regions*, Bull. Seismol. Soc. Am. *84*, 1410–1442.

HERAK, M., and HERAK, D. (1993), *Distance Dependence of M_s and Calibrating Function for 20 Second Rayleigh Waves*, Bull. Seismol. Soc. Am. *83*, 1881–1892.

HERRIN, E., and GOFORTH, T. (1977), *Phase-matched Filtering: Application to the Study of Rayleigh Waves*, Bull. Seismol. Soc. Am. *67*, 1259–1275.

KENNETT, B. L. N., *Seismic Wave Propagation in Stratified Media* (Cambridge University Press, Cambridge, UK, 1983).

LEVSHIN, A. L., RITZWOLLER, M. H., and SMITH, S. S. (1996), *Group Velocity Variations Across Eurasia.* In *Proceed. 18th Annual Seismic Res. Symp. on Monitoring A Comprehensive Test Ban Treaty*, 4–6 September, 1996, Phillips Laboratory Report PL-TR-96-2153, July, pp. 70–79, ADA313692.

MARSHALL, P. D., and BASHAM, P. W. (1972), *Discrimination Between Earthquakes and Underground Nuclear Explosions Employing an Improved M_s Scale*, Geophys. J. R. Astr. Soc. *28*, 431–458.

MCGARR, A. (1969), *Amplitude Variations of Rayleigh Waves – Propagation Across a Continental Margin*, Bull. Seismol. Soc. Am. *59*, 1281–1305.

MCGARR, A., and ALSOP, L. E. (1967), *Transmission and Reflection of Rayleigh Waves at Vertical Boundaries*, J. Geophys. Res. *72*, 2169–2180.

MCLAUGHLIN, K. L. (1988), *Maximum-likelihood Event Magnitude Estimation with Bootstrapping for Uncertainty Estimation*, Bull. Seismol. Soc. Am. *78*, 855–862.

MITCHELL, B. J., CONG, L., and XIE, J. (1996), *Seismic Attenuation Studies in the Middle East and Southern Asia*, St. Louis University Scientific Report No. 1, PL-TR-96-2154, ADA317387.

MOONEY, W., LASKE, G., and MASTERS, G. (1998), *Crust 5.1: A Global Crustal Model at 5 × 5 Degrees*, J. Geophys. Res. *103* (B1), 727–747.

MURPHY, J. R., and BARKER, B. W. (1996), *A Preliminary Evaluation of Seismic Magnitude Determination at the International Data Center (IDC)*, EOS Transact. Am. Geophys. Union, November, P. F7.

NOLET, G., *Seismic wave propagation and seismic tomography.* In *Seismic Tomography with Applications in Global Seismology and Exploration Geophysics* (ed. NOLET, G.) (D. Reidel Publishing, Dordrecht, Holland, 1987).

OKAL, E. A., and TALANDIER, J. (1987), *M_m: Theory of a Variable-period Mantle Magnitude*, Geophys. Res. Lett. *14*, 836–839.

REZAPOUR, M., and PEARCE, R. G. (1998), *Bias in Surface-wave Magnitude M_s due to Inadequate Distance Corrections*, Bull. Seismol. Soc. Am. *88*, 43–61.

RITZWOLLER, M. H., LEVSHIN, A. L., RATNIKOVA, L. I., and TREMBLAY, D. M. (1996), *High resolution group velocity variations across Central Asia.* In *Proceed. 18th Annual Seismic Res. Symp. on Monitoring a Comprehensive Test Ban Treaty*, 4–6 September, 1996, Phillips Laboratory Report PL-TR-96-2153, July, pp. 98–107, ADA313692.

RITZWOLLER, M. H., VDOVIN, O. Y., and LEVSHIN, A. L. (1999), *Surface Wave Dispersion across Antarctica: A First Look*, Antarctic J. US, in press.

STEVENS, J. L. (1986), *Estimation of Scalar Moments From Explosion-generated Surface Waves*, Bull. Seismol. Soc. Am. *76*, 123–151.

STEVENS, J. L., and DAY, S. M. (1985), *The Physical Basis of the m_b:M_s and Variable Frequency Magnitude Methods for Earthquake/Explosion Discrimination*, J. Geophys. Res. *90*, 3009–3020.

STEVENS, J. L., and MCLAUGHLIN, K. L. (1996), *Regionalized Maximum Likelihood Surface Wave Analysis*, Maxwell Technologies Technical Report submitted to Phillips Laboratory, PL-TR-96-2273, SSS-DTR-96-15562, September, ADA321813.

STEVENS, J. L., and MCLAUGHLIN, K. L. (1997), *Improved Methods for Regionalized Surface Wave Analysis*, Maxwell Technologies Final Report submitted to Phillips Laboratory, MFD-TR-97-15887, September.

STEVENS, J. L., and MCLAUGHLIN, K. L. (1988), *Analysis of Surface Waves from the Novaya Zemlya, Mururoa, and Amchitka Test Sites, and Maximum Likelihood Estimation of Scalar Moments from Earthquakes and Explosions*, S-CUBED Technical Report submitted to Air Force Technical Applications Center, SSS-TR-89-9953, September.

SATO, R. (1967), *Attenuation of seismic waves*, J. Phys. Earth *15*, 32–61.

TAKEUCHI, H., and SAITO, M., *Seismic surface waves*, In *Methods of Computational Physics* (ed. Bolt, B. A.) v. 11 (Academic Press, New York, 1972) pp. 217–295.

THOMAS, J. H., MARSHALL, P. D., and DOUGLAS, A. (1978), *Rayleigh-wave Amplitudes from Earthquakes in the Range 0–150 Degrees*, Geophys. J. R. Astr. Soc. 53, 191–200.

VDOVIN, O. Y., RIAL, J. A., RITZWOLLER, M. H., and LEVSHIN, A. L., (1999), *Group-velocity Tomography of South America and the Surrounding Oceans*, Geophys. J. Int. *136*, 324–330.

VON SEGGERN, D. H. (1975a), *Q for 20-Second Rayleigh Waves from Complete Great-Circle Paths*, Teledyne Geotech Report SDAC-TR-75-3 submitted to the Defense Advanced Research Projects Agency, February.

VON SEGGERN, D. H. (1975b), *Distance-Amplitude Relationships for Long-Period P, S, and LR from Measurements on Recordings of the Long-Period Experimental Stations*, Teledyne Geotech Report SDAC-TR-75-15 submitted to the Defense Advanced Research Projects Agency, September.

(Received June 24, 1999, revised December 20, 1999, accepted January 15, 2000)

 To access this journal online:
http://www.birkhauser.ch